This textbook presents a modern account of turbulence, one of the greatest challenges in physics. The state-of-the-art is put into historical perspective five centuries after the first studies of Leonardo and half a century after the first attempt by A. N. Kolmogorov to predict the properties of flow at very high Reynolds numbers. Such "fully developed turbulence" is ubiquitous in both cosmical and natural environments, in engineering applications and in everyday life.

First, a qualitative introduction is given to bring out the need for a probabilistic description of what is in essence a deterministic system. Kolmogorov's 1941 theory is presented in a novel fashion with emphasis on symmetries (including scaling transformations) which are broken by the mechanisms producing the turbulence and restored by the chaotic character of the cascade to small scales. Considerable material is devoted to intermittency, the clumpiness of small-scale activity, which has led to the development of fractal and multifractal models. Such models, pioneered by B. Mandelbrot, have applications in numerous fields besides turbulence (diffusion limited aggregation, solid-earth geophysics, attractors of dynamical systems, etc). The final chapter contains an introduction to analytic theories of the sort pioneered by R. Kraichnan, to the modern theory of eddy transport and renormalization and to recent developments in the statistical theory of two-dimensional turbulence. The book concludes with a guide to further reading.

The intended readership for the book ranges from first-year graduate students in mathematics, physics, astrophysics, geosciences and engineering, to professional scientists and engineers. Elementary presentations of dynamical systems ideas, of probabilistic methods (including the theory of large deviations) and of fractal geometry make this a self-contained textbook.

TURBULENCE

THE LEGACY OF A.N. KOLMOGOROV

TURBULENCE

THE LEGACY OF A.N. KOLMOGOROV

URIEL FRISCH
Observatoire de la Côte d'Azur

CAMBRIDGE UNIVERSITY PRESS
Cambridge, New York, Melbourne, Madrid, Cape Town, Singapore, São Paulo

Cambridge University Press
The Edinburgh Building, Cambridge CB2 2RU, UK

Published in the United States of America by Cambridge University Press, New York

www.cambridge.org
Information on this title: www.cambridge.org/9780521451031

First published 1995
Reprinted 1996, 1998, 2001, 2004

A catalogue record for this publication is available from the British Library

Library of Congress Cataloguing in Publication data
Frisch, U. (Uriel), 1940–
Turbulence : the legacy of A. N. Kolmogorov / Uriel Frisch.
p. cm.
Includes bibliographical references (p. –) and index.
ISBN 0-521-45103-5 (hc). – ISBN 0-521-45713-0 (pb)
1. Turbulence. I. Kolmogorov, A. N. (Andreĭ Nikolaevich), 1903–1987. II. Title.
QA913.F74 1996
532′.0527–dc20 95-12140 CIP

ISBN-13 978-0-521-45103-1 hardback
ISBN-10 0-521-45103-5 hardback

ISBN-13 978-0-521-45713-2 paperback
ISBN-10 0-521-45713-0 paperback

Transferred to digital printing 2006

Contents

CHAPTER 4
Probabilistic tools: a survey

CHAPTER 5
Two experimental laws of fully developed turbulence

CHAPTER 6
The Kolmogorov 1941 theory

CHAPTER 7
Phenomenology of turbulence in the sense of Kolmogorov 1941

CHAPTER 8
Intermittency

CHAPTER 9
Further reading: a guided tour

Preface

Andrei Nikolaevich Kolmogorov's work in 1941 remains a major source of inspiration for turbulence research. Great classics, when revisited in the light of new developments, may reveal hidden pearls, as is the case with Kolmogorov's very brief third 1941 paper 'Dissipation of energy in locally isotropic turbulence' (Kolmogorov 1941c). It contains one of the very few exact and nontrivial results in the field, as well as very modern ideas on scaling, ideas which cannot be refuted by the argument Lev Landau used to criticize the universality assumptions of the first 1941 paper.

Revisiting Kolmogorov's fifty-year-old work on turbulence was one goal of the lectures on which this book is based. The lectures were intended for first-year graduate students in 'Turbulence and Dynamical Systems' at the University of Nice–Sophia–Antipolis. My presentation deliberately emphasizes concepts which are central in dynamical systems studies, such as symmetry-breaking and deterministic chaos. The students had some knowledge of fluid dynamics, but little or no training in modern probability theory. I have therefore included a significant amount of background material. The presentation uses a physicist's viewpoint with more emphasis on systematic arguments than on mathematical rigor. Also, I have a marked preference for working in coordinate space rather than in Fourier space, whenever possible.

Modern work on turbulence focuses to a large extent on trying to understand the reasons for the partial failure of the 1941 theory. This 'intermittency' problem has received here considerable coverage. Kolmogorov himself became a pioneer in this line of investigation in 1961, following the work of his collaborator A.M. Obukhov (Kolmogorov 1961). Although some of their suggestions can be criticized as mathematically or physically inconsistent, their 1961 work has been and remains a major source of inspiration. For pedagogical reasons, I have chosen to discuss historical aspects only after presentation of more recent work on 'fractal' and 'multifractal' models of turbulence.

Some of the material on Kolmogorov presented here has appeared in a special issue of the *Proceedings of the Royal Society* 'Kolmogorov's ideas 50 years on', which also contains a whole range of alternative views on Kolmogorov and on what matters for turbulence research (Frisch 1991). Other useful references on Kolmogorov are the selected works (Tikhomirov 1991), the obituary (Kendall 1990), the review of the turbulence work of one of his close collaborators (Yaglom 1994) and the personal recollections concerned more with the mathematician and the man (Arnold 1994).

In an introductory course on turbulence, of about thirty hours of lecturing, many aspects had to be left out. I have included at the end of this book a guided tour to further reading as a partial remedy. It is also intended to convey briefly my — possibly very biased — views of what matters. No attempt has been made to present a balanced historical perspective of a subject now at least five centuries old (see p. 112); the reader will nevertheless find a number of historical sections and remarks and may discover for example that the concept of eddy viscosity was introduced in the middle of the nineteenth century (see p. 223).

More information on the organization of this book may be found in Section 1.2 (see p. 11).

The intended readership for the book ranges from first-year graduate students in mathematics, physics, astrophysics, geophysics and engineering, to professional scientists and engineers. Primarily, it is intended for those interested in learning about the basics of turbulence or wanting to take a fresh look at the subject. Much of the material on probabilistic background, on fractals and multifractals also has applications beyond fluid mechanics, for instance, to solid-earth geophysics.

I am deeply grateful to J.P. Rivet who in many respects has given life to this book and I am particularly indebted to A.M. Yaglom for numerous discussions and comments. Very useful remarks and suggestions were received from V.I. Arnold, G. Barenblatt, G.K. Batchelor, L. Biferale, M. Blank, M.E. Brachet, G. Eyink, H. Frisch, H.L. Grant, M. Hénon, J. Jiménez, R. Kraichnan, B. Legras, A. Migdal, G.M. Molchan, A. Noullez, K. Ohkitani, S.A. Orszag, A. Praskovsky, A. Pumir, Z.S. She, Ya. Sinai, J. Sommeria, P.L. Sulem, M. Vergassola, E. Villermaux and B. Villone. M.C. Vergne has realized some of the figures. I also wish to thank the students of the 'DEA Turbulence et

Systèmes Dynamiques' of the University of Nice–Sophia–Antipolis who have helped me with their questions, since I started teaching this material as a graduate course in 1990.

Part of the work for this book was done while I was visiting Princeton University (Center for Fluid Dynamics Research). Significant support was received from the 'Direction des Recherches et Moyens Techniques', from various programs of the European Union and from the 'Fondation des Treilles'.

I would like to dedicate this new printing to Giovanni Paladin who died in a mountaineering accident on June 29, 1996.

Finally, it was a pleasure and a privilege to work in close collaboration with Alison, Maureen, Simon and Stephanie at Cambridge University Press.

Nice, France U. Frisch
July 1995

1
Introduction

1.1 Turbulence and symmetries

In Chapter 41 of his *Lectures on Physics*, devoted to hydrodynamics and turbulence, Richard Feynman (1964) observes this:

Often, people in some unjustified fear of physics say you can't write an equation for life. Well, perhaps we can. As a matter of fact, we very possibly already have the equation to a sufficient approximation when we write the equation of quantum mechanics:

$$H\psi = -\frac{\hbar}{i}\frac{\partial \psi}{\partial t}. \tag{1.1}$$

Of course, if we only had this equation, without detailed observation of biological phenomena, we would be unable to reconstruct them. Feynman believes, and this author shares his viewpoint, that an analogous situation prevails in *turbulent* flow of an incompressible fluid. The equation, generally referred to as the Navier–Stokes equation, has been known since Navier (1823):

$$\partial_t \boldsymbol{v} + \boldsymbol{v} \cdot \nabla \boldsymbol{v} = -\nabla p + \nu \nabla^2 \boldsymbol{v}, \tag{1.2}$$

$$\nabla \cdot \boldsymbol{v} = 0. \tag{1.3}$$

It must be supplemented by initial and boundary conditions (such as the vanishing of $\boldsymbol{v}$ at rigid walls). We shall come back later to the choice of notation.

The Navier–Stokes equation probably contains all of turbulence. Yet it would be foolish to try to guess what its consequences are without looking at experimental facts. The phenomena are almost as varied as in the realm of life.

A good way to make contact with the rich world of turbulence phenomena is through the book of Van Dyke (1982) *An Album of Fluid Motion*. To communicate a first impression of the experimental facet of turbulence, we shall mainly use pictures from this book.

In this Introduction, we have chosen to stress the ideas of *broken symmetries* and of *restored symmetries*. Symmetry consideration are indeed central to the study of both *transition phenomena* and *fully developed turbulence*. For the time being we shall leave aside the quantitative aspects of experimental data with the exception of the control parameter, the Reynolds number, which is defined as

$$R = \frac{LV}{\nu}, \tag{1.4}$$

L and V being respectively a characteristic scale and velocity of the flow, and ν its (kinematic) viscosity.[1] Remember a consequence of the *similarity principle* for incompressible flow: for a given geometrical shape of the boundaries, the Reynolds number is the only control parameter of the flow.

With this in mind, let us observe what happens when increasing the Reynolds number in flow past a cylinder. We have chosen a cylinder in order to ensure some degree of symmetry, while selecting an *external* flow. External flow is more difficult to control and to study but has more life than internal flow which is confined by its boundaries, such as Rayleigh–Bénard convection or Taylor–Couette flow.

As shown in Fig. 1.1, we consider a flow of uniform velocity $V = (V, 0, 0)$ (at infinity), parallel to the x-axis, incident from the left on an infinite cylinder, of circular cross-section with diameter L, the axis being along the z-direction.

Fig. 1.2 is a visualization of the flow at $R = 0.16$. At first, the flow appears to possess the following symmetries:

- *Left–right* (x-reversal),
- *Up–down* (y-reversal),
- *Time-translation* (t-invariance),
- *Space-translation* parallel to the axis of the cylinder (z-invariance).

All these symmetries, except the first, are consistent with the Navier–Stokes equation and the boundary conditions. Let us be a little bit more

[1] In c.g.s. units the kinematic viscosity is about one-seventh for air and one-hundredth for water.

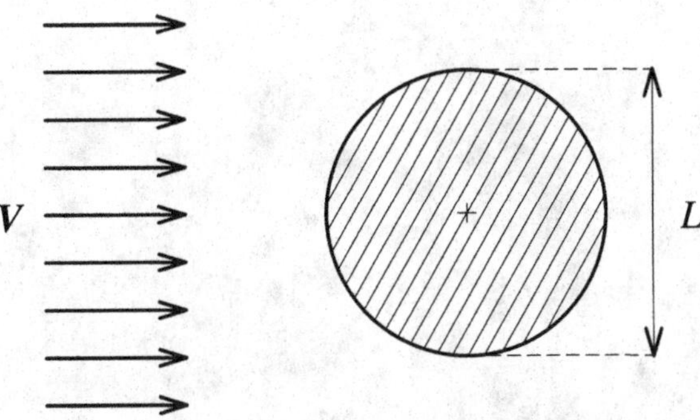

Fig. 1.1. Uniform flow with velocity V, incident on a cylinder of diameter L.

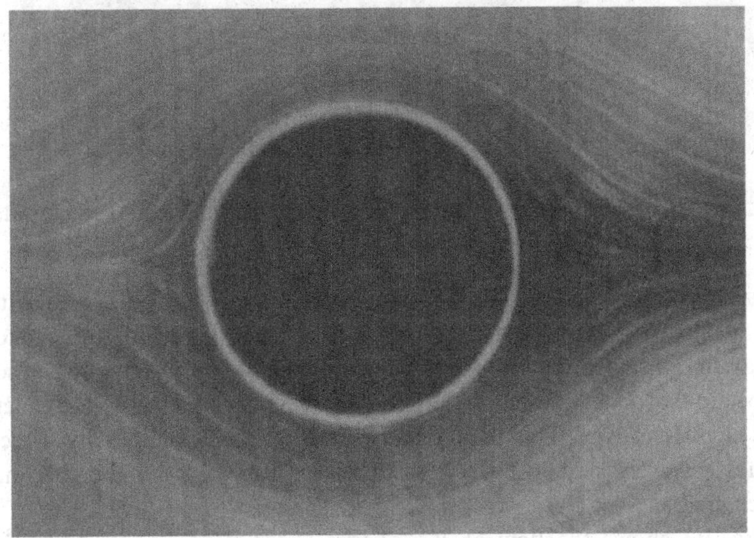

Fig. 1.2. Uniform flow past a cylinder at $R = 0.16$ (Van Dyke 1982). Photograph S. Taneda.

specific. We denote by (u, v, w) the components of the velocity. The left–right symmetry is

$$(x, y, z) \to (-x, y, z), \qquad (u, v, w) \to (u, -v, -w). \tag{1.5}$$

The up–down symmetry is

$$(x, y, z) \to (x, -y, z), \qquad (u, v, w) \to (u, -v, w). \tag{1.6}$$

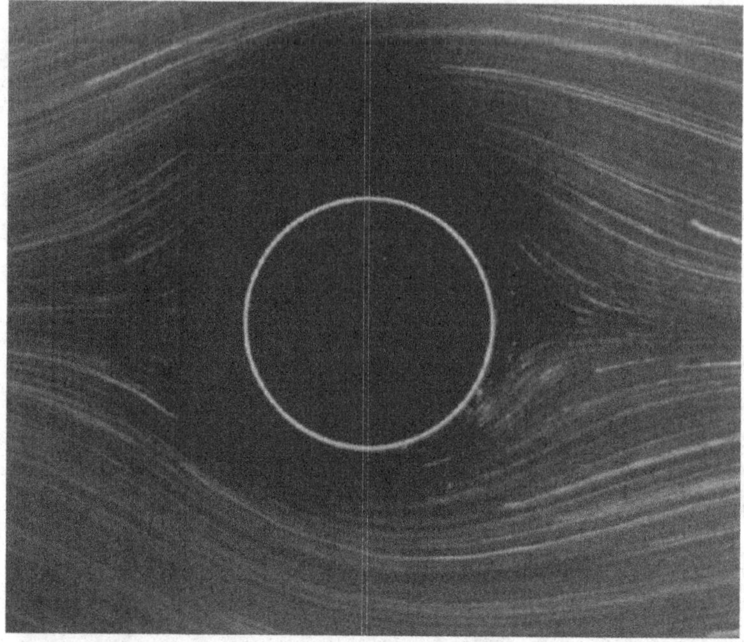

Fig. 1.3. Circular cylinder at $R = 1.54$ (Van Dyke 1982). Photograph S. Taneda.

It is easily checked that the left–right symmetry is not consistent with the Navier–Stokes equation, although it is consistent with the Stokes equation, obtained by dropping the nonlinear term. Actually, closer inspection of Fig. 1.2 shows that the left–right symmetry is not exact: it is slightly *broken*. This is an effect of the residual nonlinearity, which would get even weaker if we were to let the Reynolds number become much smaller.

Fig. 1.3 shows the flow at $R = 1.54$. There is now a marked left–right asymmetry. Around $R = 5$ the flow begins to separate behind the cylinder. Although no symmetry-breaking occurs, there is a change in the topology of the flow associated with the formation of recirculating standing eddies, shown in Fig. 1.4 for various values of R from 9.6 to 26.

Around $R = 40$ the first true loss of symmetry occurs by an Andronov–Hopf bifurcation which makes the flow time-periodic; in other words, the continuous t-invariance is broken in favor of a discrete t-invariance. The flow in the immediate neighborhood of the bifurcation point is shown in Fig. 1.5. At higher values of R, such as shown in Figs. 1.6, 1.7 and

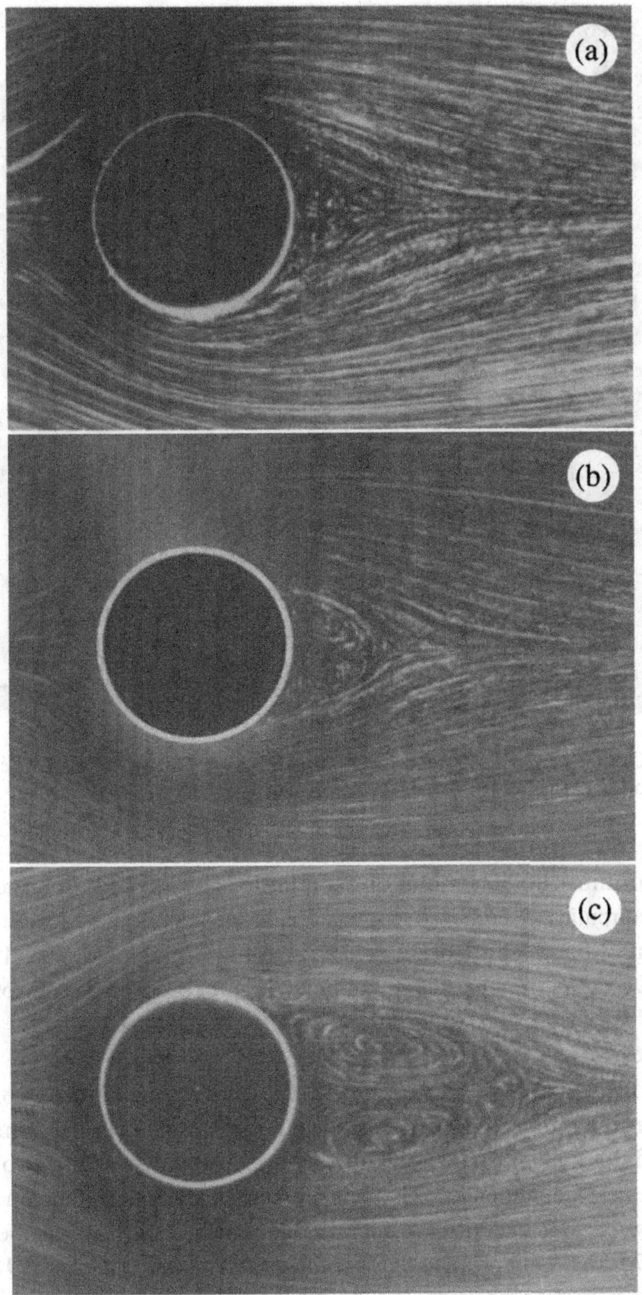

Fig. 1.4. Circular cylinder at $R = 9.6$ (a), $R = 13.1$ (b) and $R = 26$ (c) (Van Dyke 1982). Photograph S. Taneda.

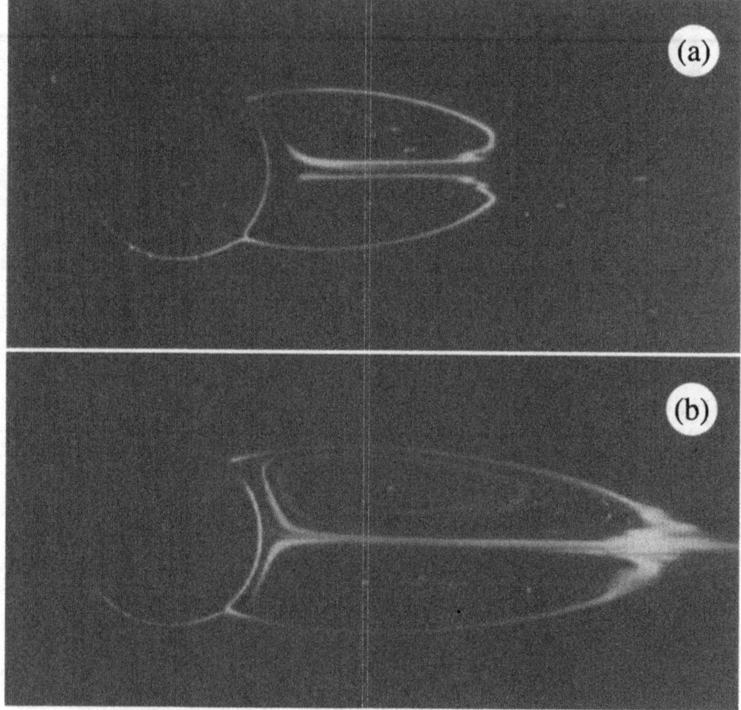

Fig. 1.5. Circular cylinder at $R = 28.4$ (a) and $R = 41.0$ (b) (Van Dyke 1982). Photograph S. Taneda.

1.8 (a two-dimensional simulation by a lattice gas method), the shedding of the recirculation eddies becomes very conspicuous and leads to the formation of the celebrated *Kármán street* of alternating vortices. It must be observed that the up–down symmetry is not really broken insofar as, after half a period, the upper eddies will be exact mirror images of the lower ones.

It is not known at what Reynolds number the z-invariance is broken. Experimentally, this cannot be easily found, since the cylinder cannot be made infinite and must be held by some device which will unavoidably introduce a z-dependence. There is numerical evidence that when the Reynolds number exceeds a critical value which is somewhere between 40 and 75 the z-invariance is *spontaneously broken* (Rivet 1991). A symmetry is said to be spontaneously broken if it is consistent with the equations of motion and the boundary conditions but is not present in the solution.

Fig. 1.6. Kármán vortex street behind a circular cylinder at $R = 140$ (Van Dyke 1982). Photograph S. Taneda.

Fig. 1.7. Kármán vortex street behind a circular cylinder at $R = 105$ (Van Dyke 1982). Photograph S. Taneda.

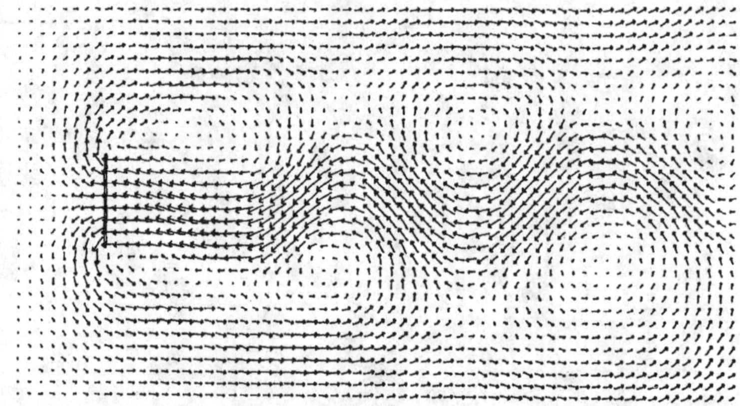

Fig. 1.8. Lattice gas simulation of a two-dimensional Kármán vortex street behind a flat plate (d'Humières, Pomeau and Lallemand 1985).

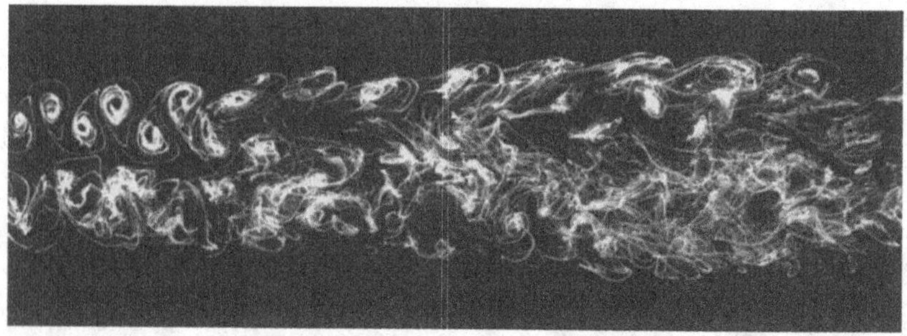

Fig. 1.9. Wake behind two identical cylinders at $R = 240$. Courtesy R. Dumas.

There is also a threshold (not accurately known) in Reynolds number beyond which the flow becomes *chaotic* in its time-dependence. A particularly manifest form of chaos is Lagrangian turbulence (also known as 'chaotic advection'), the erratic movement of marked fluid particles which can be observed in Fig. 1.9 for $R = 240$. There are now two cylinders instead of just one, but this does not change the basic symmetries. Fig. 1.10 shows the same setup with $R = 1800$. The Kármán streets behind the two cylinders display only about two distinct eddies before merging into a quasi-uniform turbulent wake.

Instead of two obstacles, one can use a large regular array of obstacles forming some kind of a grid. Figs. 1.11 and 1.12 show turbulent flows generated by grids. Far enough behind the grid (say, 10–20 meshes)

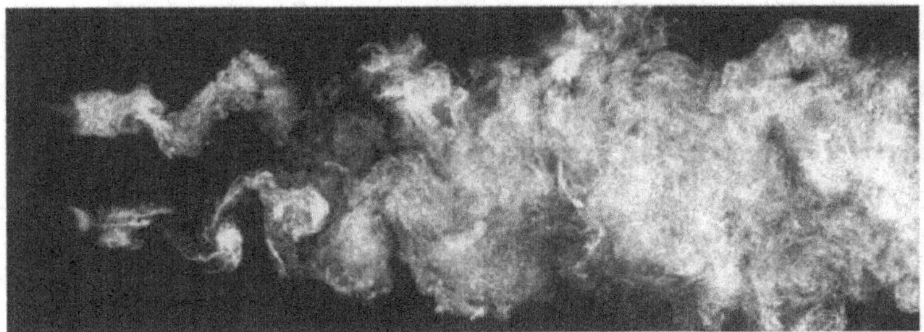

Fig. 1.10. Wake behind two identical cylinders at $R = 1800$. Courtesy R. Dumas.

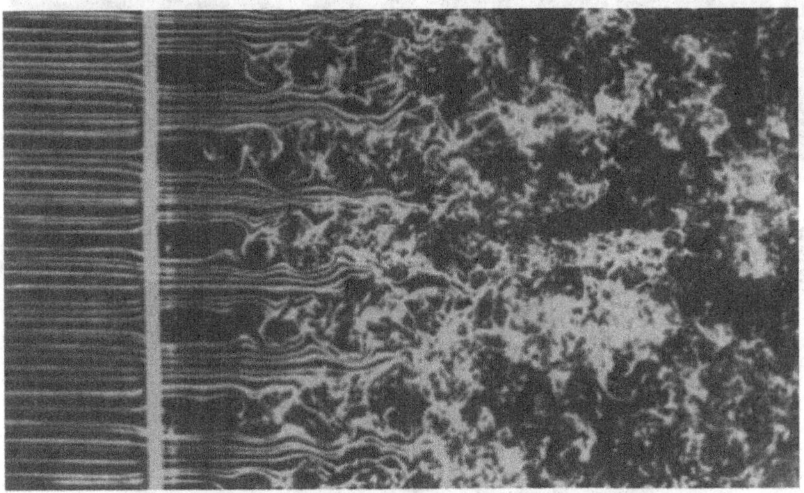

Fig. 1.11. Homogeneous turbulence behind a grid. Photograph T. Corke and H. Nagib.

the flow displays a form of spatial disorder known since Lord Kelvin (1887) as *homogeneous isotropic turbulence* because its overall aspect seems not to change under translations and rotations. This, of course, can only be a statistical statement which will be made more precise later. Fig. 1.13 illustrates another aspect of the homogeneity and isotropy of grid turbulence.

Finally, Fig. 1.14 shows a turbulent jet at $R = 2300$. It is used to illustrate the presence of eddy-motion at all scales, suggesting that some form of (statistical) scale-invariance may be present.

Let us summarize what we have observed. As the Reynolds number is increased, the various symmetries permitted by the equations (and the

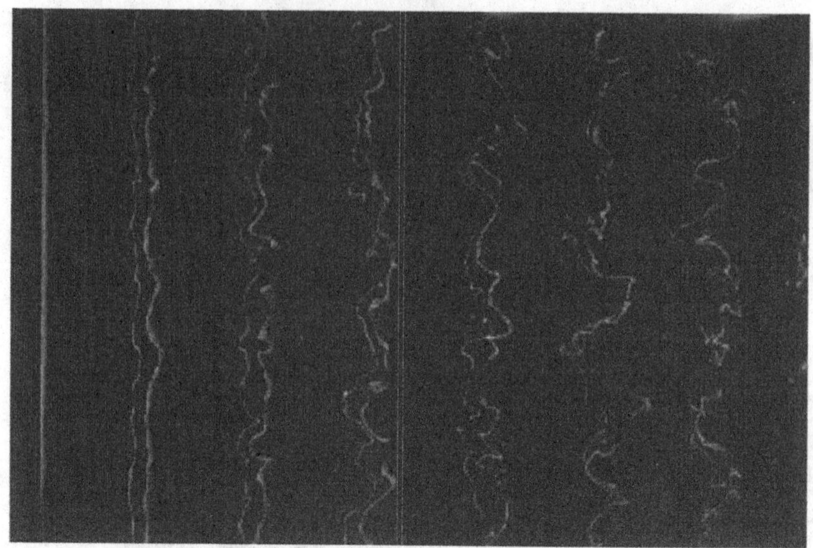

Fig. 1.12. Growth of material lines in isotropic turbulence. $R = 1360$, based on grid rod diameter (Van Dyke 1982). Photograph S. Corrsin and M. Karweit.

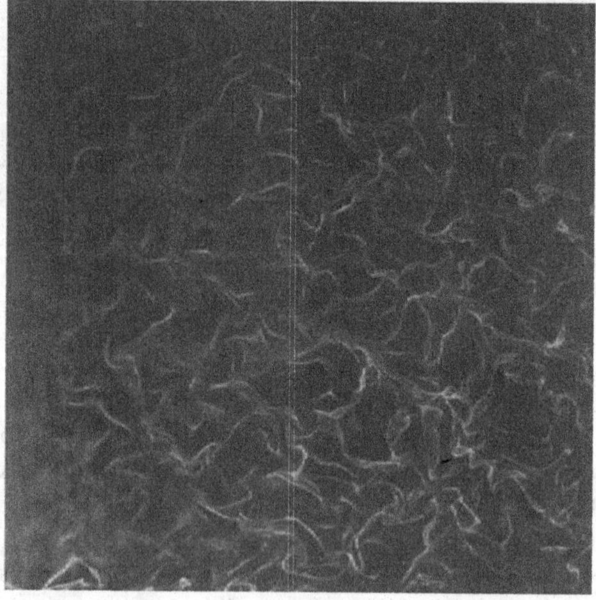

Fig. 1.13. Wrinkling of a fluid surface in isotropic turbulence (Van Dyke 1982). Photograph M. Karweit.

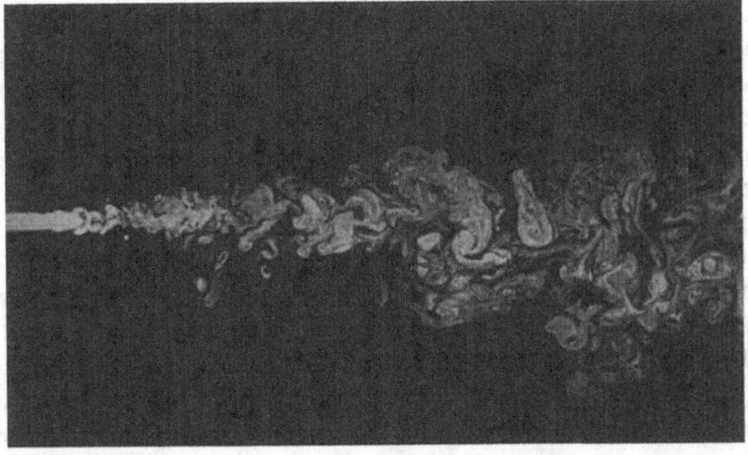

Fig. 1.14. Turbulent water jet (Van Dyke 1982). Photograph P. Dimotakis, R. Lye and D. Papantoniou.

boundary conditions) are successively broken. However, at very high Reynolds number, there appears a tendency to *restore* the symmetries in a statistical sense far from the boundaries.[2]

Turbulence at very high Reynolds numbers, when all or some of the possible symmetries are restored in a statistical sense, is known as *fully developed turbulence.* For this it is necessary that the flow should not be subject to any constraint, such as a strong large-scale shear, which would prevent it from 'accepting' all possible symmetries. Fully developed turbulence will be the central topic of this book.

1.2 Outline of the book

We now explain the organization of the book. This includes comments on 'nonstandard' choices we have made in the presentation; such comments are intended for readers previously exposed to other lectures or textbooks. Note that most of the chapters have their own introductions.

The qualitative material just presented naturally leads to a presentation of the basic symmetries of the incompressible[3] Navier–Stokes equation (Chapter 2). There is no systematic presentation of the fundamentals of fluid mechanics, which can be found in many textbooks, for example

[2] Actually, it is known that chaotic dynamical can possess *symmetry-increasing* bifurcations (Chossat and Golubitsky 1988).

[3] Compressible turbulence will not be discussed here.

in Landau and Lifshitz (1987), Batchelor (1970) or Tritton (1988). In Chapter 2, we shall also discuss the basic conservation laws (energy, helicity, etc.). This includes a 'scale-by-scale energy budget equation' (Section 2.4) which allows us to make sense of the idea of transfer of energy among different scales of motion without requiring at this stage a probabilistic formalism.

In Chapter 3 we address the question of why a *probabilistic* description of turbulence is appropriate. The presentation is made in the spirit of modern (but elementary) dynamical systems theory. In Chapter 4 we present some of the basic tools of probability theory, which are frequently used in turbulence. We mostly follow Kolmogorov's (1933) somewhat abstract viewpoint, but avoid delicate measure-theoretic issues which are discussed in many textbooks for the more mathematically minded reader.

Then, we turn to two of the most basic experimental laws of fully developed turbulence (Chapter 5) which provide direct motivation for Kolmogorov's 1941 theory. This theory (here often abbreviated to 'K41') is presented in Chapter 6 in an unusual way: instead of beginning with the first 1941 turbulence paper with its (now) somewhat questionable hypotheses, we start with a different set of hypotheses linked to the basic symmetries of the Navier–Stokes equation and to the experimental laws reported in the previous chapter. This includes the assumption that the solutions are (statistically) scale-invariant, but no assumption about the value of the scaling exponent. The value of the latter (1/3) is obtained without further assumptions, from Kolmogorov's third 1941 turbulence paper. This is the 'hidden pearl' we were referring to in the Preface. It therefore deserves a detailed presentation (Section 6.2). We then discuss Landau's objection to one aspect of the 1941 theory. A surprising conclusion is that a possible answer to Landau's objection is contained in the third 1941 turbulence paper and that our 'revised set of hypotheses' are actually faithful to Kolmogorov.

Phenomenology of turbulence is presented in Chapter 7. To avoid a 'black magic' impression, we present phenomenology only *after* more systematic theory for which it is essentially a shorthand system. Both the power of standard phenomenology and some of its shortcomings are illustrated.

Intermittency, a particular case of the break-down of the 1941 theory, is discussed at length in Chapter 8 (see the introduction to that chapter). Kolmogorov's (1961, 1962) work is presented only after the discussion of various more recent intermittency models and in the light of exact results about admissible deviations to his 1941 theory.

There are two sections with historical material: Section 6.5 is devoted to the early Kolmogorov 1941 theory and Section 8.8 to intermittency.

Chapter 9, as stated in the Preface, is a guided tour of further reading on turbulence; it also contains additional historical material.

2
Symmetries and conservation laws

Let us return to the Navier–Stokes equation (1.2). We rewrite (1.2)–(1.3) as

$$\partial_t v_i + v_j \partial_j v_i = -\partial_i p + \nu \partial_{jj} v_i, \tag{2.1}$$

$$\partial_i v_i = 0. \tag{2.2}$$

p and ν are here called the pressure and the viscosity, respectively. Actually, they are obtained by dividing the true pressure and the dynamical viscosity by the mean density ρ_0. In addition to the standard rule of summation over repeated indices, we use the notation:

$$\partial_t \equiv \frac{\partial}{\partial t}, \qquad \partial_j \equiv \frac{\partial}{\partial x_j}, \qquad \partial_{ij} \equiv \frac{\partial^2}{\partial x_i \partial x_j}. \tag{2.3}$$

2.1 Periodic boundary conditions

In order to achieve maximum symmetry, it is advantageous not to have *any* boundaries. We could thus assume that the fluid fills all of the space $\mathbb{R}^3$. The unboundedness of the space does, however, lead to some mathematical difficulties. We shall therefore often assume *periodic boundary conditions* in the space variable $\boldsymbol{r} = (x, y, z)$:

$$\boldsymbol{v}(x + nL, y + mL, z + qL) = \boldsymbol{v}(x, y, z), \tag{2.4}$$

for all x, y, z and all signed integers n, m, q. The positive real number L is called the period. It is then obviously enough to consider the restriction of the flow to a *periodicity box* such as B_L: $0 \le x < L$, $0 \le y < L$, $0 \le z < L$ (Fig. 2.1). Later, we shall recover the case of a fluid in the unbounded space $\mathbb{R}^3$ by letting $L \to \infty$.

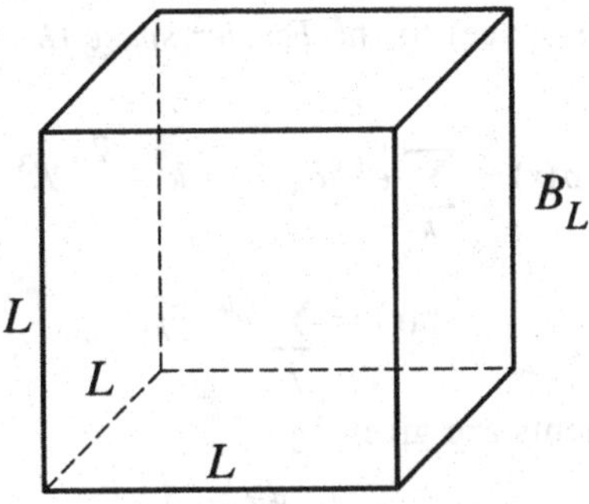

Fig. 2.1. The periodicity box.

The space of L-periodic functions $\boldsymbol{v}(\boldsymbol{r})$ satisfying $\nabla \cdot \boldsymbol{v} = 0$ will be denoted by $\mathscr{H}$.

In principle, the definition of $\mathscr{H}$ should be supplemented by prescribing a suitable norm. We shall refrain from this because our purpose is not to derive fully rigorous results: the present state of the mathematics for the three-dimensional Navier–Stokes equation, which will not be reviewed here (see, e.g., Rose and Sulem 1978; Constantin 1991, 1994; Gallavotti 1993; and Section 9.3) makes it unreasonable to set higher standards.

With periodic boundary conditions, it is easy to eliminate the pressure from the Navier–Stokes equation, as we now show. This is a rather elementary exercise which gives us an opportunity to introduce useful notation. Taking the divergence of (2.1) and using (2.2), we obtain

$$\partial_i \left(v_j \partial_j v_i\right) = \partial_{ij} \left(v_i v_j\right) = -\partial_{ii} p = -\nabla^2 p. \tag{2.5}$$

Eq. (2.5) is an instance of the *Poisson equation*:

$$\nabla^2 p = \sigma. \tag{2.6}$$

The Poisson equation can be solved within the class of L-periodic functions provided that $\sigma(\boldsymbol{r})$ has a vanishing average:[1]

$$\langle \sigma \rangle = \frac{1}{L^3} \int_{B_L} \sigma(\boldsymbol{r}) d\boldsymbol{r} = 0. \tag{2.7}$$

Obviously, the function $\sigma = -\partial_{ij}(v_i v_j)$, being made of space-derivatives of periodic functions, possesses the solvability property (2.7).

[1] Angular brackets will denote *space averages* until further notice.

The solution of the Poisson equation is readily obtained by going from the *physical space* ($\boldsymbol{r}$-space) to the *Fourier space* ($\boldsymbol{k}$-space), using Fourier series. We write

$$\sigma(\boldsymbol{r}) = \sum_{\boldsymbol{k}} e^{i\boldsymbol{k}\cdot\boldsymbol{r}}\hat{\sigma}_{\boldsymbol{k}}, \qquad \boldsymbol{k} \in \frac{2\pi}{L}\mathbf{Z}^3, \tag{2.8}$$

$$p(\boldsymbol{r}) = \sum_{\boldsymbol{k}} e^{i\boldsymbol{k}\cdot\boldsymbol{r}}\hat{p}_{\boldsymbol{k}}. \tag{2.9}$$

The Fourier coefficients are given by

$$\hat{\sigma}_{\boldsymbol{k}} = \langle e^{-i\boldsymbol{k}\cdot\boldsymbol{r}}\sigma(\boldsymbol{r})\rangle, \tag{2.10}$$

$$\hat{p}_{\boldsymbol{k}} = \langle e^{-i\boldsymbol{k}\cdot\boldsymbol{r}}p(\boldsymbol{r})\rangle. \tag{2.11}$$

Notice that by (2.7), $\hat{\sigma}_0$ vanishes. It follows from (2.6) that

$$\hat{p}_{\boldsymbol{k}} = -\frac{\hat{\sigma}_{\boldsymbol{k}}}{k^2}, \qquad \boldsymbol{k} \neq 0, \tag{2.12}$$

where k is the modulus of the *wavevector* $\boldsymbol{k}$. The coefficient $\hat{p}_0$ is arbitrary. Indeed, the solution of the Poisson equation is defined up to an additive constant. But, adding a constant to the pressure does not change the Navier–Stokes equation. This (not quite unique) solution will be denoted in the physical space as $\nabla^{-2}\sigma$. Note that, in the physical space, its is a non-local operator (its explicit expression involves a convolution). After the pressure has been eliminated by solving the Poisson equation, the Navier–Stokes equation may be rewritten as

$$\partial_t v_i + \left(\delta_{i\ell} - \partial_{i\ell}\nabla^{-2}\right)\partial_j\left(v_j v_\ell\right) = \nu\nabla^2 v_i. \tag{2.13}$$

It is now sufficient to impose the divergence condition $\partial_j v_j = 0$ at $t = 0$, since (2.13) will propagate this condition to all times.

An alternative way to eliminate the pressure is to work with the *vorticity*

$$\boldsymbol{\omega} = \nabla \wedge \boldsymbol{v}. \tag{2.14}$$

Taking the curl of the Navier–Stokes equation (1.2), and using the identity $\nabla v^2 = 2\boldsymbol{v}\cdot\nabla\boldsymbol{v} + 2\boldsymbol{v}\wedge(\nabla\wedge\boldsymbol{v})$, we obtain the *vorticity equation*:

$$\partial_t\boldsymbol{\omega} = \nabla\wedge(\boldsymbol{v}\wedge\boldsymbol{\omega}) + \nu\nabla^2\boldsymbol{\omega}. \tag{2.15}$$

If we try to rewrite (2.15) in terms, solely, of the vorticity field, we first must solve (2.14) for the velocity. This is done by taking the curl of (2.14) and solving the resulting Poisson equation. Hence, the same non-local operator ∇^{-2} appears as in the velocity formalism of (2.13).

2.2 Symmetries

Theoretical physicists are used to designating as 'symmetries' any discrete or continuous invariance groups of a dynamical theory. We shall also make use of this extended meaning and often use the term symmetry for invariance group. Let $\mathbf{G}$ denote a group of transformations acting on space-time functions $\boldsymbol{v}(t,\boldsymbol{r})$, which are spatially periodic and divergence-less. $\mathbf{G}$ is said to be a *symmetry group* of the Navier–Stokes equation if, for all $\boldsymbol{v}$s which are solutions of the Navier–Stokes equation, and all $g \in \mathbf{G}$, the function $g\boldsymbol{v}$ is also a solution.

Hereafter we give a list of known symmetries of the Navier–Stokes equation.

- *Space-translations* $g_{\boldsymbol{\rho}}^{\text{space}}: t,\boldsymbol{r},\boldsymbol{v} \mapsto t,\boldsymbol{r}+\boldsymbol{\rho},\boldsymbol{v}, \qquad \boldsymbol{\rho} \in \mathbb{R}^3.$
- *Time-translations* $g_{\tau}^{\text{time}}: t,\boldsymbol{r},\boldsymbol{v} \mapsto t+\tau,\boldsymbol{r},\boldsymbol{v}, \qquad \tau \in \mathbb{R}.$
- *Galilean transformations* $g_{\boldsymbol{U}}^{\text{Gal}}: t,\boldsymbol{r},\boldsymbol{v} \mapsto t,\boldsymbol{r}+\boldsymbol{U}t,\boldsymbol{v}+\boldsymbol{U}, \ \boldsymbol{U} \in \mathbb{R}^3.$
- *Parity* $\boldsymbol{P}: t,\boldsymbol{r},\boldsymbol{v} \mapsto t,-\boldsymbol{r},-\boldsymbol{v}.$
- *Rotations*[2] $g_{A}^{\text{rot}}: t,\boldsymbol{r},\boldsymbol{v} \mapsto t,A\boldsymbol{r},A\boldsymbol{v}, \qquad A \in \mathrm{SO}(\mathbb{R}^3).$
- *Scaling*[3] $g_{\lambda}^{\text{scal}}: t,\boldsymbol{r},\boldsymbol{v} \mapsto \lambda^{1-h}t,\lambda\boldsymbol{r},\lambda^{h}\boldsymbol{v}, \qquad \lambda \in \mathbb{R}_{+},\ h \in \mathbb{R}.$

Concerning the notation used in the list, let us observe that it is simpler to write, for example, $t,\boldsymbol{r},\boldsymbol{v} \mapsto t,A\boldsymbol{r},A\boldsymbol{v}$ than the equivalent statement $\boldsymbol{v}(t,\boldsymbol{r}) \mapsto A\boldsymbol{v}(t,A^{-1}\boldsymbol{r})$. We did not write the transformations for the pressure p because the latter can be eliminated from the Navier–Stokes equation. In view of (2.5) the pressure transforms as v^2.

Proofs and comments. The space- and time-translation symmetries are obvious. As for Galilean transformations,[4] we observe that when we substitute $\boldsymbol{v}(t,\boldsymbol{r}-\boldsymbol{U}t)+\boldsymbol{U}$ for $\boldsymbol{v}(t,\boldsymbol{r})$, there is a cancellation of terms between $\partial_t\boldsymbol{v}$ and $\boldsymbol{v}\cdot\nabla\boldsymbol{v}$.

Under parity, all the terms in the Navier–Stokes equation change sign (in particular $\nabla \mapsto -\nabla$). Observe also that the symmetry $\boldsymbol{v} \mapsto -\boldsymbol{v}$ is not consistent with the equations, except when the nonlinear term is negligible. Arbitrary (continuous) rotational invariance is not consistent with periodic boundary conditions, since the latter single out certain directions, so that only a discrete subset of rotations is permitted. As for the scaling transformations, when t is changed into $\lambda^{1-h}t$, $\boldsymbol{r}$ into $\lambda\boldsymbol{r}$, and $\boldsymbol{v}$ into $\lambda^{h}\boldsymbol{v}$, all the terms in the Navier–Stokes equation are multiplied

[2] Only in the limit $L \to \infty$.

[3] Only for $v = 0$.

[4] There is a variant of Galilean invariance in which the velocity $\boldsymbol{U}$ is random and isotropically distributed (Kraichnan 1964, 1965, 1968a). This *random Galilean invariance* will be discussed in Sections 6.2.5, 7.3 and 9.5.

by λ^{2h-1}, except the viscous term which is multiplied by λ^{h-2}. Thus, for finite viscosity, only $h = -1$ is permitted. The corresponding symmetry is then equivalent to the well-known *similarity principle* of fluid dynamics, because the scaling transformations are then seen to keep the Reynolds number unchanged. If we ignore the viscous term or merely let it tend to zero, as may be justified at very high Reynolds numbers (we shall come back to this later), then we find that there are *infinitely many* scaling groups, labeled by their *scaling exponent* h, which can be any real number.

We finally observe that all the listed symmetries, except for the scaling symmetries, are just macroscopic consequences of the basic symmetries of Newton's equations governing microscopic molecular motion (in the classical approximation).

2.3 Conservation laws

It is customary in mechanics to discuss conservation laws together with symmetries. For *conservative* systems describable by a Lagrangian function there is a theorem by Noether (1918) which gives a rationale for this association (see also Goldstein 1980). This theorem states that for each symmetry there is a corresponding conservation law. For instance, momentum conservation corresponds to the invariance of the Lagrangian under space-translations. Such results are not directly relevant for turbulence since the Navier–Stokes equation is *dissipative*.[5] Still, we find it useful to discuss conservation laws at this point. We shall only discuss *global* conservation laws involving an integration over the whole volume occupied by the fluid. Other more local conservation laws, such as the conservation of circulation (see, e.g., Lamb 1932) may be even more important but have found surprisingly few applications to turbulence (so far).

Periodic boundary conditions are assumed as in the previous sections. Angular brackets are used to denote averages over the fundamental periodicity box:

$$\langle f \rangle \equiv \frac{1}{L^3} \int_{B_L} f(\boldsymbol{r}) d\boldsymbol{r}, \tag{2.16}$$

where $f(\boldsymbol{r})$ is an arbitrary periodic function. We list hereafter some useful identities which are readily proved by performing integrations by parts.

[5] The Euler equation, obtained by setting $\nu = 0$, is conservative and possesses various Lagrangian formulations.

All functions are periodic.

$$\langle \partial_i f \rangle = 0. \tag{2.17}$$

$$\langle (\partial_i f) g \rangle = -\langle f \partial_i g \rangle. \tag{2.18}$$

$$\langle (\nabla^2 f) g \rangle = -\langle (\partial_i f)(\partial_i g) \rangle. \tag{2.19}$$

$$\langle \boldsymbol{u} \cdot (\nabla \wedge \boldsymbol{v}) \rangle = \langle (\nabla \wedge \boldsymbol{u}) \cdot \boldsymbol{v} \rangle. \tag{2.20}$$

$$\langle \boldsymbol{u} \cdot \nabla^2 \boldsymbol{v} \rangle = -\langle (\nabla \wedge \boldsymbol{u}) \cdot (\nabla \wedge \boldsymbol{v}) \rangle, \qquad \text{if } \nabla \cdot \boldsymbol{v} = 0. \tag{2.21}$$

We now list the main known conservation laws. We include relations, such as the energy balance equation, which become conservation laws only when the viscosity is set equal to zero.

- Conservation of momentum

$$\frac{d}{dt} \langle \boldsymbol{v} \rangle = 0. \tag{2.22}$$

- Conservation of energy

$$\frac{d}{dt} \left\langle \frac{1}{2} v^2 \right\rangle = -\frac{1}{2} \nu \left\langle \sum_{ij} (\partial_i v_j + \partial_j v_i)^2 \right\rangle = -\nu \langle |\boldsymbol{\omega}|^2 \rangle, \tag{2.23}$$

where $\boldsymbol{\omega} = \nabla \wedge \boldsymbol{v}$.

- Conservation of helicity

$$\frac{d}{dt} \left\langle \frac{1}{2} \boldsymbol{v} \cdot \boldsymbol{\omega} \right\rangle = -\nu \langle \boldsymbol{\omega} \cdot \nabla \wedge \boldsymbol{\omega} \rangle. \tag{2.24}$$

Proofs. Momentum conservation is proved by observing that the advection term $v_j \partial_j v_i$ in (2.1) can be rewritten, using (2.2), as $\partial_j(v_j v_i)$. Thus in the Navier–Stokes equation, all the terms other than $\partial_t v_i$ are spatial derivatives of periodic functions. Hence, (2.22) follows from (2.17). For the energy balance relation (2.23), we multiply (2.1) by v_i and use (2.2) to obtain

$$\partial_t \frac{v_i v_i}{2} + \partial_j \frac{v_j v_i v_i}{2} = -v_i \partial_i p + \nu v_i \nabla^2 v_i, \tag{2.25}$$

from which (2.23) follows by use of (2.2), (2.17), (2.18) and (2.21). For the helicity balance relation (2.24), we start from the vorticity equation (2.15) and take the scalar product with $\boldsymbol{v}$, average and observe that, by (2.20),

$$\frac{d}{dt} \langle \boldsymbol{v} \cdot \boldsymbol{\omega} \rangle = 2 \langle \boldsymbol{v} \cdot (\partial_t \boldsymbol{\omega}) \rangle. \tag{2.26}$$

Thus we obtain

$$\frac{d}{dt}\left\langle \frac{\boldsymbol{v}\cdot\boldsymbol{\omega}}{2}\right\rangle = \langle \boldsymbol{v}\cdot\nabla\wedge(\boldsymbol{v}\wedge\boldsymbol{\omega})\rangle + \nu\langle \boldsymbol{v}\cdot\nabla^2\boldsymbol{\omega}\rangle, \tag{2.27}$$

from which the helicity relation (2.24) follows by use of (2.20) and (2.21). *QED.*

We introduce now some important notation:

$$\left.\begin{aligned} E \equiv \left\langle \frac{1}{2}|\boldsymbol{v}|^2\right\rangle, \quad \Omega \equiv \left\langle \frac{1}{2}|\boldsymbol{\omega}|^2\right\rangle, \\ H \equiv \left\langle \frac{1}{2}\boldsymbol{v}\cdot\boldsymbol{\omega}\right\rangle, \quad H_\omega \equiv \left\langle \frac{1}{2}\boldsymbol{\omega}\cdot\nabla\wedge\boldsymbol{\omega}\right\rangle. \end{aligned}\right\} \tag{2.28}$$

The energy and helicity balance equations may thus be written:

$$\frac{d}{dt}E = -2\nu\Omega, \qquad \frac{d}{dt}H = -2\nu H_\omega. \tag{2.29}$$

It is standard usage to call E the *mean energy*,[6] H the *mean helicity* and Ω the *mean enstrophy*.[7] (The word 'mean' is often omitted.) As for the quantity H_ω, it might be called the *mean vortical helicity*. The *mean energy dissipation* (per unit mass)

$$\varepsilon \equiv -\frac{dE}{dt}, \tag{2.30}$$

is one of the most frequently used quantities in turbulence.

Remarks

- In deriving the conservation laws, we assumed that the velocity and the pressure fields were sufficiently smooth to permit all necessary manipulations, such as integrations by parts, derivatives of products, etc. This sort of smoothness is generally conjectured to hold for any finite positive viscosity. For the solutions of the Euler equation ($\nu = 0$) it may not hold and energy conservation may break down, as first observed by Onsager (1949). Increasingly weak smoothness conditions ensuring energy conservation for the Euler equation have been obtained by Sulem and Frisch (1975), Eyink (1994a) and Constantin, E and Titi (1994).

[6] It is actually the mean energy per unit mass, but in an incompressible fluid with a constant density this distinction is not important.

[7] The term *enstrophy* was coined by C. Leith by analogy with en–ergy. Observe that in modern Greek στρωφη designates the curl operation.

- Note that $\nu|\omega|^2$ and the *local dissipation*,

$$\varepsilon_{\text{loc}} = \frac{1}{2}\nu\sum_{ij}\left(\partial_i v_j + \partial_j v_i\right)^2, \tag{2.31}$$

have the same space average. They actually differ by a term proportional to the Laplacian of pressure. Indeed, from (2.5), we obtain

$$\nabla^2 p = \frac{1}{4}\sum_{ij}\left(\partial_i v_j - \partial_j v_i\right)^2 - \frac{1}{4}\sum_{ij}\left(\partial_i v_j + \partial_j v_i\right)^2. \tag{2.32}$$

Only the quantity ε_{loc}, which involves the rate-of-strain tensor (the symmetric part of the velocity gradient), deserves to be called a local dissipation. Indeed, in a region of quasi-uniform vorticity there is an almost solid rotation of the fluid and hence no dissipation.

- The energy balance equation plays a crucial role in proving the existence 'in the large' (for all times) for the three-dimensional Navier–Stokes equation. Unfortunately, uniqueness in the large is proven only in two dimensions. In two dimensions, there is indeed an additional balance equation for the enstrophy:

$$\frac{d}{dt}\Omega = -2\nu P, \qquad P \equiv \left\langle \frac{1}{2}|\nabla \wedge \omega|^2 \right\rangle. \tag{2.33}$$

The quantity P is called the *mean palinstrophy.* Eq. (2.33) also has important consequences for *two-dimensional turbulence*, a subject which is mostly outside the scope of the present book (see Section 9.7).

- The conservation of helicity (for $\nu = 0$) was discovered by Moreau (1961); its potential for fluid dynamics was recognized by Moffatt (1969). More recently it was found by Kuz'min (1983) and Oseledets (1989) that there is an associated *material invariant*, i.e. a quantity which is conserved along any fluid particle trajectory (see also Gama and Frisch 1993). The influence of helicity on magnetic field generation was recognized by Steenbeck, Krause and Rädler (1966). The possibility of helicity cascades analogous to energy cascades was discussed by Brissaud, Frisch, Léorat, Lesieur and Mazure (1973). Further references on helicity may be found in Moffatt and Tsinober (1992).

2.4 Energy budget scale-by-scale

The energy balance equation (2.23) does not contain any contribution from the nonlinear term in the Navier–Stokes equation. Actually, the same relation would hold if we had started from the (vector) heat equation. What is then the role of nonlinearities in relation to the energy? We shall now show that *the nonlinear term redistributes energy among the various scales of motion* without affecting the global energy budget.

We need a definition for the concept of 'scale'. For this, let us consider again Fig. 1.14. If this is shown using an overhead projector somewhat out of focus, the finest details will be blurred. Defocusing amounts approximately to a linear filtering which removes or attenuates high harmonics in the spatial Fourier decomposition of the image above a cutoff K which depends on the defocusing. There is an associated scale $\ell \sim K^{-1}$ over which there is smoothing. Let us now formalize this idea, restricting ourselves to L-periodic functions. Given a function f and its Fourier series

$$f(\boldsymbol{r}) = \sum_{\boldsymbol{k}} \hat{f}_{\boldsymbol{k}} e^{i\boldsymbol{k}\cdot\boldsymbol{r}}, \qquad \boldsymbol{k} \in \frac{2\pi}{L}\mathbf{Z}^3, \tag{2.34}$$

we define two families of functions depending on $\boldsymbol{r}$ and on the additional parameter $K > 0$. The *low-pass filtered* function is

$$f_K^<(\boldsymbol{r}) \equiv \sum_{k \le K} \hat{f}_{\boldsymbol{k}} e^{i\boldsymbol{k}\cdot\boldsymbol{r}}, \tag{2.35}$$

and the *high-pass filtered* function is

$$f_K^>(\boldsymbol{r}) \equiv \sum_{k > K} \hat{f}_{\boldsymbol{k}} e^{i\boldsymbol{k}\cdot\boldsymbol{r}}. \tag{2.36}$$

The length $\ell = K^{-1}$ will be called the *scale* of the filtering. Obviously,

$$f(\boldsymbol{r}) = f_K^<(\boldsymbol{r}) + f_K^>(\boldsymbol{r}). \tag{2.37}$$

Note that $f^<$ and $f^>$ are pronounced f 'lesser' and f 'greater', respectively. The decomposition (2.37) was used for the first time by Obukhov (1941b).

To illustrate the idea of low/high-pass filtering, let us consider the example of the one-dimensional function shown in Fig. 2.2(a). We have here deliberately chosen a function which possesses structures on two very different scales: a small scale (of the order of a few millimeters) and a large scale (of the order of a few centimeters). Let us choose $\ell = K^{-1}$ to be intermediate, say, about 1 cm. The corresponding low- and high-pass filtered functions are shown in Figs. 2.2(b) and (c).

A word of warning may be needed here. The functions $f_K^<(\boldsymbol{r})$ and $f_K^>(\boldsymbol{r})$ are not Fourier transforms of $f(\boldsymbol{r})$: they still depend on the same space variable $\boldsymbol{r}$ as $f(\boldsymbol{r})$, but they also depend on an additional scale variable.

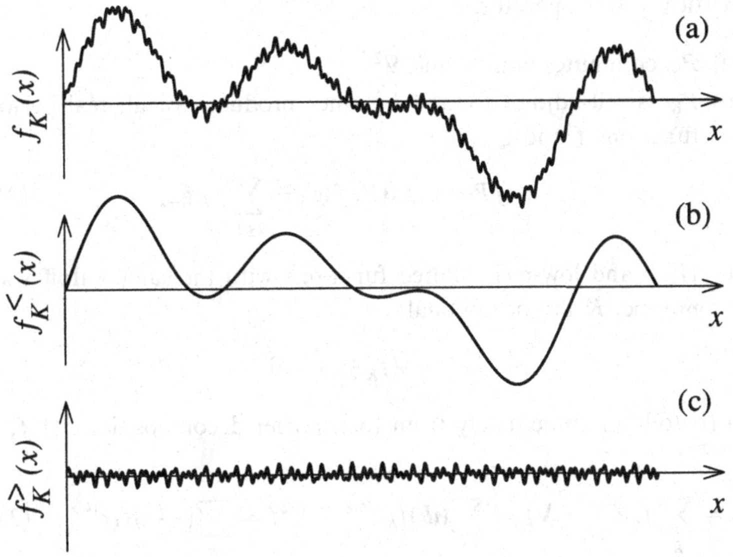

Fig. 2.2. Signal (a) subject to low-pass filtering (b) and high-pass filtering (c).

The passage from f to $f^>$ may be generalized to filters of arbitrary shape. It is then known as a *wavelet transform*.[8] For the purpose of the present book, we shall not need more than the sharp (low/high-pass) filters.

When this concept of filtering is applied to a three-dimensional turbulent velocity field, we obtain two functions $\boldsymbol{v}_K^<(\boldsymbol{r})$ and $\boldsymbol{v}_K^>(\boldsymbol{r})$. The former is conveniently identified as *eddies of scale larger than* ℓ and the latter as *eddies of scale less than* ℓ.[9]

Before turning to the scale-by-scale energy budget, we need a few technical results.

We define the low-pass filtering operator

$$P_K : \quad f(\boldsymbol{r}) \mapsto f_K^<(\boldsymbol{r}). \tag{2.38}$$

This operator sets to zero all Fourier components with wavenumber greater than K. Clearly, P_K is a projector: $P_K^2 = P_K$. We list some useful

8 For the use of wavelets in turbulence, see Farge (1992) and references therein.

9 Standard usage is to speak of small and large 'eddies' as a loose phenomenological concept. Here, we find it convenient actually to provide a definition.

properties of this operator.

(i) P_K commutes with ∇ and ∇^2.

(ii) P_K is self-adjoint for the L^2 inner product: for all real periodic functions f and g,

$$\langle f P_K g \rangle = \langle (P_K f) g \rangle = \sum_{k \le K} \hat{f}_k \, \hat{g}_{-k}. \tag{2.39}$$

(iii) High and low-pass filtered functions with the same cutoff wavenumber K are orthogonal:

$$\langle f_K^> \, g_K^< \rangle = 0. \tag{2.40}$$

Item (i) follows immediately from the Fourier decompositions of f, ∇f and $\nabla^2 f$:

$$f = \sum_k \hat{f}_k e^{i\boldsymbol{k}\cdot\boldsymbol{r}}, \quad \nabla f = \sum_k (i\boldsymbol{k}) \hat{f}_k e^{i\boldsymbol{k}\cdot\boldsymbol{r}}, \quad \nabla^2 f = \sum_k (-k^2) \hat{f}_k e^{i\boldsymbol{k}\cdot\boldsymbol{r}}. \tag{2.41}$$

Items (ii) and (iii) are consequences of Parseval's identity:

$$\langle f g \rangle = \sum_k \hat{f}_k \, \hat{g}_{-k}. \tag{2.42}$$

We return now to the Navier–Stokes equation and write it in a slightly more general form, including a forcing term:[10]

$$\partial_t \boldsymbol{v} + \boldsymbol{v} \cdot \nabla \boldsymbol{v} = -\nabla p + \nu \nabla^2 \boldsymbol{v} + \boldsymbol{f}, \tag{2.43}$$

$$\nabla \cdot \boldsymbol{v} = 0. \tag{2.44}$$

The force $\boldsymbol{f}$ is assumed to be periodic in the space variable, it may depend also on the time and the velocity.[11]

We apply P_K to (2.43), use (2.37) and item (i), to obtain

$$\left.\begin{aligned} \partial_t \boldsymbol{v}_K^< + P_K \left(\boldsymbol{v}_K^< + \boldsymbol{v}_K^>\right) \cdot \nabla \left(\boldsymbol{v}_K^< + \boldsymbol{v}_K^>\right) & \\ = -\nabla p_K^< + \nu \nabla^2 \boldsymbol{v}_K^< + \boldsymbol{f}_K^<, & \\ \nabla \cdot \boldsymbol{v}_K^< = 0. & \end{aligned}\right\} \tag{2.45}$$

Now we take the scalar product of (2.45) with $\boldsymbol{v}_K^<$, average and use

[10] The rationale for introducing a driving force into the Navier–Stokes equation will be discussed in Section 6.2.1.

[11] The simplest two-dimensional example was introduced by Kolmogorov in the late 1950s; the force is $\boldsymbol{f} = (0, \sin x_1)$ and the flow is known as the 'Kolmogorov flow'. It was studied first by Meshalkin and Sinai (1961); see also Section 9.6.3.2.

items (i), (ii) and (iii) to obtain an energy budget of the low-pass filtered velocity:

$$\partial_t \left\langle \frac{|\boldsymbol{v}_K^<|^2}{2} \right\rangle + \left\langle \boldsymbol{v}_K^< \cdot \left[\left(\boldsymbol{v}_K^< + \boldsymbol{v}_K^> \right) \cdot \nabla \left(\boldsymbol{v}_K^< + \boldsymbol{v}_K^> \right) \right] \right\rangle = - \left\langle \boldsymbol{v}_K^< \cdot \nabla p_K^< \right\rangle + \nu \left\langle \boldsymbol{v}^< \cdot \nabla^2 \boldsymbol{v}_K^< \right\rangle + \left\langle \boldsymbol{v}_K^< \cdot \boldsymbol{f}_K^< \right\rangle. \tag{2.46}$$

Eq. (2.46) can be simplified using some of the same transformations as for the energy balance equation (Section 2.3). For example, we have $\langle \boldsymbol{v}^< \cdot \nabla p_K^< \rangle = 0$ and $\nu \langle \boldsymbol{v}_K^< \cdot \nabla^2 \boldsymbol{v}_K^< \rangle = -\nu \langle |\boldsymbol{\omega}_K^<|^2 \rangle$. The main difference is that now the contribution from the nonlinear term in the Navier–Stokes equation (the second term on the l.h.s.) does not vanish. When this term is expanded, it produces four terms, of which two are identically vanishing:

$$\langle \boldsymbol{v}_K^< \cdot (\boldsymbol{v}_K^< \cdot \nabla \boldsymbol{v}_K^<) \rangle = \langle \boldsymbol{v}_K^< \cdot (\boldsymbol{v}_K^> \cdot \nabla \boldsymbol{v}_K^<) \rangle = 0. \tag{2.47}$$

Eq. (2.47) is proved using incompressibility, (2.17) and (2.18). The vanishing of the leftmost side means that interactions among 'lesser' scales cannot change the energy content of the lessers. Similarly, the vanishing of the middle term means that advection of lessers by greaters does not change the energy content of the lessers. Collecting the remaining terms from (2.46), we obtain the *scale-by-scale energy budget equation*:

$$\partial_t \mathscr{E}_K + \Pi_K = -2\nu \Omega_K + \mathscr{F}_K. \tag{2.48}$$

Here, we have introduced the *cumulative energy* between wavenumber 0 and K:

$$\mathscr{E}_K \equiv \frac{1}{2} \langle |\boldsymbol{v}_K^<|^2 \rangle = \frac{1}{2} \sum_{k \le K} |\hat{\boldsymbol{v}}_k|^2, \tag{2.49}$$

the *cumulative enstrophy*:

$$\Omega_K \equiv \frac{1}{2} \langle |\boldsymbol{\omega}_K^<|^2 \rangle = \frac{1}{2} \sum_{k \le K} k^2 |\hat{\boldsymbol{v}}_k|^2, \tag{2.50}$$

the *cumulative energy injection* (by the force):

$$\mathscr{F}_K \equiv \langle \boldsymbol{f}_K^< \cdot \boldsymbol{v}_K^< \rangle = \sum_{k \le K} \hat{\boldsymbol{f}}_k \cdot \hat{\boldsymbol{v}}_{-k}, \tag{2.51}$$

and the *energy flux*[12] through wavenumber K:

$$\Pi_K \equiv \langle \boldsymbol{v}_K^< \cdot (\boldsymbol{v}_K^< \cdot \nabla \boldsymbol{v}_K^>) \rangle + \langle \boldsymbol{v}_K^< \cdot (\boldsymbol{v}_K^> \cdot \nabla \boldsymbol{v}_K^>) \rangle. \tag{2.52}$$

[12] An alternative expression for the energy flux will be given in Section 6.2.2.

Eq. (2.48) can be interpreted as follows: the rate of change of the energy at scales down to $\ell = K^{-1}$ is equal to the energy injected at such scales by the force ($\mathscr{F}_K$) minus the energy dissipated at such scales ($2\nu\Omega_K$) minus the flux of energy (Π_K) to smaller scales due to nonlinear interactions. As we shall see later, at high Reynolds numbers it is typical to have the energy injection confined to large scales ($O(\ell_0)$) and the energy dissipation confined to small scales ($O(\ell_d)$) with $\ell_d \ll \ell_0$.

We finally observe that it is traditional in turbulence theory to derive an energy budget equation in Fourier space under the (restrictive) assumptions of statistical homogeneity and isotropy (see Chapter 3 for these notions). In contrast, the scale-by-scale energy budget equation derived above makes no use of probabilistic tools. It is therefore applicable to a much wider range of situations.

3
Why a probabilistic description of turbulence?

3.1 There is something predictable in a turbulent signal

In Chapter 1 we presented some pictures chosen to prompt the study of the symmetries of the Navier–Stokes equation. However important flow visualizations may be, experimental data on turbulence also include a considerable body of quantitative results. Velocimetry, the measurement of the flow velocity (or one component thereof) at a given point as a function of time, is by far the most common way of getting quantitative information. There are many different techniques of velocimetry which we shall not review here.

Let us turn directly to an example. Fig. 3.1(a) shows a one-second signal obtained from a hot-wire probe placed in the very large wind tunnel S1 of ONERA.[1] The signal is the 'streamwise' velocity (component parallel to the mean flow). It is sampled five thousand times per second (5 kHz). The mean flow has been subtracted so that the signal appears to fluctuate around zero.

What strikes us when looking at this signal?

(i) The signal appears highly *disorganized* and presents structures on all scales.
(ii) The signal appears *unpredictable* in its detailed behavior.
(iii) Some properties of the signal are quite *reproducible.*

Regarding item (i), we observe that in contrast to the signal shown in Fig. 2.2 which had only two scales present, the signal shown here displays structures on all scales: the eye directly perceives structures with time-scales of the order of one second, of one-tenth of a second, of one-hundredth of a second, and possibly smaller.

[1] We shall come back later to some of the characteristics of this wind tunnel.

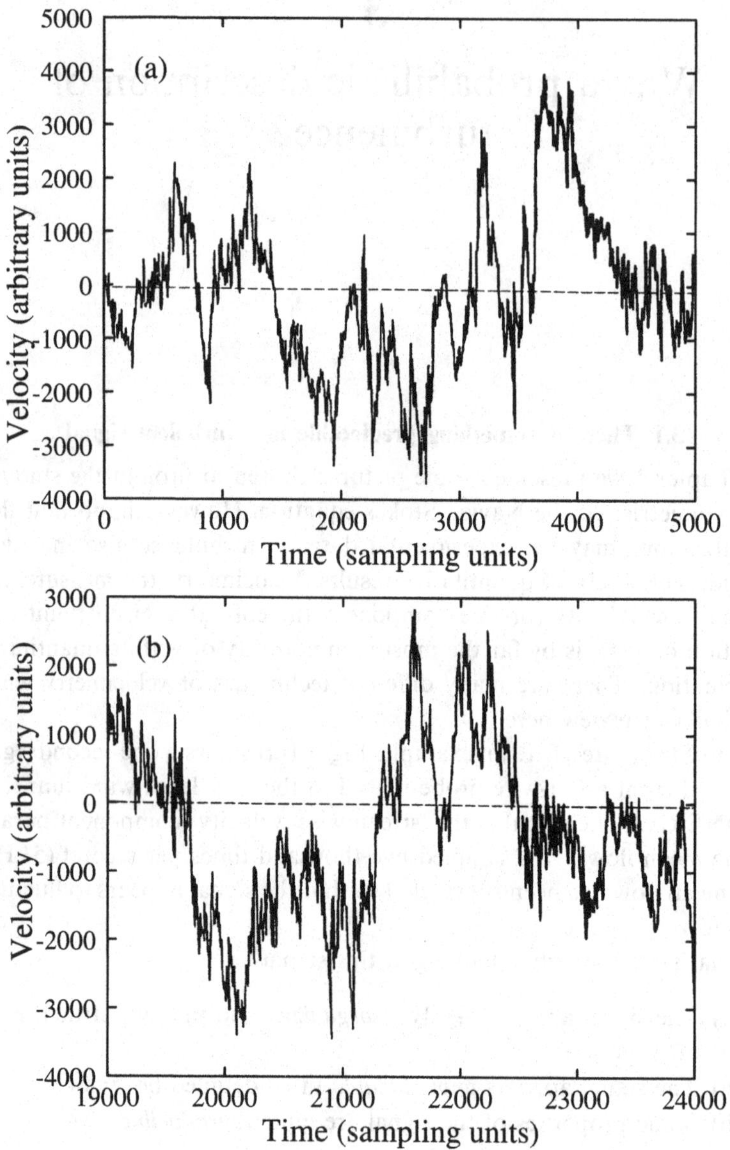

Fig. 3.1. One second of a signal recorded by a hot-wire (sampled at 5 kHz) in the S1 wind tunnel of ONERA (a); same signal, about four seconds later (b). Courtesy Y. Gagne and E. Hopfinger.

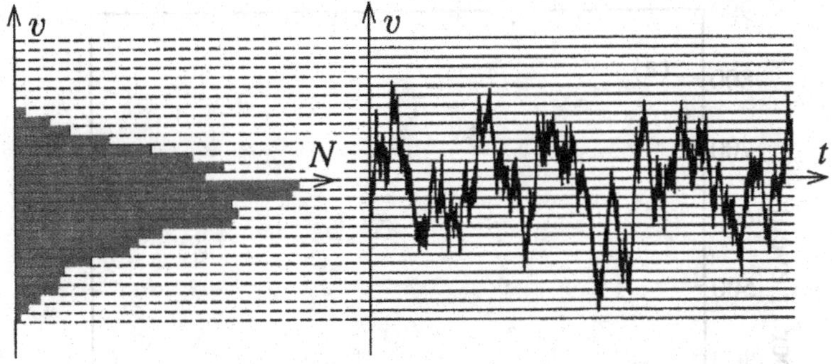

Fig. 3.2. Construction of the histogram of a signal by binning.

Regarding item (ii), let us look at a sample of the same duration taken about four seconds later (Fig. 3.1(b)). The general aspect is the same but all the details are different and could not have been predicted from looking at the previous figure.

Regarding item (iii), one instance of a reproducible property is the *histogram* of the signal. As shown in Fig. 3.2, let us take a finite record of the (discretely sampled) signal and divide the v-axis into a large number N of equal small bins centered around values v_i $(i = 1, \cdots, N)$. The histogram is defined as the function N_i giving the number of times the ith bin is visited. Let us apply this procedure to the signal from the S1 wind tunnel. Fig. 3.3(a) shows the histogram obtained from a record of duration 150 s sampled 5000 times with 100 bins.[2] Fig. 3.3(b) shows the same sort of histogram taken from a record of the same duration but several minutes later. (Several hours would work equally well.) We see that the two histograms are essentially identical.

We can summarize our findings by saying that although the detailed properties of the signal appear not to be predictable, its *statistical properties* are reproducible. Such observations, which have been known for a long time, have induced theoreticians to look for a *probabilistic description* of turbulence (Taylor 1935, 1938). However, we know that the basic equation (Navier–Stokes) is *deterministic*: Although there is no rigorous proof of this, it is widely conjectured that for a given initial condition there is a unique solution for all times. How can *chance* or *chaos* arise in

[2] Why the record has to be much longer than before will become clear in Section 4.4 once we have introduced the concept of integral time scale.

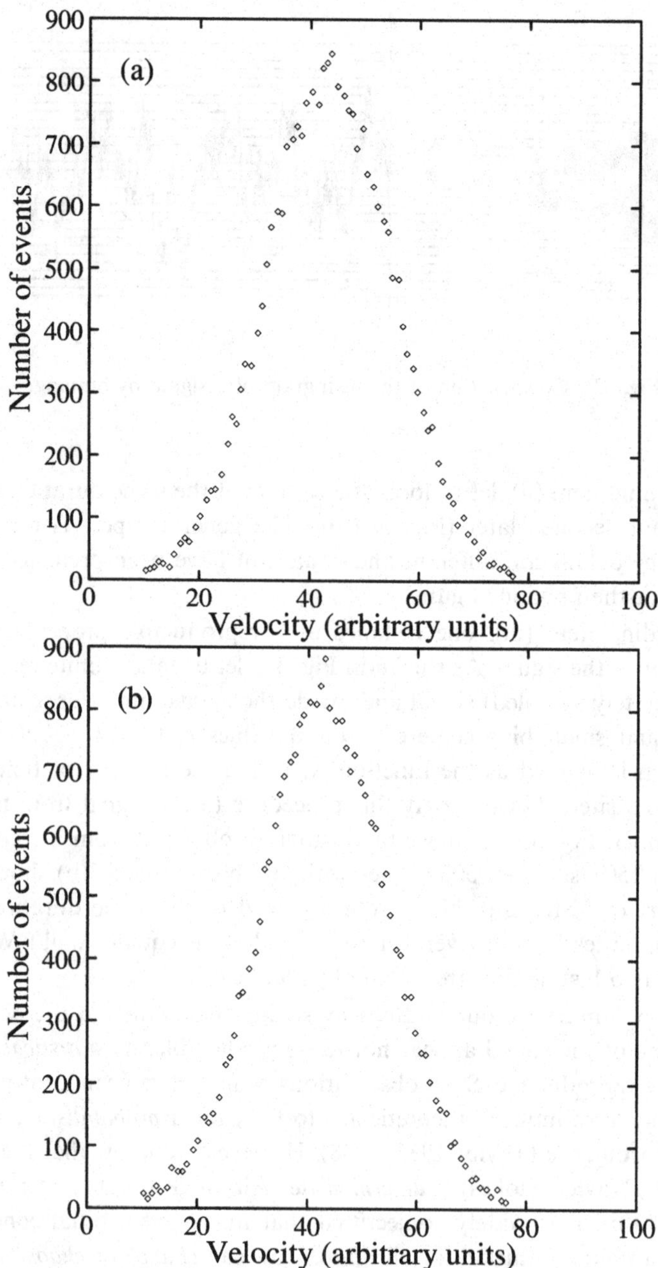

Fig. 3.3. Histogram for same signal as in Fig. 3.1(a), sampled 5000 times over a time-span of 150 seconds (a); same histogram, a few minutes later (b).

a purely deterministic context?[3] To give some insight into this question in a context which keeps a flavor of the Navier–Stokes equation, we shall now discuss a toy model.

3.2 A model for deterministic chaos

In this section we shall study the following discrete map:

$$v_{t+1} = 1 - 2v_t^2, \qquad v_0 = \varpi, \quad t = 0, 1, 2, \ldots. \tag{3.1}$$

Here, v_t is a real number between -1 and $+1$ and the time t is discrete. The tth iterate starting from the initial value ϖ is denoted by $v(t, \varpi)$. The set of iterates of a given initial value is known as its *orbit*. The map (3.1) is an instance of the logistic map $v \mapsto v - av^2$. Here, we could as well call it the *poor man's Navier–Stokes equation.* Let us indeed rewrite the map in a way paralleling the Navier–Stokes equation written directly underneath:

$$\left.\begin{array}{rcccccc} v_{t+1} - v_t = & -2v_t^2 & - & v_t & + & 1 \\ \partial_t \boldsymbol{v} & = -(\boldsymbol{v} \cdot \nabla \boldsymbol{v} + \nabla p) & + & \nu \nabla^2 \boldsymbol{v} & + & \boldsymbol{f}. \end{array}\right\} \tag{3.2}$$

Written in this way, our logistic map has the equivalent of the nonlinear term, the viscous term and the force term. Of course, the simple map has no spatial structure whatsoever.

We now define:

$$G : v_t \mapsto v_{t+1}, \tag{3.3}$$

and

$$G_\tau \equiv G^\tau : v_t \mapsto v_{t+\tau}. \tag{3.4}$$

Thus,

$$v_t(\varpi) = G_t\, \varpi. \tag{3.5}$$

With the poor man's Navier–Stokes equation we can repeat the same sort of experiment as performed with the wind-tunnel data. We choose an arbitrary initial condition ϖ (between -1 and $+1$) and iterate many times, say 5000. From these iterates we can then construct the histogram which is shown in Fig. 3.4. If we repeat the process with 5 000 consecutive iterates taken much later (say iterate numbers 20 000–25 000), we obtain again essentially the same histogram.

The reason we chose the particular map defined by (3.1) is that it

[3] 'Chance' (*Le hasard*) is the word used by Henri Poincaré in the introduction to his 'Calcul des probabilités'; nowadays, in deterministic situations, we say 'chaos'.

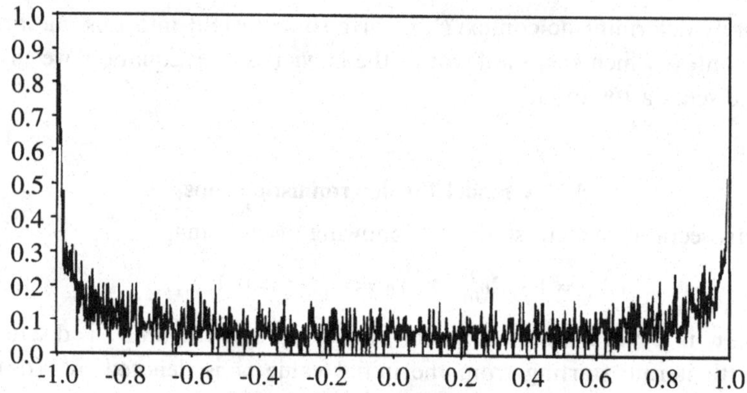

Fig. 3.4. Normalized histogram of the values of v obtained by iterating (3.1) (Ruelle 1989).

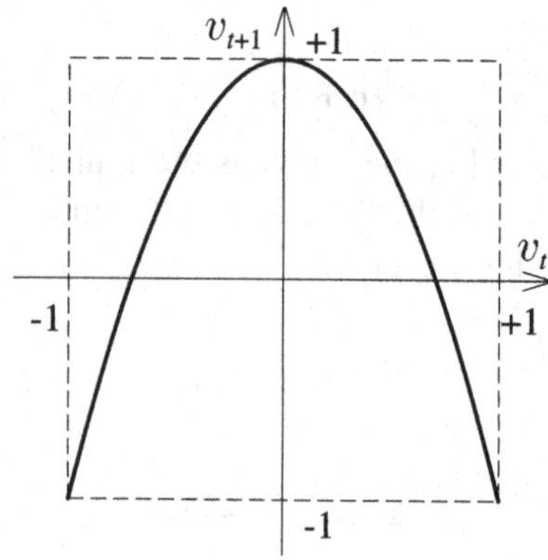

Fig. 3.5. The graph of the 'poor man's Navier–Stokes' map (3.1).

is now possible to understand the reason for this reproducibility and thereby get an insight into the behavior of a large class of nonlinear deterministic systems.

First, we shall relate the map (3.1) to a simpler map. In Fig. 3.5 we have drawn the graph of the map. Observe that it falls within a square of side two, centered at the origin. Let us make the following change of variable:

$$v_t = \sin\left(\pi x_t - \frac{\pi}{2}\right), \qquad 0 \le x_t \le 1, \tag{3.6}$$

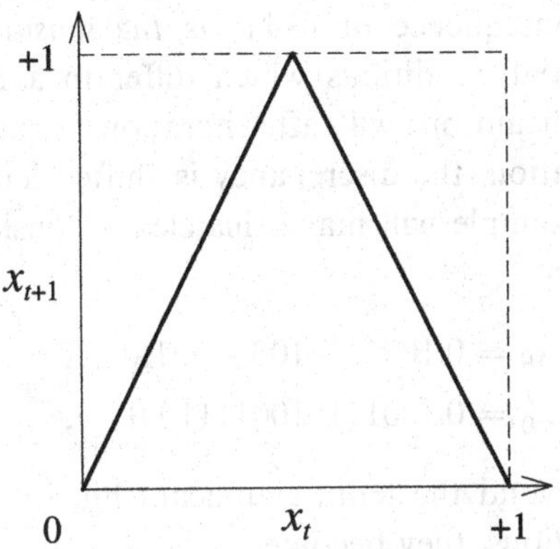

Fig. 3.6. The tent map (3.8).

and similarly:

$$v_{t+1} = \sin\left(\pi x_{t+1} - \frac{\pi}{2}\right). \tag{3.7}$$

An elementary calculation then gives

$$x_{t+1} = \begin{cases} 2x_t & \text{for } 0 \le x_t \le \frac{1}{2} \\ 2 - 2x_t & \text{for } \frac{1}{2} \le x_t \le 1. \end{cases} \tag{3.8}$$

We shall denote by B the map $x_t \mapsto x_{t+1}$. This is known as the *tent map*, because of the shape of its graph (shown in Fig. 3.6). Thus the map (3.1) and the tent map are *conjugate*: if we know how to iterate one of them, the iterates of the other are readily obtained from (3.6).

Actually, it is quite easy to iterate the tent map. For this we use the binary decomposition of real numbers between 0 and 1. Let α_1, α_2,... denote binary digits taking the values 0 or 1 and let N denote the *negation* which interchanges 0 and 1. It is then a simple exercise to check that if x has the binary decomposition

$$x = 0.\alpha_1\alpha_2\alpha_3\ldots = \alpha_1 2^{-1} + \alpha_2 2^{-2} + \alpha_3 2^{-3} + \ldots, \tag{3.9}$$

then its image Bx by the tent map has the decomposition

$$Bx = 0.(N^{\alpha_1}\alpha_2)(N^{\alpha_1}\alpha_3)(N^{\alpha_1}\alpha_4)\ldots. \tag{3.10}$$

This relation is easily iterated to give

$$\left.\begin{aligned} B^t x &= 0.(P\alpha_{t+1})(P\alpha_{t+2})(P\alpha_{t+3})\ldots, \\ P &= N^{\alpha_1+\alpha_2+\cdots+\alpha_t}. \end{aligned}\right\} \tag{3.11}$$

An immediate consequence of (3.11) is *the sensitivity to the initial conditions.* Two initial conditions which differ in a minute way (say, beyond the nth significant bit) will, after iterations, separate very quickly. Indeed, at each iteration, the discrepancy is shifted left and thus grows by a factor 2. An example will make this clear. Consider the following two initial conditions:

$$x_0 = 0.10011101001011010\ldots, \tag{3.12}$$

$$x_0' = 0.10011101001111001\ldots, \tag{3.13}$$

which differ only beyond the tenth significant bit, i.e. by about $2^{-10} \simeq 10^{-3}$. After ten iterations, they become

$$x_{10} = 0.0100101\ldots, \tag{3.14}$$

$$x_{10}' = 0.0000110\ldots. \tag{3.15}$$

The orbits have now completely separated. It is this sensitivity to initial conditions which is often loosely referred to as *chaos.*

Another important property of the tent map is the existence of an *invariant measure.* Suppose that we select x_0 at random in the interval $[0,1]$ with a uniform distribution; then all the iterates will also have a uniform distribution.

It is obviously enough to prove this assertion for the first iterate $x = Bx_0$. The statement that x_0 is uniformly distributed is tantamount to

$$\text{Prob}\{x_0 \in [a,b]\} = b - a, \qquad \forall\, 0 \le a \le b \le 1, \tag{3.16}$$

where $\text{Prob}\{\cdot\}$ denotes the probability of an event. In other words the probability measure is just the Lebesgue measure dx_0. To find how this probability measure transforms under the tent map B we must use the relation

$$\text{Prob}\{Bx_0 \in [a,b]\} = \text{Prob}\{x_0 \in B^{-1}[a,b]\}, \tag{3.17}$$

which expresses the conservation of probability. In (3.17) $B^{-1}[a,b]$ denotes the preimage under the tent map of the interval $[a,b]$, i.e. the set of points which are mapped into $[a,b]$. To understand this preimage it is useful to draw a picture (Fig. 3.7). It is seen that the preimage of $[a,b]$ is made of two disjoint intervals, each half the length of the original interval. This immediately implies the invariance of the uniform measure. In other words, the Lebesgue measure is an invariant measure for the tent map.

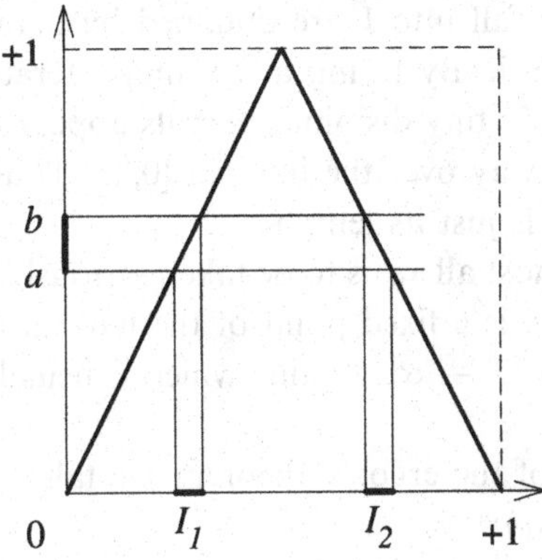

Fig. 3.7. Construction of the preimages of an interval $[a, b]$ for the tent map.

At this point, the reader may have the feeling that probabilities were introduced through the back door. Given a purely deterministic system such as the tent map, why should we decide to resort to a probabilistic description with the initial value x_0 *selected at random*? The answer is that it (almost surely) does not matter if the initial value is deterministic or random. Indeed there is an important result called Birkhoff's ergodic theorem which states the following (roughly).[4]

Ergodic theorem for the tent map. *Let* $f(x)$ *be an integrable function defined in the interval* $[0, 1]$. *For almost all* x_0

$$\lim_{T\to\infty} \frac{1}{T} \sum_{t=0}^{t=T} f\left(B^t x_0\right) = \int_0^1 f(x)dx. \tag{3.18}$$

The ergodic theorem states that the 'time average' of $f(\cdot)$ along the orbit of almost all initial x_0 is equal to its *ensemble average* calculated with the invariant measure (the uniform measure). Thus the deterministic tent map behaves in an essentially probabilistic way.

For the proof of the ergodic theorem we refer the reader to textbooks on ergodic theory such as Halmos (1956). A feeling for why the theorem holds can be gotten as follows. Since (3.18) is linear in $f(\cdot)$, we can without loss of generality suppose that f is equal to 1 in a small interval $I = [a, b]$ and zero outside. The l.h.s. of (3.18) is then just the average fraction of the time the orbit of x_0 visits the interval I. The points x_0 such

[4] Since a reader interested in turbulence may not necessarily be familiar with measure-theoretic jargon, we shall generally water down our statements using, e.g., 'integrable' instead of 'measurable'. We apologize for this to the more mathematically minded reader.

that their tth iterates fall into I are obtained by iterating the preimage construction of Fig. 3.7. By trying a few more iterates, the reader will see that the resulting 2^t tiny disjoint intervals appear to be spread in an increasingly uniform way over the interval $[0, 1]$. Thus the average time the orbit spends in I is just its length.

The restriction 'almost all x_0' is to be taken seriously. Actually, suppose we take $x_0 = 0$, which is a fixed point of the tent map, then the l.h.s. of (3.18) tends to $f(0)$ as $T \to \infty$, a value which is usually not equal to the r.h.s..

As an illustration of the ergodic theorem, we take $f(x) = x^n$. We then have (for almost all x_0):

$$\lim_{T\to\infty} \frac{1}{T} \sum_{0}^{T} (x_t)^n = \int_0^1 x^n dx = \frac{1}{n+1}. \tag{3.19}$$

We are now in a position to return to the original poor man's Navier–Stokes equation, the map (3.1). Since it is conjugated to the tent map by the transformation

$$x \mapsto \sin\left(\pi x - \frac{\pi}{2}\right), \tag{3.20}$$

it will also possess an invariant measure, which is the image of the uniform measure by (3.20). An elementary calculation leads to the invariant measure $P(v)dv$ with the *probability density function* (p.d.f.) $P(v)$ given by

$$P(v) = \frac{1}{\pi\sqrt{1 - v^2}}. \tag{3.21}$$

Except for a normalization factor (the integral of a p.d.f. is 1), this is the same as the histogram shown in Fig. 3.4.

3.3 Dynamical systems

Birkhoff's theorem, which allows us to replace time averages over one orbit by ensemble averages, is valid in a much broader context than the two maps considered in Section 3.2. The appropriate framework is that of *dynamical systems.* We here introduce the following definition:

Definition. *A dynamical system is a quadruplet* $(\Omega, \mathscr{A}, P, G_t)$. *The set* Ω *is called the* probability space.[5] $\mathscr{A}$ *is a family of subsets*[6] *of* Ω. P, *the*

[5] In this and the next chapter Ω has its standard probabilistic meaning; elsewhere Ω will denote the enstrophy.

[6] Actually, a σ-algebra but, as we stated before, we do not intend to go into measure-theoretic fine points.

probability measure, *maps $\mathscr{A}$ to the real numbers between 0 and 1 and satisfies*

$$P(A) \geq 0 \quad \forall A \in \mathscr{A}, \qquad P\left(\cup_i A_i\right) = \sum_i P(A_i), \qquad P(\Omega) = 1, \tag{3.22}$$

where A_i is any enumerable set of disjoint sets $\in \mathscr{A}$. The time-shifts, G_t, *are a family of operators depending on a variable $t \geq 0$ which can be either continuous or discrete. The G_ts satisfy the semi-group property*

$$G_0 = I, \qquad G_t G_{t'} = G_{t+t'} \tag{3.23}$$

and conserve the probability:

$$P\left(G_t^{-1} A\right) = P(A), \qquad \forall t \geq 0, \quad \forall A \in \mathscr{A}. \tag{3.24}$$

The special case of the tent map corresponds to the following choices: $\Omega = [0, 1]$, P is the Lebesgue measure dx and G_t is the tth iterate of the tent map.

In the general framework of dynamical systems, *Birkhoff's ergodic theorem* states (roughly) the following. The basic hypothesis is that the only sets in $\mathscr{A}$ which are globally invariant under the time shifts G_t are those of measure zero and one (e.g., the empty set or the entire set Ω).[7] It then follows that for any integrable function f defined on Ω and for almost all $\varpi' \in \Omega$,

$$\lim_{T\to\infty} \frac{1}{T} \int_0^T f\left(G_t \varpi'\right) dt = \int_\Omega f(\varpi)\, dP \equiv \langle f \rangle. \tag{3.25}$$

Eq. (3.25) is seen to be a generalization of (3.18).

3.4 The Navier–Stokes equation as a dynamical system

We can now return to the flow of an incompressible fluid governed by the Navier–Stokes equation and formulate it as a dynamical system. The equation is written as

$$\left.\begin{aligned} &\partial_t \boldsymbol{v} + \boldsymbol{v} \cdot \nabla \boldsymbol{v} = -\nabla p + \nu \nabla^2 \boldsymbol{v} + \boldsymbol{f}, \\ &\nabla \cdot \boldsymbol{v} = 0, \\ &\boldsymbol{v}_0 \equiv \boldsymbol{v}(t = 0) = \varpi \quad \text{(plus boundary conditions).} \end{aligned}\right\} \tag{3.26}$$

The initial condition, denoted ϖ, is chosen in a suitable space $\mathscr{H}$ of functions satisfying the boundary conditions and the incompressibility

[7] This assumption is known as 'metric transitivity' and may be very hard to prove for a given dynamical system.

constraint.[8] The force $\boldsymbol{f}$ is assumed independent of the time.[9] The space Ω is now simply the space $\mathscr{H}$ of all possible initial conditions ϖ. The time-shift G_t is the map

$$G_t : \varpi \mapsto \boldsymbol{v}(t). \tag{3.27}$$

(Since $\boldsymbol{f}$ does not depend on the time, G_t also maps $\boldsymbol{v}(s)$ into $\boldsymbol{v}(s+t)$.) As for P, it is a probability measure on Ω, invariant under the time shift.

The existence of the time-shift G_t and an invariant measure P are, in general, only conjectures. In three dimensions, we do not even have a theorem guaranteeing the existence and the uniqueness of the solution to the Navier–Stokes equation. The existence of invariant measures is an even harder problem (see Ruelle 1989; Vishik and Fursikov 1988). For chaotic systems, rigorous proofs are available only for very simple finite-dimensional models. This is perhaps the place to warn the reader that the poor man's Navier–Stokes equation (3.1) is indeed a poor model. It is pathological in at least two ways.

First its invariant measure fills all of the available space $[0, 1]$. In contrast, it is typical for dissipative systems in finite dimensions to have their invariant measure concentrated on an *attractor* with zero Lebesgue measure and with a fractal structure (see. e.g., Ruelle 1989, 1991). A well-known instance is the Hénon (1976) map $(x, y) \mapsto (y + 1 - ax^2, bx)$.

Second, it is typical for dissipative dynamical systems to have more than one attractor and therefore more than one invariant measure.[10] Each attractor has an associated basin. The statistical properties of the solution will then depend on to which basin the initial condition belongs. Thus, not only may the detailed behavior of orbits be unpredictable (because of the sensitivity to the initial conditions), but even their *statistical properties may be unpredictable*, insofar as it may be impossible to determine to which basin the initial condition belongs. Translated into meteorological vocabulary, this is equivalent to stating that not only the weather but also the climate may be unpredictable.

To conclude this chapter, we observe that at the present stage of development of the theory of dynamical systems there has been little quantitative impact on the understanding of high Reynolds number flow. We shall come back to such matters in Section 9.4. For the moment,

[8] At this point there is no need to restrict ourselves to periodic boundary conditions.
[9] The formalism can be readily extended to periodic time-dependence.
[10] Similarly, observe that it is typical for a solid resting on a table to have more than one stable equilibrium position.

the (partial) understanding of chaos in deterministic systems gives us confidence that a *probabilistic description* of turbulence is justified.[11] In the next chapter we shall review some of the basic tools of probability theory.

[11] This statement does not in any way imply that it would be justified to describe turbulence with a finite number of averaged quantities.

4
Probabilistic tools: a survey

We shall now introduce various probabilistic concepts frequently used in turbulence. For a more complete treatment the reader is referred to Feller (1968a,b), Wax (1954), Kac (1959) and Papoulis (1991).

We start from the abstract dynamical system structure as defined in the previous chapter, namely a quadruplet $(\Omega, \mathscr{A}, P, G_t)$. The time shift G_t, being relevant only for random functions, we may ignore it, at first, while defining random variables. We shall also 'forget' about $\mathscr{A}$ in order not to discuss measure-theoretic issues. (This is not really legitimate, just a lesser evil.)

4.1 Random variables

Definition. *A random variable is a map*

$$v : \Omega \to \mathbb{R}, \qquad \varpi \mapsto v(\varpi). \tag{4.1}$$

An example is the x-component of the velocity of a turbulent fluid at a given point and a given time. (The velocity is then still a function of the initial condition ϖ.)

Definition. *The probability measure of the random variable v is the image of the measure P by the map v.*

It is customary to define the *cumulative probability* as

$$\begin{aligned} F(x) &\equiv \text{Prob}\{v(\varpi) < x\} \\ &\equiv P\left(v^{-1}(]-\infty, x[)\right), \end{aligned} \tag{4.2}$$

where $v^{-1}(I)$ denotes the set of ϖs which are mapped into the interval I by v. Obviously $F(x)$ is a nondecreasing function. Its derivative $p(x) = dF(x)/dx$ (which may be a function or a distribution) is therefore nonnegative and is called the *probability density function* of the random variable. Loosely expressed, $p(x)dx$ is the probability of finding $v(\varpi)$ between x and $x+dx$. The p.d.f. is normalized:

$$\int_{\mathbb{R}} p(x)dx = 1. \tag{4.3}$$

Definition. *The mean value (or expectation value) of the random variable v is given by*

$$\langle v\rangle \equiv \int_{\Omega} v(\varpi)dP = \int_{\mathbb{R}} x\,p(x)dx, \tag{4.4}$$

which may be infinite.

Note that the operation of taking the mean value is linear. Mean values such as $\langle v\rangle$ are also referred to as *ensemble averages* to distinguish them from time averages.

Definition. *The random variable v is said to be centered*[1] *if $\langle v\rangle = 0$.*

Definition. *The moment of the* m*th order of the random variable v is given by*

$$\langle v^m\rangle \equiv \int_{\mathbb{R}} x^m p(x)dx, \quad m \in \mathbb{N}. \tag{4.5}$$

If v is centered, $\langle v^2\rangle$ is called the *variance*, $S = \langle v^3\rangle/(\langle v^2\rangle)^{3/2}$ is called the *skewness* and $F = \langle v^4\rangle/(\langle v^2\rangle)^2$ is called the *flatness*.[2]

Definition. *The characteristic function of the random variable v is the function of the real variable z, given by*

$$K(z) \equiv \langle e^{izv}\rangle = \int_{\mathbb{R}} e^{izx}p(x)dx. \tag{4.6}$$

$K(z)$ is thus the Fourier transform of the p.d.f. $p(x)$. Fourier transforms of positive functions are said to be of *positive type.*

The main reason for using characteristic functions is that the characteristic function of the *sum of two independent random variables* is the product of their individual characteristic functions (whereas, for p.d.f.s a convolution is needed).

Definition. *The centered random variable v is said to be Gaussian*[3] *if*

$$K(z) = \langle e^{izv}\rangle = e^{-\frac{1}{2}\sigma^2 z^2}, \qquad \sigma^2 = \langle v^2\rangle. \tag{4.7}$$

[1] In the following we shall mostly work with centered variables.

[2] Sometimes called 'kurtosis', although the correct definition for the latter is $F-3$.

[3] We shall sometimes call such random variables 'scalar Gaussian' to distinguish them from vector Gaussian variables to be defined later.

A simple calculation shows then that

$$p(x) = \frac{1}{\left(2\pi\sigma^2\right)^{1/2}} e^{-\frac{x^2}{2\sigma^2}}. \tag{4.8}$$

In manipulating Gaussian variables, it is often simpler to work with (4.7) rather than with (4.8).

We now turn to the multidimensional generalization of random variables. Substituting $\mathbb{R}^n$ for $\mathbb{R}$ in the definition of a random variable, we obtain a *vector-valued random variable*, $\boldsymbol{v}(\varpi) = (v_i(\varpi),\, i = 1,\ldots,n)$. We shall just informally upgrade some of our previous definitions. When the upgrade is obvious (such as for the p.d.f.), we shall omit it. The moments are now tensors of the form

$$\langle v_{i_1} v_{i_2} \cdots v_{i_m} \rangle. \tag{4.9}$$

When $\boldsymbol{v}$ is centered (zero mean value), its *covariance tensor* is defined by

$$\Gamma_{ij} \equiv \langle v_i v_j \rangle. \tag{4.10}$$

The characteristic function of $\boldsymbol{v}$ is defined by

$$K(\boldsymbol{z}) \equiv \langle e^{i\boldsymbol{z}\cdot\boldsymbol{v}} \rangle. \tag{4.11}$$

Thus, the characteristic function is the n-dimensional Fourier transform of the p.d.f. Characteristic functions always exist, whereas moments can be infinite. When finite, they are given in terms of the derivative at 0 of the characteristic function by the relation

$$\langle v_{i_1} v_{i_2} \cdots v_{i_m} \rangle = \left(\frac{1}{i}\right)^m \frac{\partial^m}{\partial z_{i_1} \partial z_{i_2} \cdots \partial z_{i_m}} K(\boldsymbol{z}) \bigg|_{\boldsymbol{z}=0}. \tag{4.12}$$

Eq. (4.12) is obtained by differentiation of (4.11) with respect to the z_i variables and then setting $\boldsymbol{z} = 0$.

Definition. *The vector-valued centered random variable $\boldsymbol{v} \in \mathbb{R}^n$ is said to be Gaussian if for all $\boldsymbol{c} \in \mathbb{R}^n$, the scalar quantity $\boldsymbol{c}\cdot\boldsymbol{v}$ is a scalar Gaussian random variable.*

This definition immediately implies that the Gaussian property is invariant under linear transformations.

Using (4.7) and (4.10) we can calculate the characteristic function of a vector-valued Gaussian random variable $\boldsymbol{v}$:

$$\begin{aligned} K(\boldsymbol{z}) &= \langle e^{i\boldsymbol{z}\cdot\boldsymbol{v}} \rangle = e^{-\frac{1}{2}\langle(\boldsymbol{z}\cdot\boldsymbol{v})^2\rangle} \\ &= e^{-\frac{1}{2} z_j z_k \langle v_j v_k \rangle} = e^{-\frac{1}{2} z_j z_k \Gamma_{jk}}. \end{aligned} \tag{4.13}$$

Thus, the characteristic function of Gaussian random variables is completely determined by the covariance tensor. By (4.12) all the moments of a vector-valued Gaussian random variable are also completely determined by the covariance tensor, i.e. by the set of second order moments. The relation between second order and higher order moments can actually be written in explicit form.

Before doing this, we state the following important result for Gaussian variables.

Gaussian integration by parts (Furutsu 1963; Donsker 1964; Novikov 1964). *Let $\boldsymbol{v} = (v_i,\, i = 1, \dots n)$ be a vector-valued centered Gaussian variable and let f be a differentiable function of n variables, then, assuming all averages exist,*

$$\langle v_i f(v_1, v_2, \dots, v_n) \rangle = \Gamma_{ij} \left\langle \frac{\partial f}{\partial v_j} \right\rangle, \tag{4.14}$$

where $\Gamma = \langle v_i v_j \rangle$.

To prove (4.14), we first observe that when the covariance tensor Γ is diagonal (independent variables), (4.14) reduces to the following relation for scalar Gaussian random variables (no summation over i):

$$\langle v_i f(v_i) \rangle = \langle v_i^2 \rangle \left\langle \frac{\partial f}{\partial v_i} \right\rangle, \tag{4.15}$$

which follows immediately from (4.8) after one integration by parts (hence, the name). We then use the invariance of Gaussian variables under linear transformations (our very definition of vector-valued Gaussian variables) and perform a linear transformation on the v_is which diagonalizes the covariance tensor. This transforms (4.14) into (4.15), completing the proof.

Let us also note that (4.14) may be rewritten in vector notation as

$$\langle \boldsymbol{v} f(\boldsymbol{v}) \rangle = \left\langle \boldsymbol{v}' \frac{\partial}{\partial \epsilon} \left(f(\boldsymbol{v} + \epsilon \boldsymbol{v}') \right) \Big|_{\epsilon=0} \right\rangle, \tag{4.16}$$

where $\boldsymbol{v}$ and $\boldsymbol{v}'$ are assumed to be independent and identically distributed vector-valued Gaussian variables. In the form (4.16), Gaussian integration by parts also applies to Gaussian operators acting in finite- or infinite-dimensional spaces.

Gaussian integration by parts has many applications in the statistical theory of turbulence. Here, we shall use it to derive the following relation.

Moment relation for Gaussian random variables (Isserlis 1918). *Let $\boldsymbol{v}$ be a centered vector-valued Gaussian random variable, then*

$$\langle v_{i_1} v_{i_2} \ldots v_{i_{2m+1}} \rangle = 0, \quad \forall\, m,\, i_1, i_2, \ldots, i_{2m+1} \tag{4.17}$$

and

$$\left.\begin{array}{c} \langle v_{i_1} v_{i_2} \ldots v_{i_{2m}} \rangle = \sum \langle v_{i_{\ell_1}} v_{i_{\ell_2}} \rangle \langle v_{i_{\ell_3}} v_{i_{\ell_4}} \rangle \cdots \langle v_{i_{\ell_{2m-1}}} v_{i_{\ell_{2m}}} \rangle, \\ \forall\, m,\, i_1, i_2, \ldots, i_{2m}, \end{array}\right\} \tag{4.18}$$

where

$$(\ell_1, \ell_2), (\ell_3, \ell_4), \cdots, (\ell_{2m-1}, \ell_{2m}) \tag{4.19}$$

is an arbitrary partition of $\{1, 2, \ldots, 2m\}$ *into pairs and the summation is over all possible partitions.*[4]

Proof. Eq. (4.17) is a consequence of the observation that changing $\boldsymbol{v}$ into $-\boldsymbol{v}$ does not change the covariance and thus leaves all the moments invariant; hence, odd order moments must vanish. The proof of (4.18) is recursive. It obviously holds for $m = 1$ because there is then a single term in the partition. Assuming it holds for the value $m - 1$, we apply Gaussian integration by parts to the l.h.s. of (4.18). For the function f, we take the product $v_{i_2} \ldots v_{i_{2m}}$. Using Leibnitz's formula, we then see that Gaussian integration by parts gives a sum of $2m - 1$ terms of the form

$$\langle v_{i_1} v_j \rangle \langle v_{i_2} \ldots \hat{v}_j \ldots v_{i_{2m}} \rangle, \tag{4.20}$$

where $\hat{v}_j$ means that the factor v_j is omitted. Partitions into pairs of $2m$ indices are obviously obtained by pairing the first index with any of the $2m - 1$ remaining ones and then writing all possible partitions of the $2m - 2$ indices left over. By the recursive hypothesis, the sum of all terms of the form (4.20) will then give exactly the right hand side of (4.18). *QED.*[5]

In practice the moment formula is often used as follows. Suppose we wish to evaluate $\langle v_i v_j v_k v_\ell \rangle$. We put four consecutive points on a line, labeled i, j, k, ℓ as shown in Fig. 4.1. We then write all possible partitions into pairs by drawing diagrams in which the paired points are connected by bridges.[6] There are three different ways to do this (shown in Fig. 4.1).

[4] Observe that the number of different partitions is $1 \times 3 \times 5 \times \cdots \times (2m - 1)$, as is easily shown by recursion.

[5] The decomposition (4.18) for the moments of Gaussian random variables can be generalized to the non-Gaussian case by using *cumulants*; see, e.g., Frisch (1968) and Section 9.5.1.

[6] Such diagrams are known as Feynman diagrams in quantum field theory and the moment formula is the equivalent of Wick's theorem.

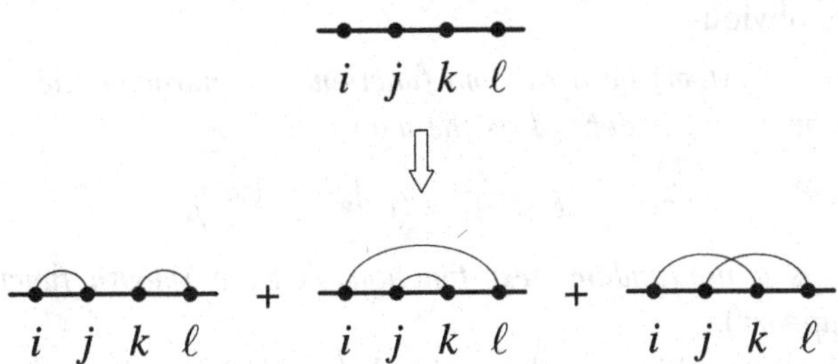

Fig. 4.1. An illustration of the pairing construction for moments of a Gaussian random function.

To each diagram we then associate the product of the covariances for the paired indices. We thus obtain

$$\langle v_i v_j v_k v_\ell \rangle = \Gamma_{ij}\Gamma_{k\ell} + \Gamma_{i\ell}\Gamma_{jk} + \Gamma_{ik}\Gamma_{j\ell}. \tag{4.21}$$

4.2 Random functions

Definition. *A random function*[7] *is a family of (scalar or vector-valued) random variables depending on one or several space or time variables.*[8]

An example is the velocity field $\boldsymbol{v}(t, \boldsymbol{r}, \varpi)$ of the solution of the Navier–Stokes equation with the initial condition ϖ.

We shall again just upgrade our previous definitions where necessary.

The moments of order n of a random function are, in general, tensors depending on n space-time variables. The moments of the random velocity field $\boldsymbol{v}$ are for example given by

$$\langle v_{i_1}(t_1, \boldsymbol{r}_1) v_{i_2}(t_2, \boldsymbol{r}_2) \cdots v_{i_m}(t_m, \boldsymbol{r}_m) \rangle. \tag{4.22}$$

When the random function is centered ($\langle \boldsymbol{v} \rangle = 0$), as we shall generally assume,

$$\Gamma_{ij}(t, \boldsymbol{r}; t', \boldsymbol{r}') \equiv \langle v_i(t, \boldsymbol{r}) v_j(t', \boldsymbol{r}') \rangle \tag{4.23}$$

is called the *correlation function.*

[7] Also often called a stochastic process.

[8] For random functions, the probabilistic labeling variable ϖ is often omitted. We shall drop it occasionally in this chapter and systematically in subsequent chapters.

Until further notice, we shall restrict ourselves to random functions which depend only upon the time and have scalar-value; generalizations are usually obvious.

Definition. *Let $v(t,\varpi)$ be a random function. Its characteristic functional (Kolmogorov 1935) is defined as the map*

$$z(t) \mapsto K[z(\cdot)] \equiv \left\langle e^{i\int_{\mathbb{R}} dt z(t)v(t,\varpi)}\right\rangle, \tag{4.24}$$

where $z(t)$ is a nonrandom 'test function' (e.g., a smooth function with compact support).

If the time variable and the integral $\int z(t)v(t)$ are discretized, (4.24) reduces to the definition (4.11) of the characteristic function for vector-valued random variables. (The components of the vector being the values of $v(t)$ at discrete times.) The relation (4.12) between moments and the characteristic function can be extended to random functions with the use of functional derivatives; this is left to the reader.

Definition. *The random function $v(t)$ is said to be Gaussian if for all test functions $z(t)$*

$$\int_{\mathbb{R}} z(t)v(t,\varpi)\,dt \tag{4.25}$$

is a Gaussian random variable.

It follows, in almost the same way as for a vector-valued Gaussian random variable, that the characteristic functional of a centered Gaussian random function is given by

$$K[z(\cdot)] = e^{-\frac{1}{2}\int\int_{\mathbb{R}^2} dtdt' z(t)z(t')\langle v(t)v(t')\rangle}. \tag{4.26}$$

4.3 Statistical symmetries

In Section 2.2 we gave a list of possible symmetries of the Navier–Stokes equation. For each of these deterministic symmetries a corresponding *statistical symmetry* may be defined. We begin with time-translations. So far we have not made use of the semi-group of time-shifts G_t which is part of the definition of a dynamical system, given in Section 3.3. We recall that the G_ts act on the probability space in such a way as to conserve the probability. The time-shifts are needed to define the concept of *stationarity.*

Definition. *A random function $v(t,\varpi)$ is said to be G_t-stationary if for all t and ϖ*

$$v(t+h,\varpi) = v(t, G_h\varpi), \qquad \forall\, h \geq 0. \tag{4.27}$$

The full solution $\boldsymbol{v}(t,\boldsymbol{r},\varpi)$ of the Navier–Stokes problem, as formulated in (3.26), is an example of a stationary random function. Stationarity is then obtained *by construction*, since G_t is defined as the time-shift for the solution and P is an invariant measure.

Some consequences of the definition of stationarity are listed below.

Propositions:

(i) If $v(t,\varpi)$ is a G_t-stationary random function and $f(\cdot)$ is a deterministic function of one variable, the random function $f(v(t,\varpi))$ is also G_t-stationary.

(ii) The sum of two G_t-stationary random functions is G_t-stationary.

(iii) The moments of a G_t-stationary random function, if they exist, are invariant under simultaneous translations of all their arguments:

$$\begin{aligned}&\langle v(t_1+h)v(t_2+h)\cdots v(t_m+h)\rangle\\&\quad=\langle v(t_1)v(t_2)\cdots v(t_m)\rangle,\quad \forall t_1,t_2,\ldots,t_m,\ h.\end{aligned}\tag{4.28}$$

Proof. Items (i) and (ii) are immediate consequences of the definition. As for item (iii), let us first assume that $h\geq 0$. We can then write the l.h.s. of (4.28) as

$$\begin{aligned}&\int_\Omega v(t_1+h,\varpi)v(t_2+h,\varpi)\cdots v(t_m+h,\varpi)\,dP\\&\quad=\int_\Omega v(t_1,G_h\varpi)v(t_2,G_h\varpi)\cdots v(t_m,G_h\varpi)\,dP.\end{aligned}\tag{4.29}$$

We make the change of variable: $\varpi'=G_h\varpi$ and use the conservation of probability. The r.h.s. of (4.29) becomes precisely the r.h.s. of (4.28). Finally, we prove the relation for negative time-shifts, by making in (4.28) the changes of variables:

$$t_1=t_1'-h,\ t_2=t_2'-h,\ \ldots,t_m=t_m'-h.\tag{4.30}$$

QED.

It follows from (4.28) that the correlation function depends only on the difference of its two time arguments

$$\langle v(t)v(t')\rangle=\Gamma(t-t'),\tag{4.31}$$

and that the variance $\langle v^2(t)\rangle=\Gamma(0)$ does not depend on the time t.

Equality in law. Another somewhat looser definition of stationarity of a random function $v(t)$ is to say that for any h, the random functions $v(t+h)$ and $v(t)$ have the 'same statistical properties', i.e. all corresponding multiple time moments and/or p.d.f.s are equal. This is called equality in law and denoted here

$$v(t+h) \stackrel{\text{law}}{=} v(t). \qquad (4.32)$$

Such notation has the advantage that there is no need to write the variable ϖ and the operators G_t. In the following this will allow us to speak of stationarity without specifying the time-shifts G_t at the cost, however, of some ambiguity. Indeed, let $v(t)$ be G_t-stationary. It is then immediately seen that $v(2t)$ is G_{2t}-stationary.[9] Both of these functions being stationary, it might be inferred that $v(t) + v(2t)$ is also stationary. Actually,

$$\begin{aligned} \langle (v(t) + v(2t))^2 \rangle &= \langle v^2(t) \rangle + \langle v^2(2t) \rangle + 2\langle v(t)v(2t) \rangle \\ &= 2\Gamma(0) + 2\Gamma(2t - t), \end{aligned} \qquad (4.33)$$

which obviously is not time-independent, as it should be if $v(t) + v(2t)$ were truly stationary.

A concept which is somewhat broader than stationarity, but which is very useful in turbulence, is contained in the following

Definition. *A random function $v(t, \varpi)$ is said to have G_t-stationary increments if for all t, t' and ϖ*

$$v(t' + h, \varpi) - v(t + h, \varpi) = v\left(t', G_h\varpi\right) - v\left(t, G_h\varpi\right), \qquad \forall\, h \geq 0. \qquad (4.34)$$

Stationarity implies stationary increments but not the converse. A well-known example of a random function with stationary increments is the *Brownian motion* function $W(t, \varpi)$ (Lévy 1965).[10] This is a Gaussian random function, defined for $t \geq 0$ and satisfying the following conditions:

$$W(0) = 0, \qquad \langle W(t) \rangle = 0, \qquad \langle W(t)W(t') \rangle = \inf(t, t'). \qquad (4.35)$$

It follows that

$$\langle (W(t') - W(t))^2 \rangle = |t' - t|. \qquad (4.36)$$

Observe that $W(t, \varpi)$ is not stationary. Indeed, from (4.35), we have $\langle W^2(t) \rangle = t$ which is not independent of the time.

[9] Observe that the semi-group G_{2t} is distinct from G_t.

[10] The letter 'W's stands for Norbert Wiener. He and Paul Lévy, pioneered the study of the mathematical properties of Brownian motion. A modern account may be found in Kahane (1985).

We now turn to other statistical symmetries. For space-translations, we have the following notion, formalized by Kampé de Fériet (1953; Section III):

Definition. *The random function $\boldsymbol{v}(t, \boldsymbol{r}, \varpi)$ is said to be homogeneous if there is a group G_ρ^{space} of 'space-shift' transformations of Ω, conserving the probability and commuting with the time-shifts G_t, such that*

$$\boldsymbol{v}(t, \boldsymbol{r} + \boldsymbol{\rho}, \varpi) = \boldsymbol{v}\left(t, \boldsymbol{r}, G_\rho^{\text{space}}\varpi\right). \tag{4.37}$$

A consequence of homogeneity is that all moments are invariant under simultaneous space-translations of their arguments. The correlation tensor of a stationary and homogeneous random velocity field has the following form:

$$\langle v_i(t, \boldsymbol{r}) v_j(t', \boldsymbol{r}') \rangle = \Gamma_{ij}(t - t', \boldsymbol{r} - \boldsymbol{r}'). \tag{4.38}$$

Just as for stationarity (Section 4.3), homogeneity can be weakened into the property of having homogeneous increments.

It is now clear that similar definitions can be given for other symmetries, such as rotations, scaling transformations, parity, etc. Statistical invariance under rotation is referred to as *isotropy*. A velocity field $\boldsymbol{v}$ which is statistically invariant under parity is called *nonhelical* or *nonchiral*. If it is invariant under one of the scaling groups of Section 2.2, it is said to be *scale-invariant*.

4.4 Ergodic results

We have seen in Section 3.3 that Birkhoff's ergodic theorem ensures that, under suitable conditions, time averages are equivalent to ensemble averages. For a stationary[11] random function $v(t, \varpi)$ the statement of the ergodic theorem is that for almost all ϖ

$$\lim_{T\to\infty} \frac{1}{T} \int_0^T v(t, \varpi)dt = \langle v \rangle. \tag{4.39}$$

This may be viewed as an extension of the well-known *strong law of large numbers*, which in its simplest form states that (with suitable restrictions) the average of N identically and independently distributed random variables tends, for $N \to \infty$, almost surely to their common mean value.

[11] From now on 'stationary' will be used to mean G_t-stationary.

Eq. (4.39) is used extensively in experiments to measure statistical quantities such as moments, moments of increments, p.d.f.s, etc. In practice, the time averages are calculated from a sample of finite length. In order to get a feeling for how long the sample should be, we shall now give a watered-down proof of the ergodic formula. Namely, we shall prove convergence in mean square rather than almost sure convergence. This has the advantage that it provides an estimate of the error. Without loss of generality, we may assume that $\langle v \rangle = 0$. (Otherwise change $v(t)$ into $v(t) - \langle v \rangle$.) The main result is now stated.

Mean square ergodic theorem. *Let $v(t, \varpi)$ be a centered and stationary random function. Let $\Gamma(t - t') = \langle v(t)v(t') \rangle$ be its correlation function. It is assumed that this function decreases sufficiently fast at infinity to ensure that*

$$\int_0^\infty |\Gamma(t)| dt < \infty. \tag{4.40}$$

It then follows that

$$\lim_{T \to \infty} \left\langle \left(\frac{1}{T} \int_0^T v(t, \varpi) dt \right)^2 \right\rangle = 0. \tag{4.41}$$

Proof. The l.h.s. of (4.41) involves the average of the square of an integral. Averaging and integration commute (because averaging is a linear operation). The presence of the square inhibits the use of this property. We can, however, transform the square of the integral into a double integral (using Fubini's theorem). Thus, using (4.31), we obtain

$$\begin{aligned} \left\langle \left(\frac{1}{T} \int_0^T v(t, \varpi) dt \right)^2 \right\rangle &= \frac{1}{T^2} \int_0^T \int_0^T dt_1 dt_2 \Gamma(t_1 - t_2) \\ &= \frac{2}{T^2} \int_0^T dt_1 \int_0^{t_1} dt_2 \Gamma(t_2) \\ &\leq \frac{2}{T} \int_0^\infty dt_2 |\Gamma(t_2)|. \end{aligned} \tag{4.42}$$

Clearly, this tends to zero as $T \to \infty$. *QED.*[12]

A more quantitative estimate can be given by introducing the *integral time scale*

$$T_{\text{int}} = \frac{\int_0^\infty dt |\langle v(t)v(0) \rangle|}{\langle v^2 \rangle} = \frac{\int_0^\infty dt |\Gamma(t)|}{\Gamma(0)}. \tag{4.43}$$

[12] Condition (4.40) is sufficient but not necessary for obtaining (4.41). A necessary and sufficient condition has been found by Slutsky (1938; see also Monin and Yaglom 1971, Section 4.7).

We can now compare the estimate for the mean square value of the time average, given by (4.42), with the mean square value of the velocity itself. It follows from (4.42) that the former will be negligible compared to the latter when

$$T \gg T_{\rm int}. \tag{4.44}$$

The ergodic theorem (in its almost sure or mean square versions) can also be applied to the evaluation of moments of $v(t,\varpi)$, since $v^n(t,\varpi)$ is also a stationary function. It is important, however, to stress that the integral time scale for v^n usually grows very rapidly with n. Thus the measurement of high order moments by time averaging requires very long samples.

It is possible to get much more information than just bounds on how time averages differ from ensemble averages. We now mention briefly two types of increasingly refined results.

The *central limit theorem* states that under suitable conditions (not given here), the difference between the l.h.s. and the r.h.s. of (4.39), when multiplied by $T^{1/2}$, tends for $T \to \infty$ to a Gaussian random variable. Loosely expressed, the fluctuations around the mean are typically $O(1/T^{1/2})$. This is consistent with the $1/T$ bound obtained in (4.42) for the mean *square* of this difference. Proving the central limit theorem for the addition of independent random variables, using the method of characteristic functions, is very elementary (see any textbook on probability theory). The proof is not elementary for random functions. At this point we must warn the reader that the importance of the central limit for *turbulence* should not be overstated. Turbulence has some near-Gaussian features, such as the distribution of the velocities at a given point. It cannot, however, be truly Gaussian. Indeed, a (centered) Gaussian velocity field has vanishing moments of all odd orders. Hence, the energy flux given by (2.52) would identically vanish (assuming that angular brackets are reinterpreted as ensemble averages). Actually, as we shall see in Chapter 8, turbulence is highly non-Gaussian, particularly at small scales.

An even more refined description of the *strong* discrepancies between ensemble averages and time averages over a long but finite time span is given by the theory of *large deviations.* Although the typical discrepancies are small, namely $O(1/T^{1/2})$, order unity discrepancies can happen, but with a very small probability which decreases exponentially with T. Such discrepancies are strongly relevant when investigating the behavior of $\langle \exp \int_0^T v(t,\varpi)\, dt \rangle$ for large Ts. It is a frequent misconception that the leading order contribution can be obtained by treating $\int_0^T v(t,\varpi)\,dt$ as if it were Gaussian (see, e.g., Lumley 1972). As we shall see in Section 8.8, Kolmogorov himself was tempted by such lax use of the central limit theorem. We shall come back to large deviations in Section 8.6.4 in connection with multifractals.

Finally, we mention that the concept of ergodicity is readily extended from the time domain into the space domain provided the spatial domain is of infinite extension in at least one direction, so that averages over increasingly large distances can be taken. For example, if $\boldsymbol{v}(x,y,z,\varpi)$ is a random homogeneous and ergodic velocity field defined in all of $\mathbb{R}^3$, we have, almost surely

$$\lim_{L\to\infty}\frac{1}{L^3}\int_0^L\int_0^L\int_0^L dx\,dy\,dz\,\boldsymbol{v}(x,y,z,\varpi)=\langle\boldsymbol{v}\rangle. \tag{4.45}$$

Actually, it is enough to take the average over one coordinate. If, however, the function is periodic in the space variables, averages over the periodicity box will give a poor approximation to the ensemble average unless the period is very large compared to the integral scale (the spatial analog of (4.43)). Similarly, there cannot be an exact ergodic result involving the rotation group because one can only rotate by a finite amount.[13]

4.5 The spectrum of stationary random functions

One of the most common practical methods for analyzing a stationary random function is to determine what electrical engineers call its *power spectrum.* We shall here give a somewhat nonstandard definition of the power spectrum, using the concept of low-pass filtering introduced in Section 2.4.

Let $v(t,\varpi)$ be a centered and stationary random function. We introduce the low-pass filtered functions in the same way as in Section 2.4, except that Fourier integrals rather than series are used. In this section, the Fourier variable is denoted by f rather than the traditional ω which could be confused with ϖ. We thus define

$$\left.\begin{aligned} v(t,\varpi)&=\int_{\mathbb{R}} e^{ift}\,\hat{v}(f,\varpi)\,df,\\ v_F^{<}(t,\varpi)&=\int_{|f|\le F} e^{ift}\,\hat{v}(f,\varpi)\,df,\quad F\ge 0.\end{aligned}\right\} \tag{4.46}$$

[13] More precisely, the rotation group is compact.

Clearly, $v_F^{\lessgtr}(t,\varpi)$ is itself stationary. Observe that the Fourier transforms of homogeneous random functions are random distributions, so that $\hat{v}(f,\varpi)$ is not an ordinary function of f, although $v_F^{\lessgtr}(t,\varpi)$ is an ordinary function of t. Before distributions became widely used this difficulty was circumvented by the use of Stieltjes integrals (see, e.g., Batchelor 1953; Monin and Yaglom 1975). Our use of low- and high-pass filtering has the advantage that only ordinary functions appear and that essentially the same formalism can be used in the deterministic case (Section 2.4) and in the random case.

We now define the *cumulative energy spectrum*

$$\mathscr{E}(F) \equiv \frac{1}{2}\langle [v_F^{<}(t)]^2 \rangle, \tag{4.47}$$

which, by the assumed stationarity, does not depend on the time variable.

The factor 1/2 has been introduced into the definition to agree with the standard definition of the kinetic energy. $\mathscr{E}(F)$ may be interpreted as the mean kinetic energy in (temporal) scales greater than $\sim F^{-1}$. It is easily shown, using Parseval's theorem, that the cumulative energy spectrum is a nondecreasing function of the cutoff frequency F.

Next we define the *energy spectrum* of the stationary random function $v(t,\varpi)$ by

$$E(f) \equiv \frac{d}{df}\mathscr{E}(f) \geq 0. \tag{4.48}$$

The positivity follows from the nondecreasing property. The energy spectrum is often referred to as just the 'spectrum'.

$E(f)df$ can thus be interpreted as the contribution to the mean kinetic energy of those Fourier harmonics which have the absolute value of their frequency between f and $f + df$.

Since the filtered velocity field reduces to the unfiltered one when $F \to \infty$, it follows from (4.47) and (4.48) that

$$\frac{1}{2}\langle v^2 \rangle = \int_0^\infty E(f)df. \tag{4.49}$$

As expected, the mean kinetic energy (one-half of the variance of the random function) is the integral of the energy spectrum over all frequencies. Similarly, observing that the Fourier transform of dv/dt is $if\,\hat{v}_f$, we obtain

$$\frac{1}{2}\left\langle \left(\frac{dv(t,\varpi)}{dt}\right)^2 \right\rangle = \int_0^\infty f^2 E(f)\,df. \tag{4.50}$$

By use of the identity

$$\int_{-\infty}^{+\infty} e^{i(f-f')t}dt = 2\pi\delta(f - f'), \tag{4.51}$$

it may also be shown that

$$E(f) = \frac{1}{2\pi}\int_{-\infty}^{+\infty} e^{ifs}\Gamma(s)ds, \tag{4.52}$$

which states that the correlation function $\Gamma(s)$ and the energy spectrum are Fourier transforms of each other. Eq. (4.52) is known as the *Wiener–Khinchin formula.*[14] It implies, for example, that the Fourier transform of the correlation function of a stationary random function must be non-negative.

Another immediate consequence of (4.52) is an expression for the *second order structure function,* defined as the mean square of the velocity increment from time t to time t'. We have

$$\langle (v(t') - v(t))^2 \rangle = 2\int_{-\infty}^{+\infty} \left(1 - e^{if(t'-t)}\right) E(f)df. \tag{4.53}$$

Here, $E(f)$ is extended to negative frequencies by $E(-f) = E(f)$.

When a random function has stationary increments without being stationary, (4.52) is inapplicable, but (4.53) remains valid. A particularly relevant case for turbulence is when the energy spectrum $E(f)$ is a power-law (Kolmogorov 1940):

$$E(f) = C|f|^{-n}, \qquad C > 0. \tag{4.54}$$

When substituted into (4.49) this give a divergent integral. The divergence is either at high frequencies (ultraviolet divergence) when $n < 1$ or at low frequencies (infrared divergence) when $n > 1$ or at both when $n = 1$. This shows that there cannot exist a stationary random function with finite variance and a power-law spectrum. If, however, we substitute (4.54) into (4.53), we find that no divergence occurs as long as $1 < n < 3$. A simple calculation shows then that

$$\left.\begin{aligned} \langle (v(t') - v(t))^2 \rangle &= CA_n|t' - t|^{n-1}, \\ A_n &= 2\int_{-\infty}^{+\infty} \left(1 - e^{ix}\right) |x|^{-n}dx. \end{aligned}\right\} \tag{4.55}$$

[14] Yaglom (1987) has found that Einstein (1914) had already used this relation.

Thus, $v(t)$ has stationary increments, at least in a mean square sense. For example, with $n = 2$, we recover Brownian motion.

Random functions with stationary increments often appear as limits of stationary random functions with an infrared cutoff, when the cutoff frequency tends to zero. Consider, for example, the stationary Gaussian random function, called the 'Ornstein–Uhlenbeck process', which has the correlation function $\Gamma(t) = f_0^{-1}e^{-f_0|t|}$ and the spectrum $E(f) = (1/\pi)(f_0^2 + f^2)^{-1}$. As $f_0 \to 0$, the Ornstein–Uhlenbeck process tends to Brownian motion.

Again, everything we have presented in this section can be extended from the time domain into the spatial domain when the latter is unbounded. For example, the cumulative (spatial) energy spectrum is defined by

$$\mathscr{E}(K) \equiv \frac{1}{2}\langle|\boldsymbol{v}_K^{<}(\boldsymbol{r})|^2\rangle, \tag{4.56}$$

where $\boldsymbol{v}_K^{<}$ is the low-pass filtered (vector) velocity field containing all harmonics with wavenumber less or equal to K. Again, the *(spatial) energy spectrum* is defined by

$$E(k) \equiv \frac{d\mathscr{E}(k)}{dk}. \tag{4.57}$$

Note that, although the space is three-dimensional, the variables K and k are *wavenumbers*, i.e. positive scalars. Thus, the mean energy is obtained from $E(k)$ by the same one-dimensional integral as (4.49) (with the variable k instead of f). One can also define a three-dimensional energy spectrum $E_{3D}(\boldsymbol{k})$ which is the three-dimensional Fourier transform of the spatial correlation function $\langle\boldsymbol{v}(\boldsymbol{r}) \cdot \boldsymbol{v}(\boldsymbol{r}')\rangle$. In the incompressible isotropic case, the Wiener–Khinchin formula reads

$$E(k) = 4\pi k^2 E_{3D}(k) = \frac{1}{\pi}\int_0^\infty k\rho\Gamma(\rho)\sin k\rho\, d\rho, \tag{4.58}$$

where

$$\Gamma(\rho) \equiv \langle\boldsymbol{v}(\boldsymbol{r}) \cdot \boldsymbol{v}(\boldsymbol{r}')\rangle, \qquad \rho = |\boldsymbol{r} - \boldsymbol{r}'|. \tag{4.59}$$

When using L-periodic boundary conditions, the cumulative energy spectrum defined by (4.56) will change discontinuously with K because the only admissible wavevectors are in $(2\pi/L)\mathbf{Z}^3$, where $\mathbf{Z}$ is the set of signed integers. Hence, the energy spectrum $E(k)$ will be a sum of δ-functions. Clearly, the continuous case can be recovered by letting $L \to \infty$.

Finally, as in the temporal domain, we find that when the energy spectrum is a power-law

$$E(k) \propto k^{-n}, \qquad 1 < n < 3, \tag{4.60}$$

then the velocity field has homogeneous increments (in the mean square sense) and the second order spatial structure function is also a power-law:

$$\langle |\boldsymbol{v}(\boldsymbol{r}') - \boldsymbol{v}(\boldsymbol{r})|^2 \rangle \propto |\boldsymbol{r}' - \boldsymbol{r}|^{n-1}. \tag{4.61}$$

5

Two experimental laws of fully developed turbulence

Experimental data will now be presented to illustrate two basic empirical laws of fully developed turbulence.

(i) **Two-thirds law.** *In a turbulent flow at very high Reynolds number, the mean square velocity increment* $\langle(\delta v(\ell))^2\rangle$ *between two points separated by a distance* ℓ *behaves approximately as the two-thirds power of the distance.*[1]

(ii) **Law of finite energy dissipation.** *If, in an experiment on turbulent flow, all the control parameters are kept the same, except for the viscosity, which is lowered as much as possible, the energy dissipation per unit mass* dE/dt *behaves in a way consistent with a finite positive limit.*

These laws seem to hold, at least approximately, for almost any turbulent flow. Let us now examine examples of such data.

5.1 The two-thirds law

Fig. 5.1 shows a log–log plot of the *second order longitudinal structure function*

$$S_2(\ell) \equiv \left\langle \left(\delta v_\parallel(\ell)\right)^2 \right\rangle. \tag{5.1}$$

The measurement was done in the S1 wind tunnel of ONERA.[2]

It is seen that there is a substantial $\ell^{2/3}$ range. Let us be somewhat more specific now. The *longitudinal velocity increment* is defined as

$$\delta v_\parallel(\boldsymbol{r}, \boldsymbol{\ell}) \equiv [\boldsymbol{v}(\boldsymbol{r}+\boldsymbol{\ell}) - \boldsymbol{v}(\boldsymbol{r})] \cdot \frac{\boldsymbol{\ell}}{\ell}, \tag{5.2}$$

[1] With restrictions on the range of variation of ℓ which will be given later.

[2] All the data from S1 reported in this book have been obtained by Y. Gagne, E. Hopfinger and M. Marchand.

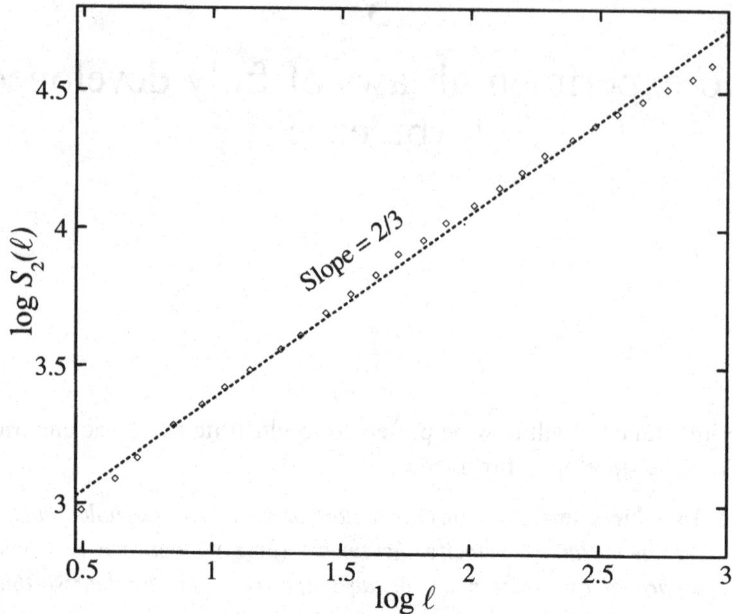

Fig. 5.1. log–log plot of the second order structure function in the time domain for data from the S1 wind tunnel of ONERA. Courtesy Y. Gagne and E. Hopfinger.

where $\ell = |\boldsymbol{\ell}|$. Thus $\delta v_{\parallel}(\boldsymbol{r}, \boldsymbol{\ell})$ is the velocity increment between two points separated by $\boldsymbol{\ell}$, projected onto the line of separation. When the turbulence is homogeneous and isotropic, we can unambiguously drop the dependence on $\boldsymbol{r}$ in the second order moment $\langle(\delta v_{\parallel}(\boldsymbol{r}, \boldsymbol{\ell}))^2\rangle$ and use ℓ instead of $\boldsymbol{\ell}$ as is done in (5.1). The wind tunnel S1 is shown on Fig. 5.2. It is over 150 m long. The widest part, known as the return duct, has a circular cross section with a diameter of 24 m. A hot wire probe was suspended near the point marked 'M'. It recorded the streamwise (parallel to the mean flow) component of the velocity. The averaging for the data shown in Fig. 5.1 is a time average, using ergodicity (Section 4.4). The mean flow velocity was 20 m/s. The Reynolds number based on this mean flow and the diameter of the duct was about 3×10^7. The Reynolds number based on the r.m.s. velocity and the integral scale (around 15 m) was about 1.5×10^6. The r.m.s. velocity fluctuations represent about 7% of the mean flow.

This relatively small ratio, the *turbulence intensity*, typical of wind tunnels, justifies the use of the *Taylor hypothesis*,[3] as now explained. Let

[3] From G.I. Taylor, the Cambridge fluid dynamicist.

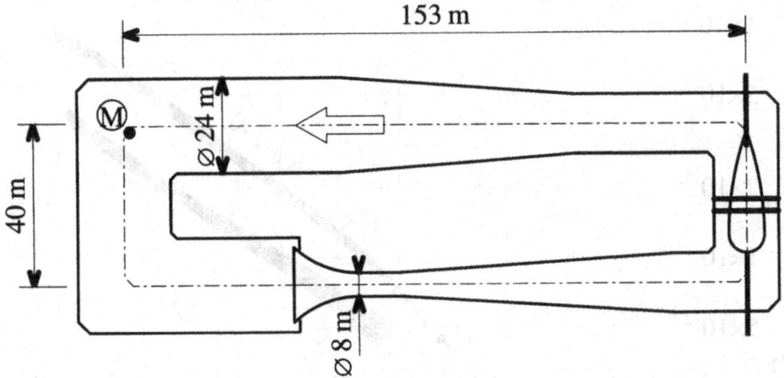

Fig. 5.2. The S1 wind tunnel of ONERA.

us denote by $v'(t,x)$ the velocity which would be measured in the frame of reference of the mean flow. The x-coordinate is along the mean flow. In the frame of reference of the laboratory the measured velocity will be

$$v(t,x) = v'(t, x - Ut) + U, \tag{5.3}$$

where U is the mean flow. If we now assume that the 'turbulence intensity' I is given by

$$I = \frac{\sqrt{\langle v'^2 \rangle}}{U} \ll 1, \tag{5.4}$$

it is easily checked that most of the time-dependence in $v(t,x)$ comes from the spatial argument $x - Ut$ of v'. Taylor's hypothesis reinterprets the temporal variation of v at a fixed spatial location as being a spatial variation of v'. The correspondence between spatial increments ℓ for v' and temporal increments τ for v is then simply[4]

$$\ell = U\tau. \tag{5.5}$$

Most experimental data on fully developed turbulence are obtained in the time domain and then recast into the space domain via the Taylor hypothesis. The time (or frequency) axis is then often relabeled as a position (or wavenumber) axis. To avoid possible confusion, we always indicate in figure captions whether the data were obtained in the time domain or in the space domain.

The question of how the Taylor hypothesis should be corrected when

[4] Attention has to be paid to the fact that increasing ts correspond to decreasing xs.

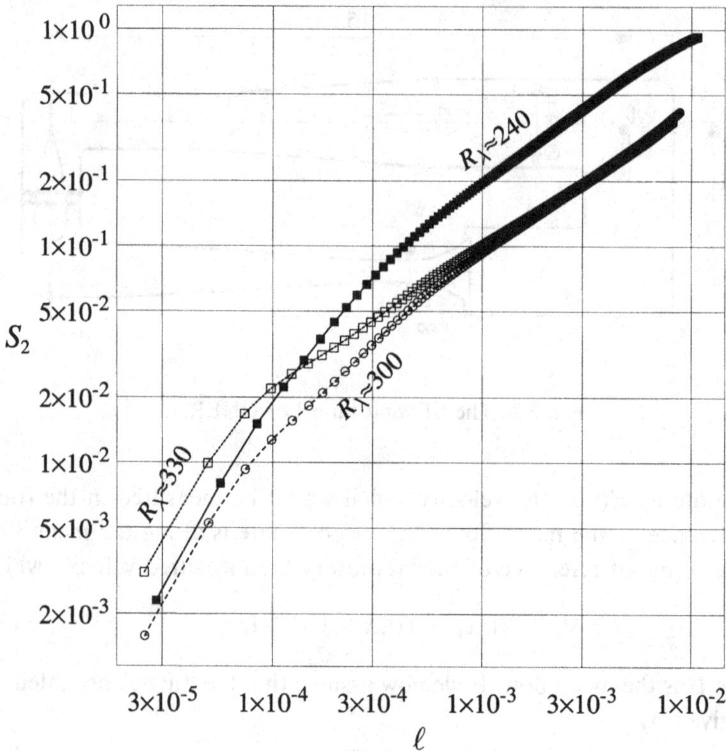

Fig. 5.3. log–log plot of the second order transverse structure function measured in the space domain by the RELIEF flow tagging technique in a turbulent jet at various R_λs, as labeled (Noullez, Wallace, Lempert, Miles and Frisch 1996).

the turbulence intensity ceases to be small has been addressed by Lumley (1965) and Pinton and Labbé (1994).

It is possible to make measurements directly in the space domain by using nonintrusive optical techniques in which a pattern, say, a straight line, is 'written' into the flow and then 'interrogated' after a very short time. From its deformation, the velocity component perpendicular to the line can be reconstructed. One promising technique is the Raman Excited Laser Induced Electronic Fluorescence (RELIEF) of Miles, Lempert, Zhang and Zhang (1991). Fig. 5.3 shows the second order *transverse* structure functions, obtained by the RELIEF technique in a turbulent jet at various Reynolds numbers. The second order transverse structure function is defined as

$$S_2^{\perp}(\ell) \equiv \langle (\delta v_{\perp}(\ell))^2 \rangle, \tag{5.6}$$

where $v_\perp$ is a particular component perpendicular to $\boldsymbol{\ell}$ of the velocity increment $\boldsymbol{v}(\boldsymbol{r}+\boldsymbol{\ell})-\boldsymbol{v}(\boldsymbol{r})$. (If the turbulence is isotropic, which particular component is chosen does not matter and the resulting structure function depends only on the modulus ℓ of $\boldsymbol{\ell}$.) It is seen that this transverse structure function, like the longitudinal one, displays a substantial $\ell^{2/3}$ range.

It is traditional for experimental and numerical data on fully developed turbulence to use not the Reynolds number based on the integral scale and the r.m.s. velocity fluctuations, but instead the so-called Taylor-scale Reynolds number, which is easier to measure. The latter is defined as

$$R_\lambda \equiv \frac{v_{\mathrm{rms}}\lambda}{\nu}. \tag{5.7}$$

Here, v_{rms} is the r.m.s. fluctuation of, say, the v_1 component of the velocity and the 'Taylor-scale' λ is defined as

$$\frac{1}{\lambda^2} \equiv \frac{\langle(\partial_1 v_1)^2\rangle}{v_{\mathrm{rms}}^2}. \tag{5.8}$$

For S1 the Taylor-scale Reynolds number is $R_\lambda \approx 2700$.

Under the assumption of isotropy, it is easily shown (Batchelor 1953) that

$$v_{\mathrm{rms}} = (2E/3)^{1/2} \qquad \langle(\partial_1 v_1)^2\rangle = (2/15)\Omega, \tag{5.9}$$

E being the mean energy and Ω the mean enstrophy,[5] defined in Section 2.3 (eq. (2.28)). Hence, for isotropic turbulence,[6]

$$\frac{1}{\lambda^2} = \frac{\Omega}{5E}, \tag{5.10}$$

and

$$R_\lambda = (10/3)^{1/2}\frac{E}{\Omega^{1/2}\nu}. \tag{5.11}$$

To illustrate further the two-thirds law we give several examples of energy spectra. In view of (4.60) and (4.61), the two-thirds law is equivalent to the statement that the energy spectrum $E(k)$ follows a $k^{-5/3}$

[5] From here on Ω recovers its hydrodynamical meaning; the notation Ω was used for the probability space in Chapters 3 and 4.

[6] Sometimes the factor 5 in the denominator is omitted; such a choice is not consistent with the standard definition used by experimentalists who often have access only to one component of the velocity.

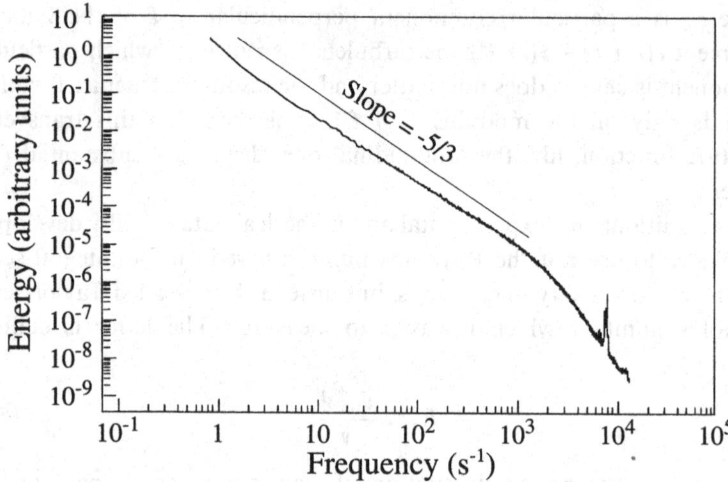

Fig. 5.4. Energy spectrum in the time domain for data from S1. Reynolds number $R_\lambda = 2720$. Courtesy Y. Gagne and M. Marchand.

law over a suitable range. The larger the Reynolds number, the wider this range.

Fig. 5.4 shows the energy spectrum for the best data obtained so far from the S1 wind tunnel. This is again a log–log plot. The horizontal axis is a frequency which can be reinterpreted as a wavenumber by use of the Taylor hypothesis. A power-law scaling k^{-n} with an exponent n close to 5/3 is observed over a very substantial range of about three decades of wavenumber. This range is called the *inertial range*, a name which will be justified in Section 6.2.5.

The features shown in Fig. 5.4 need some comment. First, there is a narrow spike in the spectrum at high wavenumbers. This is probably due to a mechanical vibration which is hard to avoid when a scientific experiment is set up temporarily in a major industrial facility such as S1. Second, it is not known if the discrepancy from a pure 5/3-law is significant (a straight line of slope $-5/3$ is shown for comparison).[7] Third, it should be noticed that the range of wavenumbers over which scaling holds for $E(k)$ is much larger than the range of distances over which scaling holds for $S_2(\ell)$. As observed by M. Nelkin (private communication, 1988), the fact that the two functions are, by (4.53), essentially Fourier transforms of each other does not imply identical spans of

[7] When a power-law is fitted to the data, the value of the exponent changes somewhat with the choice of the fitting region.

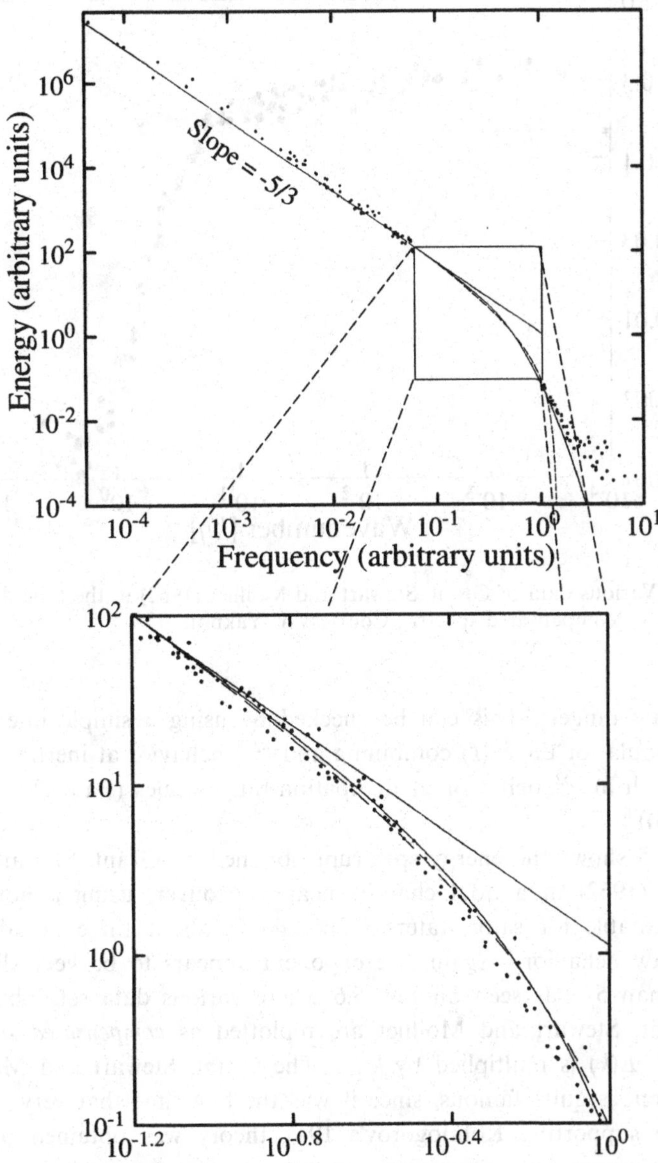

Fig. 5.5. log–log plot of the energy spectrum in the time domain and enlargement of the beginning of the dissipation range for tidal channel data (Grant, Stewart and Moilliet 1962).

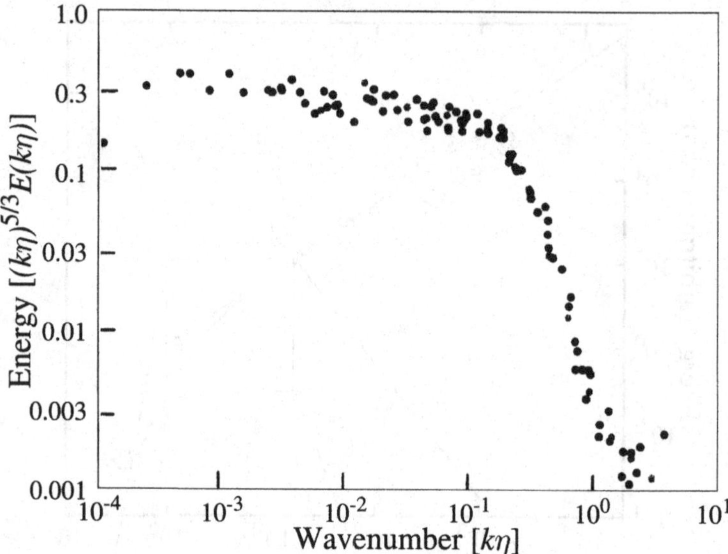

Fig. 5.6. Various data of Grant, Stewart and Moilliet (1962) in the time domain replotted as 'compensated spectra'. Courtesy V. Yakhot.

power-law ranges. This can be checked by using a simple interpolation formula for the $S_2(\ell)$ combining and $\ell^{2/3}$ behavior at inertial-range scales with an ℓ^2 behavior at dissipation-range scales (Batchelor 1951, eq. (7.10)).[8]

Fig. 5.5 shows the energy spectrum obtained by Grant, Stewart and Moilliet (1962) in a tidal channel near Vancouver, using a hot film probe suitable for salty waters. This shows about three decades of power-law behavior. Again, the exponent appears to be very slightly larger than 5/3 as seen on Fig. 5.6 where various data sets obtained by Grant, Stewart and Moilliet are replotted as *compensated spectra* in which $E(k)$ is multiplied by $k^{5/3}$. The Grant, Stewart and Moilliet experiment is quite famous, since it was the first time that very strong evidence supporting Kolmogorov's 1941 theory was obtained over a sufficient range of wavenumbers to permit simultaneous measurement of the dissipation rate (see later in this section).[9] Many other geometries give

[8] See also Nelkin (1994, Section 3.3 and references therein).

[9] Before 1961, a number of measurements at the Institute of Atmospheric Physics (Moscow), by A.S. Gurvich, L.R. Tsvang and S.L. Zubkovsky had already confirmed the existence of the $k^{-5/3}$ law (see Monin and Yaglom 1975).

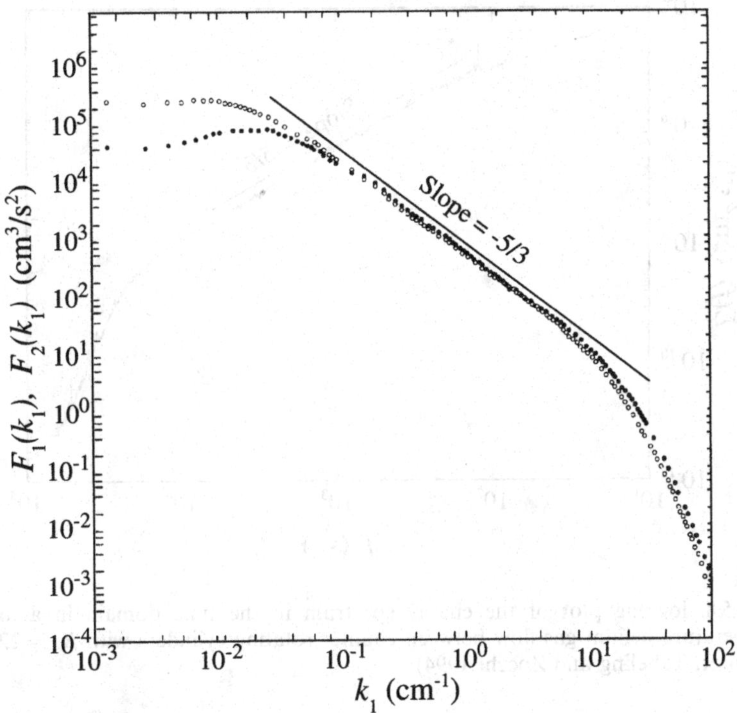

Fig. 5.7. log–log plot of the energy spectra of the streamwise component (white circles) and lateral component (black circles) of the velocity fluctuations in the time domain in a jet with $R_\lambda = 626$ (Champagne 1978).

similar energy spectra, provided the Reynolds number is high enough. Fig. 5.7 shows spectra of streamwise and lateral components of the velocity fluctuations in an axisymmetric jet with $R_\lambda = 626$. About two decades of an approximately $k^{-5/3}$ range are observed. Fig. 5.8 also displays about two decades of $k^{-5/3}$ range. It was obtained in a low temperature helium gas flow between two counter-rotating disks[10] at $R_\lambda = 1200$. Such helium facilities look very promising since they can achieve R_λs comparable to those of large industrial facilities and still fit on a table, thanks to the very low kinematic viscosity of helium, about 10^{-3} in c.g.s. units. A major challenge is to develop probes (or nonintrusive optical techniques) suitable for the very small scales (of the order of a micrometer) encountered in such experiments (Castaing,

[10] This flow geometry is sometimes called 'von Kármán swirling flow' although von Kármán (1921) considered a single infinite rotating disk.

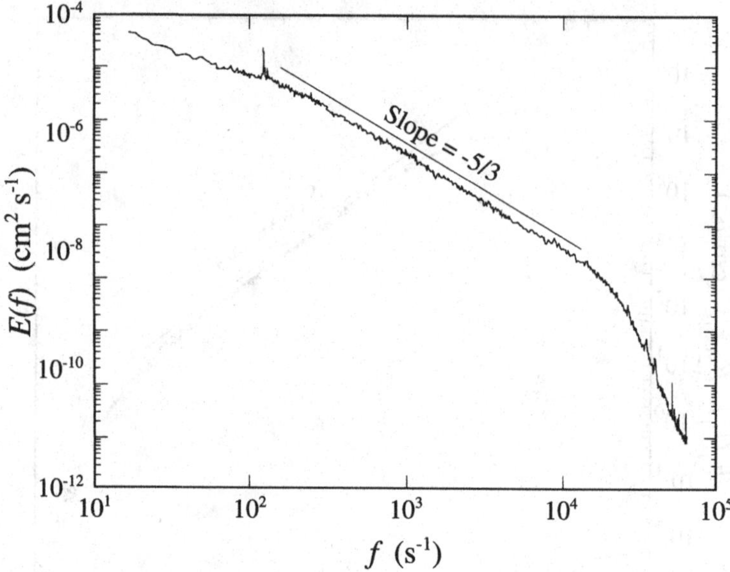

Fig. 5.8. log–log plot of the energy spectrum in the time domain in a low-temperature helium gas flow between counter-rotating cylinders with $R_\lambda = 1200$ (Maurer, Tabeling and Zocchi 1994).

Chabaud and Hébral 1992). Another difficulty, also found in jets, is that the turbulence intensity I (given by (5.4)) is about 0.25–0.3; this is too large to make the use of the Taylor hypothesis safe, but the difficulty may be obviated by a resampling technique (Pinton and Labbé 1994).

Fig. 5.9 shows various energy spectra obtained by Sanada and Ishii (private communication; see also Sanada 1991) from direct numerical simulations of the Navier–Stokes equation on a supercomputer. The simulations used 256^3 grid points and had Reynolds numbers of several thousands.[11] The spectra are genuine spatial spectra. They display an inertial range $\propto k^m$ with an exponent $m \approx -5/3$ over about one decade of wavenumbers.[12]

[11] Somewhat higher Reynolds numbers can be achieved for flows with special symmetries.

[12] Simulations of Borue and Orszag (1995) display a wider and steeper inertial range with an exponent $m \approx 1.85 \pm 0.05$. The discrepancy could be caused by a modification of the dissipation term: instead of an ordinary Laplacian it involves the eighth power thereof. According to Lévêque and She (1995) this can significantly affect the value of inertial-range exponents.

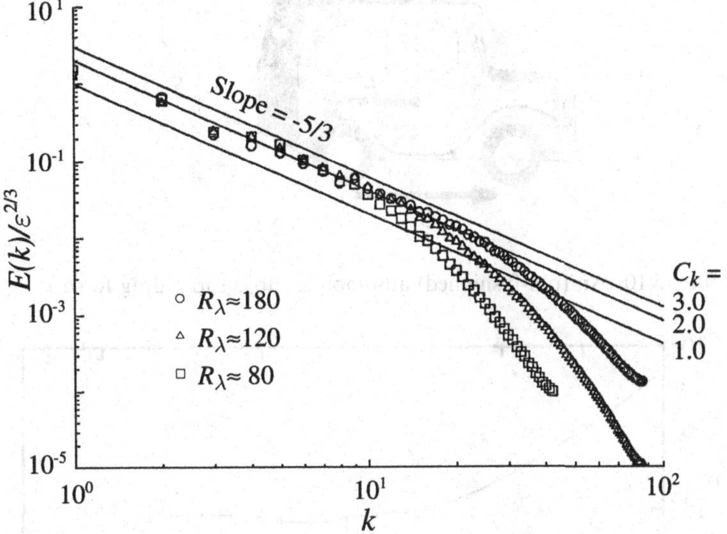

Fig. 5.9. Energy spectra in the space domain for three values of R_λ as indicated, obtained from a 256^3 computer simulation. Three different values for the Kolmogorov constant C_k are tried. Courtesy T. Sanada and K. Ishii.

5.2 The energy dissipation law

We now turn to the second law concerning the energy dissipation. Anybody with an interest in automobiles will have noticed that car manufacturers wishing to advertize the aerodynamic qualities of their products often quote the drag coefficients. The fact that such a number exists irrespective of the speed of the car is actually a confirmation of the energy dissipation law. Let us explain this. Fig. 5.10 shows an (old-fashioned) automobile moving with speed U.

The car is subject to a drag force of strength given by

$$F = \frac{1}{2} C_D \rho S U^2, \tag{5.12}$$

where S is the area of the cross-section, C_D is the (dimensionless) drag coefficient and ρ is the density of the air. A simple interpretation of this formula follows. The quantity

$$p = \rho S U^2 \tau \tag{5.13}$$

is the momentum of a cylinder of air with cross-section S, moving with speed U and of length $U\tau$. If we assume that this momentum is

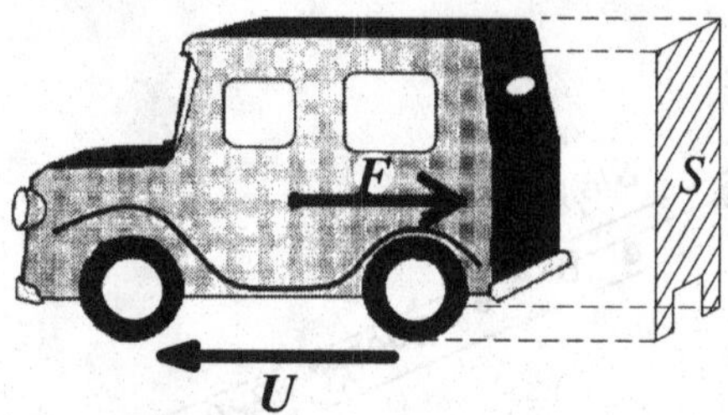

Fig. 5.10. An (old-fashioned) automobile subject to a drag force F.

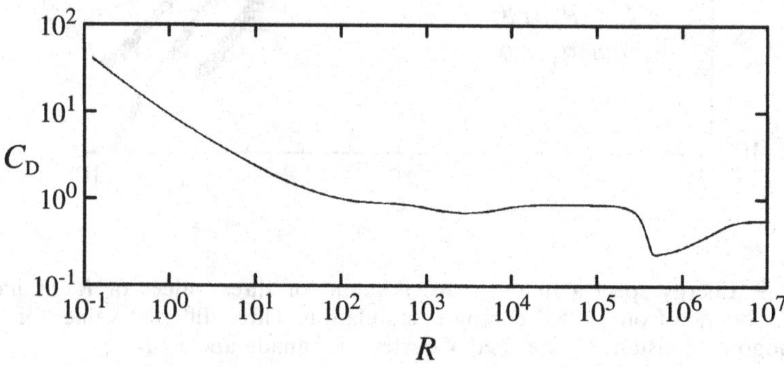

Fig. 5.11. Variation of drag coefficient with Reynolds number for circular cylinders.

completely transferred from the air to the car in a time τ, we obtain a force $f = dp/d\tau = \rho S U^2$. The presence in (5.12) of the extra factor $C_D/2 < 1$ suggests that, actually, only a fraction of this momentum is transferred. Proceeding now in a more systematic way, we can use the similarity principle for incompressible flow to show that a formula of the type of (5.12) holds, but with

$$C_D = C_D(R), \qquad R = \frac{LU}{\nu}, \tag{5.14}$$

i.e. a drag coefficient which is a function of the Reynolds number (the reference length L can be taken to be $S^{1/2}$). The similarity principle implies that $C_D(R)$ is the same for two bodies of the same shape but of different sizes. Careful measurements of $C_D(R)$ have been made for a number of shapes. Fig. 5.11 shows $C_D(R)$ for circular cylinders and is based on experimental data quoted in Tritton (1988). We observe that for very small Reynolds numbers the drag coefficient goes as R^{-1}. (Try finding out why.) At high Reynolds numbers the drag coefficient stays approximately constant, except for an accident which occurs around Reynolds numbers of a few hundred thousands.[13] Thus, taking $C_D(R)$ to be constant (at least piecewise) is a reasonable approximation. Taking this as the experimental input, let us now calculate the amount of kinetic energy dissipated (per unit time). This is equal to the work performed in moving the object (say, the car) with a speed U against the force F:

$$W = FU = \frac{1}{2} C_D \rho L^2 U^3. \tag{5.15}$$

Thus the (kinetic) energy dissipated per unit mass is[14]

$$\varepsilon = \frac{W}{\rho L^3} = \frac{1}{2} C_D \frac{U^3}{L}. \tag{5.16}$$

Insofar as the drag coefficient does not depend on the Reynolds number, the expression for ε has the remarkable property that *it does not involve the viscosity* and thus will indeed have a finite limit for $\nu \to 0$, as stated by the energy dissipation law. This law has, of course, the same limitations as the statement that the drag coefficient is independent of the Reynolds number.

Direct experimental evidence for the existence of a finite limit to the energy dissipation has been obtained by Sreenivasan (1984). Using homogeneous turbulence generated by a square-mesh grid, he measured $\varepsilon \ell_0 / v_0^3$, where ε is the energy dissipation, v_0 the r.m.s. turbulent velocity fluctuation and ℓ_0 the integral scale. He found that this nondimensionalized dissipation is independent of the Reynolds number (with about a 20% scatter) for $50 < R_\lambda < 500$. For other geometries of the grid (e.g., parallel rods) there is a larger scatter of values.

Reasonable support for the constancy of the energy dissipation also comes from computer simulations. Fig. 5.12 gives the temporal evolution of the mean energy dissipation $2\nu\Omega$ for different Reynolds numbers. The unit of time corresponds to a circulation time. The Navier–Stokes equation with periodic boundary conditions was integrated using the

13 This *drag crisis* is related to a transition in the shear layers coming from laminar separation (Schewe 1983).

14 The notation ε for the mean energy dissipation has been traditional since the work of Kolmogorov and has become a 'sacred cow'. The notation ϵ will be reserved for small parameters.

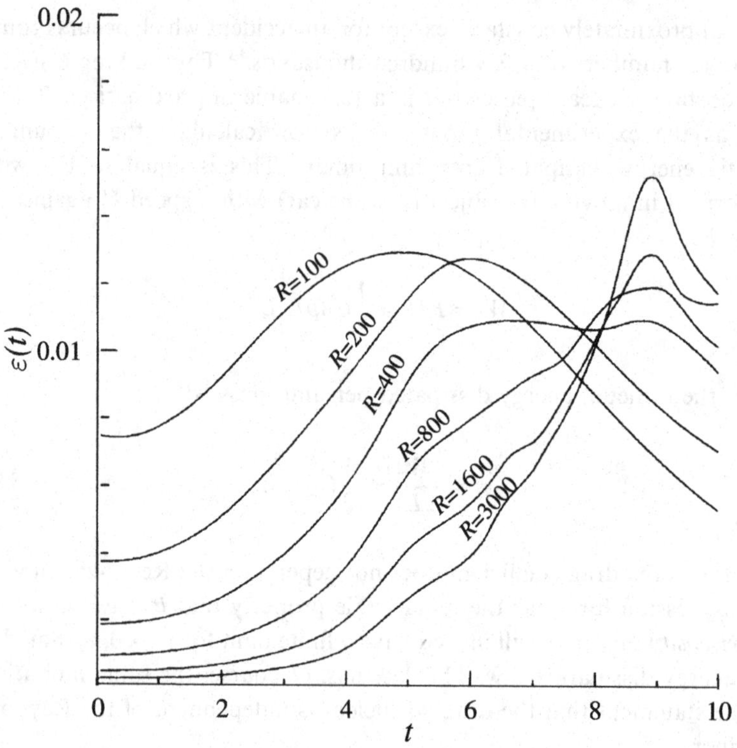

Fig. 5.12. Evolution of the energy dissipation for the Taylor–Green vortex at various Reynolds numbers (Brachet *et al.* 1983)

Taylor–Green initial condition:

$$\left.\begin{aligned} v_1 &= \sin x_1 \cos x_2 \cos x_3, \\ v_2 &= -\cos x_1 \sin x_2 \cos x_3, \\ v_3 &= 0. \end{aligned}\right\} \tag{5.17}$$

Orszag (1974) showed that one can take advantage of the symmetries of the flow to do *spectral*[15] simulations at a spatial resolution in each

[15] The spectral method (see, e.g., Gottlieb and Orszag 1977) calculates multiplications in the physical space and derivatives and their inverses in the Fourier space, shuttling back and forth by (discrete) fast Fourier transforms. Because the Fourier transform of an analytic function decreases exponentially at high wavenumbers, the spectral method can achieve an accuracy which varies exponentially with the resolution and has become the most reliable tool for 'numerical experiments' on turbulent flows at Reynolds numbers up to a few thousand.

direction roughly four times larger than would otherwise be feasible. The present state-of-the-art is 864^3 (Brachet 1990, 1991) and even 1024^3 for flows having even more symmetries than the Taylor–Green flow (Yamada, Kida and Ohkitani 1993; Boratav and Pelz 1994). Fig. 5.12 shows the energy dissipation for a range of Reynolds numbers from $R = 100$ to $R = 3000$. A similar figure (Brachet 1990, Fig. 1, not shown here) gives the energy dissipation for $R = 5000$ and is almost indistinguishable from the $R = 3000$ curve. It is seen that the energy dissipation rises to a peak value which is quite insensitive to the Reynolds number (defined here as the inverse viscosity). For example, when the Reynolds number changes by more than a factor 10 (from 400 to 5000), the peak dissipation changes by less than 40%.

In spite of such evidence, one should feel free to question the law of the energy dissipation. At the level of principles no contradiction is known to occur if one assumes that it is slightly violated, i.e. by having the energy dissipation displaying a logarithmic dependence on the viscosity or varying as a power-law with a small exponent (Saffman 1968).

Whether exact or approximate, the energy dissipation law has an important consequence. Referring to (5.16),we observe that

$$\frac{1}{2}\frac{U^3}{L} = \frac{\frac{1}{2}U^2}{L/U} = \frac{\text{kinetic energy per mass}}{\text{circulation time}}, \tag{5.18}$$

where the *circulation time* L/U is the time it takes for the fluid to move by a distance equal to the reference length L. Thus, it appears that in about one circulation time, a finite fraction of the kinetic energy carried by the moving fluid is transferred by nonlinear interactions to scales sufficiently small for viscosity to be able to remove it into heat.

Recalling that the mean energy dissipation may also be written as $2\nu\Omega$, where Ω is the mean enstrophy (2.29), we infer that at high Reynolds numbers (small ν), the enstrophy must become very large, i.e. considerable and rapid *vortex stretching* must be taking place.

6
The Kolmogorov 1941 theory

There is presently no fully *deductive* theory which starts from the Navier–Stokes equation and leads to the two basic experimental laws reported in Chapter 5. Still, it is possible to formulate *hypotheses*, compatible with these laws and leading to additional *predictions*. This was the purpose of the celebrated Kolmogorov 1941 theory (in short K41). It will here be reformulated rather freely. In Section 6.1 we shall present a modern viewpoint with emphasis on postulated *symmetries* rather than on postulated *universality*, i.e. independence on the particular mechanism by the turbulence is generated. We thereby obtain a scaling theory with an undetermined scaling exponent. The latter is determined in Section 6.2 from the 'four-fifths' law, an exact relation derived by Kolmogorov, also in 1941. The main results of the K41 theory are presented in Section 6.3.

6.1 Kolmogorov 1941 and symmetries

In Section 2.2 we made a list of known symmetries for the Navier–Stokes equation (time- and space-translations, rotations, Galilean transformations, scaling transformations, etc). What are their implications for turbulence?

Let us begin with *time-translations*. At low Reynolds numbers, if the boundary conditions and any external driving force are time-independent, the flow is steady and thus does not break the time-invariance symmetry. When the Reynolds number is increased, an Andronov–Hopf bifurcation may occur. This makes the flow time-periodic and turns the continuous time-invariance symmetry into a discrete one. When the Reynolds number is increased further the flow will usually, at some point, become chaotic. As we learned in Chapter 3, the continuous time-

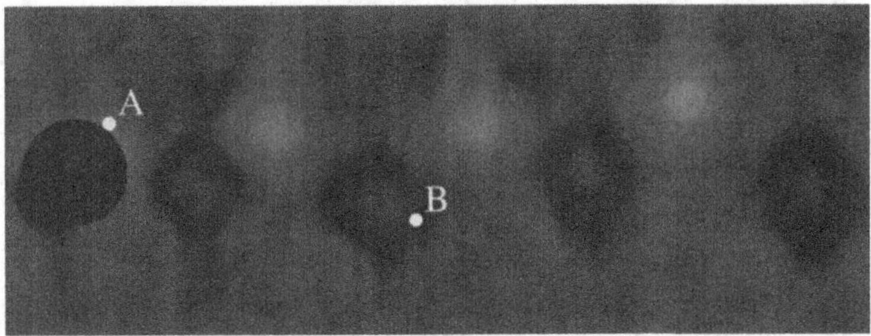

Fig. 6.1. The turbulent fluctuations cannot be the same at A, near the surface of the cylinder and at B, in the wake. Courtesy J.P. Rivet.

invariance symmetry is then restored, not for individual solutions, but at the level of the invariant measure of the dynamical system. In other words, the statistical properties of the solution become invariant under time-translations (stationary).

It is natural to try to extend this result to other symmetries of the Navier–Stokes equation. Let us consider, for example, the invariance under space-translations. Here, we meet a difficulty: let us suppose, as is indeed often the case, that the turbulence is generated by flow around a rigid body, say a cylinder as shown in Fig. 6.1. The presence of this body will trivially break the translation symmetry. For example the r.m.s. velocity fluctuations at point A very close to the cylinder cannot be the same as at point B somewhere in the wake, since the velocity and its fluctuations must vanish at rigid boundaries. Such a turbulent flow can therefore never be strictly homogeneous (i.e. statistically invariant under space-translations). However, discrete translation-invariance is conceivable if the bodies generating the turbulence are arranged in a spatially periodic fashion, such as for the flow past the grid shown in Fig. 6.2. Translations parallel to the grid by a multiple of the mesh preserve the geometry of the flow and thus leave the flow invariant as long as the Reynolds number is sufficiently low. At higher Reynolds numbers, when the flow becomes turbulent, its statistical properties will be invariant under such translations.

Similar remarks can be made about all the other symmetries of the Navier–Stokes equation: *the mechanisms responsible for the generation of the turbulent flow are usually not consistent with most of the possible symmetries* (listed in Section 2.2).

However, we remember from Chapter 1 that the qualitative aspect of many turbulent flows suggests some form of homogeneity, isotropy and,

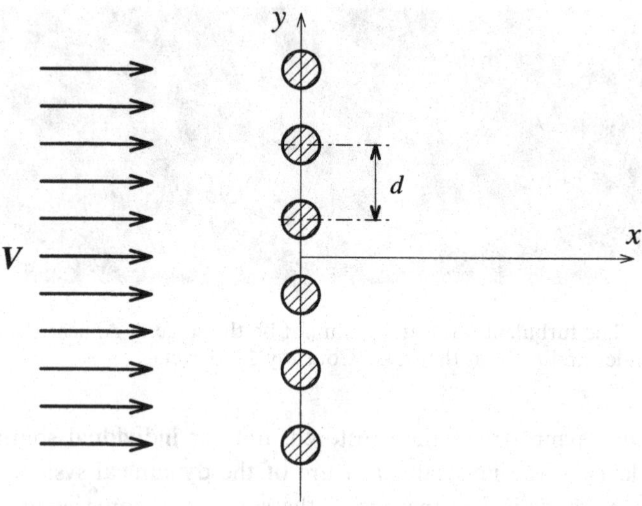

Fig. 6.2. A periodic array of cylinders forming a grid.

possibly, scale-invariance. Furthermore, the power-law behavior of the second order structure function (Section 5.1) is clearly indicative of some form of scale-invariance.

Such conflicting aspects can be reconciled through the following hypothesis.[1]

H1 *In the limit of infinite Reynolds numbers, all the possible symmetries of the Navier–Stokes equation, usually broken by the mechanisms producing the turbulent flow, are restored in a statistical sense at small scales and away from boundaries.*

By 'small scales' we understand scales $\ell \ll \ell_0$ where ℓ_0, the *integral scale*, is characteristic of the production of turbulence, for example, the diameter of the cylinder in the example of Fig. 6.1. Small-scale homogeneity is defined as the property of having homogeneous increments (Section 4.3), i.e. in terms of velocity increments:[2]

$$\delta \boldsymbol{v}(\boldsymbol{r}, \boldsymbol{\ell}) \equiv \boldsymbol{v}(\boldsymbol{r} + \boldsymbol{\ell}) - \boldsymbol{v}(\boldsymbol{r}). \tag{6.1}$$

[1] Because our hypotheses are formulated in a way rather different from Kolmogorov's, the numbering is not the same as in his first 1941 paper.

[2] High-pass filtered velocities defined in Section 2.4 with $K\ell_0 \gg 1$ could be used as well as increments.

Specifically, it is assumed that

$$\delta \boldsymbol{v}(\boldsymbol{r}+\boldsymbol{\rho}, \boldsymbol{\ell}) \stackrel{\text{law}}{=} \delta \boldsymbol{v}(\boldsymbol{r}, \boldsymbol{\ell}), \tag{6.2}$$

for all increments $\boldsymbol{\ell}$ and all displacements $\boldsymbol{\rho}$ which are small compared to the integral scale. (Equality in law is defined in Section 4.3.)·

Similarly, *isotropy* means, in the present context, that the statistical properties of velocity increments are invariant under simultaneous rotations of $\boldsymbol{\ell}$ and $\delta \boldsymbol{v}$. For *parity*, $\boldsymbol{\ell}$ and $\delta \boldsymbol{v}$ are to be reversed simultaneously.

As for the scale-invariance, since there are infinitely many scaling groups, which depend on the choice of the scaling exponent h, we need an additional hypothesis.

H2 *Under the same assumptions as in H1, the turbulent flow is self-similar at small scales, i.e. it possesses a unique scaling exponent h.*

Thus, there exists a scaling exponent $h \in \mathbb{R}$ such that

$$\delta \boldsymbol{v}(\boldsymbol{r}, \lambda \boldsymbol{\ell}) \stackrel{\text{law}}{=} \lambda^h \, \delta \boldsymbol{v}(\boldsymbol{r}, \boldsymbol{\ell}), \qquad \forall \lambda \in \mathbb{R}_+, \tag{6.3}$$

for all $\boldsymbol{r}$ and all increments $\boldsymbol{\ell}$ and $\lambda \boldsymbol{\ell}$ small compared to the integral scale.

As we shall see in Section 6.2, the unique scaling exponent h is equal to $1/3$ and is determined by postulating the energy dissipation law introduced in Section 5.2. This law constitutes the third hypothesis.

H3 *Under the same assumptions as in H1, the turbulent flow has a finite nonvanishing mean rate of dissipation ε per unit mass.*

For H3, we must keep the integral scale ℓ_0 and the r.m.s. velocity fluctuations v_0 fixed, and let $\nu \to 0$. Otherwise, ε must be nondimensionalized through division by v_0^3/ℓ_0.

At this point, we should mention that in his first 1941 paper, Kolmogorov was actually postulating something quite different from the hypotheses listed so far:

Kolmogorov's second universality assumption[3] (not used here). *In the limit of infinite Reynolds number, all the small-scale statistical properties are uniquely and universally determined by the scale ℓ and the mean energy dissipation rate ε.*

The first universality assumption of Kolmogorov will be given in Section 6.3.2.

As an illustration of the universality assumption, consider the second order structure function $\langle(\delta v(\ell))^2\rangle$. Straightforward dimensional analysis shows that this quantity has dimensions $[L]^2[T]^{-2}$, where $[L]$ and $[T]$ are

[3] Called by Kolmogorov the 'second hypothesis of similarity' and recast here in slightly different language.

units of length and time, respectively. Since the mean energy dissipation rate per unit mass, ε, obviously has the dimensions $[L]^2[T]^{-3}$, it follows from the universality assumption that

$$\langle(\delta \boldsymbol{v}(\ell))^2\rangle = C\varepsilon^{2/3}\ell^{2/3}, \tag{6.4}$$

with a universal dimensionless constant C. On the other hand, by the hypothesis H2, the second order structure function should be proportional to ℓ^{2h}. (We shall go through this kind of argument in more detail in Section 6.3.1.) Thus $h = 1/3$ is the only consistent value.

The trouble is that Landau gave an argument indicating that constants such as C should not be universal (Section 6.4). We shall therefore refrain from using the universality assumption and derive the value $h = 1/3$ by other means.

6.2 Kolmogorov's four-fifths law

In his third 1941 turbulence paper Kolmogorov (1941c) found that an *exact* relation can be derived for the third order longitudinal structure function, the average of the cube of the longitudinal velocity increment. He assumed homogeneity, isotropy and hypothesis H3 about the finiteness of the energy dissipation. Without any further assumptions he derived the following result from the Navier–Stokes equation:

Four-fifths law. *In the limit of infinite Reynolds number, the third order (longitudinal) structure function of homogeneous isotropic turbulence, evaluated for increments ℓ small compared to the integral scale, is given in terms of the mean energy dissipation per unit mass ε (assumed to remain finite and nonvanishing) by*

$$\langle(\delta v_\parallel(\boldsymbol{r}, \boldsymbol{\ell}))^3\rangle = -\frac{4}{5}\varepsilon\ell. \tag{6.5}$$

This is one of the most important results in fully developed turbulence because it is both exact and nontrivial. It thus constitutes a kind of 'boundary condition' on theories of turbulence: such theories, to be acceptable, must either satisfy the four-fifths law, or explicitly violate the assumptions made in deriving it. We have thus been led to give a rather detailed derivation of this law. Actually, Kolmogorov did not include much detail in his 1941 derivation of the four-fifths law. He was using a previously derived relation of Kármán and Howarth and some very simple arguments (less than 20 lines) which appear plausible[4] but which can be made more systematic.[5]

6.2.1 The Kármán–Howarth–Monin relation for anisotropic turbulence

In his 1941 paper on the four-fifths law, Kolmogorov assumed a freely decaying turbulent flow. 'Realistic turbulence' is usually maintained by such mechanisms as the interaction of an incoming flow with boundaries, thermal convective instability, etc. The inhomogeneities induced by the maintaining mechanism may be weak enough to be partially ignored at small scales and far from boundaries. However, something is needed to replenish the energy dissipated by viscosity. A simple device which achieves this function, already used in Section 2.4, is to add a forcing term $\boldsymbol{f}(t, \boldsymbol{r})$ in the Navier–Stokes equation:

$$\left.\begin{aligned} \partial_t \boldsymbol{v} + \boldsymbol{v} \cdot \nabla \boldsymbol{v} &= -\nabla p + \nu \nabla^2 \boldsymbol{v} + \boldsymbol{f}, \\ \nabla \cdot \boldsymbol{v} &= 0. \end{aligned}\right\} \tag{6.6}$$

We assume that this 'stirring force' is active only at large scales, so as to model the mechanism of production of turbulence which often involves some large-scale instability (Edwards 1964, 1965). In deriving the four-fifths law, there is considerable freedom in the choice of the stirring force because only the mean energy input/dissipation will turn out to be relevant. We assume that the random force $\boldsymbol{f}(t, \boldsymbol{r})$ is stationary, homogeneous, i.e. its statistical properties are invariant under translations in time and space.[6] So far, we do not assume isotropy. We also assume that the solution of the Navier–Stokes equation is homogeneous, but not necessarily stationary, so as to be able to cover instances where the external force vanishes. We finally assume that all moments required in subsequent manipulations are finite (as long as $\nu > 0$).

We define

$$\varepsilon(\boldsymbol{\ell}) \equiv -\partial_t \frac{1}{2} \langle \boldsymbol{v}(\boldsymbol{r}) \cdot \boldsymbol{v}(\boldsymbol{r} + \boldsymbol{\ell}) \rangle \Big|_{\mathrm{NL}}, \tag{6.7}$$

[4] Just before writing the equivalent of (6.5), Kolmogorov (1941c) uses the following words: *i.e. we may thus assume that ....* This can be misread as an *additional* hypothesis, when actually it is just cautious language meant to bring out that his derivation is not fully rigorous.

[5] As observed by G. Eyink (private communication), our derivation resembles the 'point-splitting regularization' method used by Schwinger (1951a) in the first derivation of the 'axial anomaly' in quantum electrodynamics.

[6] Edwards (1965) furthermore assumed the force to be delta-correlated in time; this has some technical advantages but, foremost, it ensures that the problem is invariant under random Galilean transformations (see Section 6.2.5).

where the notation $\partial_t\,(\cdot)|_{\rm NL}$ stands for 'contribution to the time-rate-of-change stemming from the nonlinear terms (advection and pressure) in the Navier–Stokes equation'. $\varepsilon(\boldsymbol{\ell})$ has the dimension of a time-rate-of-change of an energy per unit mass and will be called the *physical-space energy flux.* In Section 6.2.2 it will be shown that the more usual *Fourier-space energy flux* is expressible in terms of the Fourier transform of $\varepsilon(\boldsymbol{\ell})$.

Kármán–Howarth–Monin relation. *Homogeneous (but not necessarily isotropic) solutions of the Navier–Stokes equation (6.6) satisfy*

$$\begin{aligned}\varepsilon(\boldsymbol{\ell}) &= -\frac{1}{4}\nabla_{\boldsymbol{\ell}}\cdot\langle|\delta\boldsymbol{v}(\boldsymbol{\ell})|^2\delta\boldsymbol{v}(\boldsymbol{\ell})\rangle\\ &= -\partial_t\frac{1}{2}\langle\boldsymbol{v}(\boldsymbol{r})\cdot\boldsymbol{v}(\boldsymbol{r}+\boldsymbol{\ell})\rangle+\left\langle\boldsymbol{v}(\boldsymbol{r})\cdot\frac{\boldsymbol{f}(\boldsymbol{r}+\boldsymbol{\ell})+\boldsymbol{f}(\boldsymbol{r}-\boldsymbol{\ell})}{2}\right\rangle\\ &\quad+\nu\nabla_{\boldsymbol{\ell}}^2\langle\boldsymbol{v}(\boldsymbol{r})\cdot\boldsymbol{v}(\boldsymbol{r}+\boldsymbol{\ell})\rangle,\end{aligned}\tag{6.8}$$

where $\nabla_{\boldsymbol{\ell}}$ denotes partial derivatives with respect to the (vector) increment $\boldsymbol{\ell}$ and

$$\langle|\delta\boldsymbol{v}(\boldsymbol{\ell})|^2\delta\boldsymbol{v}(\boldsymbol{\ell})\rangle\equiv\langle|\delta\boldsymbol{v}(\boldsymbol{r},\boldsymbol{\ell})|^2\delta\boldsymbol{v}(\boldsymbol{r},\boldsymbol{\ell})\rangle,\tag{6.9}$$

the velocity increment $\delta\boldsymbol{v}(\boldsymbol{r},\boldsymbol{\ell})$ being defined in (6.1). (Observe that, after averaging, no dependence on $\boldsymbol{r}$ is left because of homogeneity.)

Proof. Hereafter, v_i, f_i, $\boldsymbol{r}'$, v_i', ∂_i, ∂_i', and ∇_{ℓ_i} denote $v_i(\boldsymbol{r})$, $f_i(\boldsymbol{r})$, $\boldsymbol{r}+\boldsymbol{\ell}$, $v_i(\boldsymbol{r}')$, $\partial/\partial x_i$, $\partial/\partial x_i'$ and $\partial/\partial\ell_i$, respectively. After averaging, we have, by homogeneity,

$$\partial_i\langle(\cdot)\rangle=-\partial_i'\langle(\cdot)\rangle=-\nabla_{\ell_i}\langle(\cdot)\rangle.\tag{6.10}$$

Starting from the Navier–Stokes equation (6.6), we obtain

$$\begin{aligned}\partial_t\frac{1}{2}\langle v_iv_i'\rangle &= -\frac{1}{2}\partial_j\langle v_iv_jv_i'\rangle-\frac{1}{2}\partial_j'\langle v_i'v_j'v_i\rangle\\ &\quad-\frac{1}{2}\langle v_i'\partial_ip\rangle-\frac{1}{2}\langle v_i\partial_i'p'\rangle\\ &\quad+\frac{1}{2}\langle v_i'f_i\rangle+\frac{1}{2}\langle v_if_i'\rangle\\ &\quad+\frac{1}{2}\nu\left(\partial_{jj}+\partial_{jj}'\right)\langle v_iv_i'\rangle.\end{aligned}\tag{6.11}$$

The terms in the second line vanish by incompressibility. The first term in the third line can be rewritten as $\langle\boldsymbol{v}(\boldsymbol{r})\cdot\boldsymbol{f}(\boldsymbol{r}-\boldsymbol{\ell})\rangle$, using the identity

$$\langle\boldsymbol{v}(\boldsymbol{r}+\boldsymbol{\ell})\cdot\boldsymbol{f}(\boldsymbol{r})\rangle=\langle\boldsymbol{v}(\boldsymbol{r})\cdot\boldsymbol{f}(\boldsymbol{r}-\boldsymbol{\ell})\rangle,\tag{6.12}$$

which follows from homogeneity. Again, by homogeneity, the terms in the fourth line of (6.11) can be transformed into $\nu\nabla^2_{\ell}\langle \boldsymbol{v}(\boldsymbol{r})\cdot\boldsymbol{v}(\boldsymbol{r}+\boldsymbol{\ell})\rangle$. Next, we observe that

$$\begin{aligned}\langle|\delta \boldsymbol{v}|^2\delta v_j\rangle &= +\langle(v_i'-v_i)(v_i'-v_i)(v_j'-v_j)\rangle \\ &= -\langle v_i'v_i'v_j\rangle+\langle v_iv_iv_j'\rangle \\ &\quad -2\langle v_iv_i'v_j'\rangle+2\langle v_iv_i'v_j\rangle. \end{aligned} \tag{6.13}$$

(The additional terms $\langle v_i'v_i'v_j'\rangle-\langle v_iv_iv_j\rangle$ cancel by homogeneity.) If we now apply the operator ∇_{ℓ_j} to the r.h.s. of (6.13), by incompressibility, only the last two terms contribute and we obtain, within a factor 4, the same as the first line on the r.h.s. of (6.11). The Kármán–Howarth–Monin relation (6.8) follows. *QED.*

What we call here the Kármán–Howarth–Monin relation is an anisotropic generalization of a relation first established by von Kármán and Howarth(1938). The anisotropic version is found in Monin and Yaglom (1975, p. 403) where it is attributed 'largely' to Monin (1959); hence the name chosen here. Our derivation of the relation is more straightforward than that given in Monin and Yaglom (1975). In the isotropic case it is usually called the Kármán–Howarth relation.

We observe that if, in the Kármán–Howarth–Monin relation, we hold the viscosity $\nu > 0$ fixed and let the separation $\boldsymbol{\ell} \to 0$, in (6.8) the term $\nabla_{\boldsymbol{\ell}}\cdot\langle|\delta\boldsymbol{v}(\boldsymbol{\ell})|^2\delta\boldsymbol{v}(\boldsymbol{\ell})\rangle$ tends to zero. Indeed, velocity increments vary linearly for very small increments (assuming smoothness for $\nu > 0$). We are thus left with

$$\partial_t\frac{1}{2}\langle v^2\rangle = \langle \boldsymbol{f}(\boldsymbol{r})\cdot\boldsymbol{v}(\boldsymbol{r})\rangle+\nu\langle\boldsymbol{v}(\boldsymbol{r})\cdot\nabla^2\boldsymbol{v}(\boldsymbol{r})\rangle, \tag{6.14}$$

an equation which just expresses that the only changes in the mean energy come from the input through the force and the viscous energy dissipation. Actually, as we shall now see, (6.8) for $\ell \neq 0$ is essentially an energy-flux relation.

6.2.2 The energy flux for homogeneous turbulence

Our starting point will be the scale-by-scale energy budget equation (2.48) which relates the (mean) cumulative energy $\mathscr{E}_K$, the (mean) cumulative enstrophy Ω_K, the (mean) cumulative energy injection $\mathscr{F}_K$ and the (mean) energy flux Π_K. Since we are now working with random homogeneous functions rather than with periodic functions, the Fourier series used in Section 2.4 to define filtering ((2.34)–(2.36)) must be replaced by Fourier integrals. For example, the low-pass filtered velocity $\boldsymbol{v}_K^<$ is now related to the velocity field $\boldsymbol{v}$ and to its Fourier transform $\hat{\boldsymbol{v}}$ by

$$\left.\begin{aligned} \boldsymbol{v}(\boldsymbol{r}) &= \int_{\mathbb{R}^3} d^3k\, e^{i\boldsymbol{k}\cdot\boldsymbol{r}}\, \hat{\boldsymbol{v}}_{\boldsymbol{k}}, \\ \hat{\boldsymbol{v}}_k &= \frac{1}{(2\pi)^3}\int_{\mathbb{R}^3} d^3r\, e^{-i\boldsymbol{k}\cdot\boldsymbol{r}}\, \boldsymbol{v}(\boldsymbol{r}), \\ \boldsymbol{v}_K^<(\boldsymbol{r}) &= \int_{|\boldsymbol{k}|\le K} d^3k\, e^{i\boldsymbol{k}\cdot\boldsymbol{r}}\, \hat{\boldsymbol{v}}_{\boldsymbol{k}}. \end{aligned}\right\} \tag{6.15}$$

Similarly, angular brackets are now to be interpreted as ensemble averages rather than spatial averages over the periodicity box. With this reinterpretation, all the results obtained in Section 2.4 remain literally true.

For convenience we rewrite the scale-by-scale energy budget relation (2.48) as

$$\partial_t \mathscr{E}_K + \Pi_K = \mathscr{F}_K - 2\nu\Omega_K. \tag{6.16}$$

In Section 2.4 we gave an expression (2.52) for the energy flux Π_K, involving the low- and high-pass filtered velocities. We shall now use the Kármán–Howarth–Monin relation (6.8) to reexpress the energy flux in terms of third order moments of velocity increments.

Expression of the energy flux for homogeneous turbulence. The energy flux through wavenumber K is expressed in terms of third order velocity moments by

$$\Pi_K = -\frac{1}{8\pi^2}\int_{\mathbb{R}^3} d^3\ell\, \frac{\sin(K\ell)}{\ell}\nabla_{\boldsymbol{\ell}}\cdot\left[\frac{\boldsymbol{\ell}}{\ell^2}\nabla_{\boldsymbol{\ell}}\cdot\langle|\delta\boldsymbol{v}(\boldsymbol{\ell})|^2\delta\boldsymbol{v}(\boldsymbol{\ell})\rangle\right]. \tag{6.17}$$

Proof of (6.17). We observe that (6.16) may be rewritten as

$$\Pi_K = -\partial_t\mathscr{E}_K\big|_{\mathrm{NL}}, \tag{6.18}$$

where $\mathscr{E}_K$ is the cumulative energy spectrum

$$\mathscr{E}_K = \frac{1}{2}\langle|\boldsymbol{v}_K^<|^2\rangle. \tag{6.19}$$

From (6.7), (6.15), (6.18), (6.19), the identity

$$\int_{\mathbb{R}^3} d^3k\, e^{i\boldsymbol{k}\cdot\boldsymbol{r}} = (2\pi)^3\delta(\boldsymbol{r}) \tag{6.20}$$

and the assumption of homogeneity, we readily obtain

$$\Pi_K = \frac{1}{(2\pi)^3}\int_{|\boldsymbol{k}|\le K} d^3k \int_{\mathbb{R}^3} d^3\ell\, e^{i\boldsymbol{k}\cdot\boldsymbol{\ell}}\, \varepsilon(\boldsymbol{\ell}). \tag{6.21}$$

Interchanging the integrations in (6.21), we see that the integration over $\boldsymbol{k}$ can be performed. It is simplest to use spherical coordinates with the polar axis along $\boldsymbol{\ell}$. After a few lines of algebra, this leads to

$$\Pi_K = \frac{1}{2\pi^2}\int_{\mathbb{R}^3} d^3\ell\, \frac{\sin(K\ell) - K\ell\cos(K\ell)}{\ell^3}\,\varepsilon(\boldsymbol{\ell}). \tag{6.22}$$

A consequence of Kolmogorov's formula (6.5) is the existence of a range of ℓs over which $\varepsilon(\boldsymbol{\ell})$ is independent of ℓ. When this assumption is made in (6.22), the integral is found to diverge at large ℓs as $\int^\infty \cos(K\ell)d\ell$. Actually, there is no divergence, because correlations tend to zero for separation ℓ much larger than the integral scale. However, it is preferable to perform an integration by parts and recast (6.22) as:

$$\Pi_K = \frac{1}{2\pi^2}\int_{\mathbb{R}^3} d^3\ell\, \frac{\sin(K\ell)}{\ell}\,\nabla_{\boldsymbol{\ell}}\cdot\left[\varepsilon(\boldsymbol{\ell})\frac{\boldsymbol{\ell}}{\ell^2}\right]. \tag{6.23}$$

Substituting the value of $\varepsilon(\boldsymbol{\ell})$ given by (6.8), we obtain (6.17). *QED.*

6.2.3 The energy flux for homogeneous isotropic turbulence

The energy flux relation (6.17) is enough to derive the value of the scaling exponent $h = 1/3$ within the K41 framework. Additional information is, however, obtained if we make use of the assumption of isotropy. It is then possible to express the energy flux in terms of third order moments of longitudinal velocity increments. These are also much simpler to measure experimentally (via the Taylor hypothesis discussed in Section 5.1).

Expression of the energy flux for homogeneous isotropic turbulence. The energy flux through wavenumber K is expressed in terms of the *third order longitudinal structure function* $S_3(\ell) = \langle(\delta v_\parallel(\boldsymbol{r},\boldsymbol{\ell}))^3\rangle$ by

$$\Pi_K = -\frac{1}{6\pi}\int_0^\infty d\ell\,\frac{\sin(K\ell)}{\ell}(1+\ell\partial_\ell)(3+\ell\partial_\ell)(5+\ell\partial_\ell)\frac{S_3(\ell)}{\ell}, \tag{6.24}$$

where $\partial_\ell \equiv \partial/\partial\ell$.

Proof of (6.24). We begin with some preliminary material, following closely the presentation of Landau and Lifshitz (1987, Section 34). Using the same notation as in the proof of the Kármán–Howarth–Monin relation (Section 6.2.1), we define

$$b_{ij,m} = \langle v_i v_j v'_m\rangle, \tag{6.25}$$

$$B_{ijm} = \langle(v'_i - v_i)(v'_j - v_j)(v'_m - v_m)\rangle. \tag{6.26}$$

Isotropy implies that the tensor $b_{ij,m}$ is expressible in terms of Kronecker deltas and the components of the unit vector

$$\ell_i^0 \equiv \frac{\ell_i}{\ell}. \tag{6.27}$$

The most general form of such a third order tensor, symmetrical in i and j, is

$$b_{ij,m} = C(\ell)\delta_{ij}\ell_m^0 + D(\ell)\left(\delta_{im}\ell_j^0 + \delta_{jm}\ell_i^0\right) + F(\ell)\ell_i^0\ell_j^0\ell_m^0. \tag{6.28}$$

Incompressibility implies

$$\partial'_m b_{ij,m} = 0, \qquad \forall i, j. \tag{6.29}$$

Using this and (6.28), by a simple calculation we obtain

$$\left[\ell^2(3C + 2D + F)\right]' = 0, \qquad C' + \frac{2}{\ell}(C + D) = 0, \tag{6.30}$$

where the primes denote derivatives with respect to ℓ. The only solution of the first equation which is consistent with the finiteness (actually, the vanishing) of $b_{ij,m}$ at $\ell = 0$ is

$$3C + 2D + F = 0. \tag{6.31}$$

Using (6.28), (6.30) and (6.31), we reexpress everything in terms of the coefficient $C(\ell)$, so that (6.28) becomes

$$b_{ij,m} = C\delta_{ij}\ell_m^0 - (C + \ell C'/2)\left(\delta_{im}\ell_j^0 + \delta_{jm}\ell_i^0\right) + (\ell C' - C)\ell_i^0\ell_j^0\ell_m^0. \tag{6.32}$$

Similarly, B_{ijm}, given by (6.26), becomes

$$B_{ijm} = -2(\ell C' + C)\left(\delta_{ij}\ell_m^0 + \delta_{im}\ell_j^0 + \delta_{jm}\ell_i^0\right) + 6(\ell C' - C)\ell_i^0\ell_j^0\ell_m^0. \tag{6.33}$$

Having established these preliminary results, we now observe that (from (6.33))

$$S_3(\ell) = \langle\left(\delta v_\parallel(\boldsymbol{r}, \boldsymbol{\ell})\right)^3\rangle = B_{ijm}\ell_i^0\ell_j^0\ell_m^0 = -12C, \tag{6.34}$$

and (from (6.33) and isotropy)

$$\langle|\delta \boldsymbol{v}(\boldsymbol{\ell})|^2\delta \boldsymbol{v}(\boldsymbol{\ell})\rangle = B_{iim}\ell_m^0\frac{\boldsymbol{\ell}}{\ell} = \left(-4C' - 16\frac{C}{\ell}\right)\boldsymbol{\ell}. \tag{6.35}$$

Hence, by (6.8) and (6.34),

$$\varepsilon(\boldsymbol{\ell}) \equiv -\frac{1}{4}\nabla_{\boldsymbol{\ell}} \cdot \langle |\delta \boldsymbol{v}(\boldsymbol{\ell})|^2 \delta \boldsymbol{v}(\boldsymbol{\ell})\rangle = -\frac{1}{12}(3+\ell\partial_\ell)(5+\ell\partial_\ell)\frac{S_3(\ell)}{\ell}. \tag{6.36}$$

Substituting this expression into (6.17), using (6.8) and performing the integration over angular variables, we obtain (6.24). *QED.*

As a corollary of (6.24), we establish the following relation.

Energy transfer relation for homogeneous isotropic turbulence. Homogeneous isotropic turbulence satisfies the following *energy transfer relation*:

$$\partial_t E(k) = T(k) + F(k) - 2\nu k^2 E(k), \tag{6.37}$$

$$T(k) \equiv -\frac{\partial}{\partial_k}\Pi_k \tag{6.38}$$

$$= \int_0^\infty \cos(k\ell)(1+\ell\partial_\ell)(3+\ell\partial_\ell)(5+\ell\partial_\ell)\frac{S_3(\ell)}{6\pi\ell}d\ell, \tag{6.39}$$

where

$$E(k) = \frac{\partial}{\partial_k}\frac{1}{2}\langle |\boldsymbol{v}_k^<|^2\rangle, \qquad F(k) = \frac{\partial}{\partial_k}\langle \boldsymbol{f}_k^< \cdot \boldsymbol{v}_k^<\rangle, \tag{6.40}$$

are the energy spectrum and the energy injection spectrum, respectively.

Proof. Eqs. (6.37)–(6.39) follow from the scale-by-scale energy budget equation (2.48) upon substitution of k for K, differentiation with respect to k, and use of (2.49)–(2.51) and expression (6.24) for the energy flux (after differentiation with respect to k under the integral). *QED.*

The function $T(k)$ which appears in (6.37)–(6.39) is called the *energy transfer*. Being the derivative with respect to the wavenumber k of minus the energy flux, it is also the time-rate-of-change per unit wavenumber of the energy spectrum, due to nonlinear interactions. Hence, by (4.58) and (6.7), the transfer is expressible in terms of the physical-space flux as

$$T(k) = -\frac{2}{\pi}\int_0^\infty k\ell \sin(k\ell)\,\varepsilon(\ell)\,d\ell. \tag{6.41}$$

Relation (6.39) is perhaps more practical, because the third order structure function is directly accessible to experiments. In this book there is more emphasis on the energy flux than on the energy transfer for two reasons: (i) the constancy of the energy flux in the inertial range makes it a physically more relevant quantity, (ii) the energy flux can still be defined when the wavevectors are discrete (e.g., for 2π-periodic non-random flows), while the energy transfer, being a k-derivative, cannot.

6.2.4 From the energy flux relation to the four-fifths law

Up to this point, we have assumed only *homogeneity and isotropy.* We shall now introduce additional assumptions specific to *fully developed turbulence.*

(i) The driving force $\boldsymbol{f}(t,\boldsymbol{r})$ is acting only at large scales. Specifically, we assume that the force has essentially no contributions coming from wavenumbers $\gg K_c \sim \ell_0^{-1}$, where ℓ_0 is the integral scale. In other words,

$$\boldsymbol{f}_K^{<}(t,\boldsymbol{r}) \simeq \boldsymbol{f}(t,\boldsymbol{r}), \qquad \text{for } K \gg K_c, \tag{6.42}$$

where the low-pass filtered force $\boldsymbol{f}_K^{<}(t,\boldsymbol{r})$ is defined in Section 2.4.

(ii) For large times, the solution of the Navier–Stokes equation tends to a statistically stationary state with a finite mean energy per unit mass.[7]

(iii) In the infinite Reynolds number limit ($\nu \to 0$), the mean energy dissipation per unit mass[8] $\varepsilon(\nu)$ tends to a finite positive limit (hypothesis H3 of Section 6.1):

$$\lim_{\nu\to 0} \varepsilon(\nu) = \varepsilon > 0. \tag{6.43}$$

(iv) Scale-invariance (hypotheses H1 and H2) *is not assumed.*

We now turn to the consequences of these assumptions. Stationarity (item (ii)) implies that the time-derivative terms may be omitted in the global energy budget equation (6.14) and in the scale-by-scale energy budget equation (6.16), which become respectively

$$\langle \boldsymbol{f} \cdot \boldsymbol{v} \rangle = -\nu \langle \boldsymbol{v} \cdot \nabla^2 \boldsymbol{v} \rangle = \varepsilon(\nu) \tag{6.44}$$

and

$$\Pi_K = \mathscr{F}_K - 2\nu \Omega_K. \tag{6.45}$$

Consider the energy injection term $\mathscr{F}_K$ for $K \gg K_c$. Using item (i) and (2.51), we obtain

$$\mathscr{F}_K = \langle \boldsymbol{f}_K^{<} \cdot \boldsymbol{v} \rangle \simeq \langle \boldsymbol{f} \cdot \boldsymbol{v} \rangle = \varepsilon(\nu). \tag{6.46}$$

[7] This assumption is not valid for two-dimensional turbulence and other problems with an 'inverse' cascade of energy, i.e. from small to large scales. The energy of the flow may then grow without bound as $t \to \infty$ (see also Section 9.7).

[8] This is the only place where it is preferable to bring out explicitly the dependence of ε on the viscosity.

Consider the energy dissipation term $2\nu\Omega_K$. We claim that, *for fixed* K,

$$\lim_{\nu\to 0} 2\nu\Omega_K = 0. \tag{6.47}$$

Indeed, we have

$$\begin{aligned} 2\nu\Omega_K &= \nu\langle|\omega_K^<|^2\rangle \le \nu K^2\langle|v_K^<|^2\rangle \\ &\le \nu K^2\langle|v|^2\rangle = 2\nu K^2 E, \end{aligned} \tag{6.48}$$

where E is the mean energy (assumed bounded by item (ii)). The first equality follows from (2.50) and the first inequality follows from the fact that the curl operator, acting on low-pass filtered vector fields with a cutoff at wavenumber K, has a norm bounded by K.

In (6.45), we take $K \gg K_c$ and let $\nu \to 0$. Using (6.43) (item (iii)), (6.46) and (6.47), we obtain

$$\lim_{\nu\to 0} \Pi_K = \varepsilon, \qquad \forall K \gg K_c. \tag{6.49}$$

What we have established is physically quite obvious: in the statistically stationary state, the energy flux is independent of the scale under consideration and equal to the energy input/dissipation, provided that there is no direct energy injection ($K \gg K_c$) and no direct dissipation ($\nu \to 0$). Combining (6.49) with the relation (6.24) for the energy flux and changing the integration variable from ℓ to $x = K\ell$, we obtain (the limit $\nu \to 0$ is understood):

$$\Pi_K = -\int_0^\infty dx \frac{\sin x}{x} F\left(\frac{x}{K}\right) = \varepsilon, \qquad \forall K \gg K_c. \tag{6.50}$$

Here,

$$F(\ell) \equiv (1 + \ell\partial_\ell)(3 + \ell\partial_\ell)(5 + \ell\partial_\ell)\frac{S_3(\ell)}{6\pi\ell}. \tag{6.51}$$

We now observe that the large-K behavior of the integral in (6.50) involves only the small-ℓ behavior of $F(\ell)$ and that we have the identity $\int_0^\infty dx\,(\sin x/x) = \pi/2$. We thus obtain that, for small ℓ,

$$F(\ell) \simeq -\frac{2}{\pi}\varepsilon. \tag{6.52}$$

After substitution into (6.51), we obtain a linear third order differential equation for $S_3(\ell)$. It is straightforward to solve this (using $\ln \ell$ as the independent variable and $S_3(\ell)/\ell$ as the dependent variable). The only solution which goes to zero with ℓ is[9]

$$S_3(\ell) = -\frac{4}{5}\varepsilon\ell. \tag{6.53}$$

This completes the derivation of the four-fifths law.

6.2.5 Remarks on Kolmogorov's four-fifths law

First, we discuss Kolmogorov's own derivation of the four-fifths law. As already stated, his derivation is very short. He essentially used the Kármán–Howarth equation, an isotropic version of (6.8). We recall that he was working with freely decaying turbulence (no driving force). He also assumed (in our notation) the existence of an inertial range of values for ℓ: (i) sufficiently small to approximate the time-rate-of-change of the velocity correlation function for separation ℓ by its value for zero separation, namely -2ε, and (ii) sufficiently large so that the dissipation term $\nu\nabla_\ell^2\langle \boldsymbol{v}(\boldsymbol{r}) \cdot \boldsymbol{v}(\boldsymbol{r}+\boldsymbol{\ell})\rangle$ may be neglected. Actually, assumption (i) was made only implicitly by Kolmogorov. Our approach was much longer, as we included the proofs of preliminary material already known to Kolmogorov and also proved the consistency of the assumptions which in our approach play the roles of (i) and (ii). For this we had to assume that the turbulence is maintained by a driving force. With our assumptions, it follows from (6.45) that consistency is established if we simultaneously have

$$K \gg K_c \sim \ell_0^{-1} \qquad \text{and} \qquad |2\nu\Omega_K| \ll \varepsilon. \tag{6.54}$$

The range of wavenumbers over which both conditions (6.54) hold is by definition the *inertial range.* It is traditionally so called because, at such wavenumbers, the dynamics is dominated by the inertia terms in the Navier–Stokes equation, i.e. all but the viscous and forcing terms. In view of (6.48), the second condition in (6.54) will be satisfied if

$$K \ll \left(\frac{\varepsilon}{2\nu E}\right)^{1/2} = \left(\frac{\Omega}{E}\right)^{1/2} = \frac{\sqrt{5}}{\lambda}, \tag{6.55}$$

where E and Ω, the mean energy and mean enstrophy, are defined in Section 2.3. The r.h.s. is the inverse of the Taylor scale λ, already defined in (5.10). For small ν, the Taylor scale $\lambda \propto \nu^{1/2}$ (since ε and E are assumed to stay finite). Thus, when the viscosity is small, we have shown

9 The vanishing of $S_3(\ell)$ as $\ell \to 0$ for $\nu > 0$ is a consequence of the postulated regularity of the flow. Here, the limit $\nu \to 0$ is taken before the limit $\ell \to 0$ and the vanishing is not guaranteed, but there is experimental evidence that it still holds.

that the inertial range extends from scales $\sim \ell_0$ down to at least scales $\sim \lambda \propto \nu^{1/2}$. Actually, we shall see that the inertial range probably extends much further, down to scales $\propto \nu^{3/4}$. This result cannot, however, be established as cleanly.

Second, we observe that (6.53) is invariant under *random Galilean transformations*. From Section 2.2, we know that, in the absence of forcing and boundaries, the Navier–Stokes equation is invariant under Galilean transformations. That is, for any vector $\boldsymbol{U}$, if $\boldsymbol{v}(t,\boldsymbol{r})$ is a solution, so is

$$\boldsymbol{v}'(t,\boldsymbol{r}) \equiv \boldsymbol{v}(t,\boldsymbol{r}-\boldsymbol{U}t)+\boldsymbol{U}. \tag{6.56}$$

If $\boldsymbol{v}(t,\boldsymbol{r})$ is homogeneous and stationary, so is $\boldsymbol{v}'(t,\boldsymbol{r})$. Yet, isotropy is not preserved, since $\boldsymbol{U}$ introduces a preferred direction. Hence, Galilean invariance cannot be easily used to test predictions of a theory of homogeneous and isotropic turbulence. To obviate this difficulty Kraichnan (1964, 1965, 1968a) proposed taking $\boldsymbol{U}$ to be random and isotropically distributed. (The distribution can, for example, be taken to be Gaussian, but this does not matter much.) Under such a random Galilean transformation, all the structure functions, and in particular $S_3(\ell)$, remain invariant. Indeed, the velocity-shift $\boldsymbol{U}$ cancels in the velocity increment and the shift $\boldsymbol{U}t$ in the spatial argument cancels by homogeneity.[10] Similarly, the mean dissipation ε is also invariant. Thus, as stated, the whole four-fifths relation (6.53) is invariant. Note also that the presence of a driving force breaks Galilean and random Galilean invariances of the Navier–Stokes equation,[11] but this does not affect the derivation of the four-fifths law since the (single-time) correlations of the velocity and force appearing in (6.8) are invariant under Galilean transformations.[12]

Third, we observe that in deriving the four-fifths law, we took several limits: (i) the limit $t \to \infty$ gives a statistical steady state; (ii) the limit $\nu \to 0$ eliminates any residual dissipation at finite scales; (iii) the limit $\ell \to 0$ eliminates the direct influence of large-scale forcing. The correct formulation of (6.53) is thus

$$\lim_{\ell\to 0}\lim_{\nu\to 0}\lim_{t\to\infty}\frac{S_3(\ell)}{\ell} = -\frac{4}{5}\varepsilon. \tag{6.57}$$

Any attempt to take the limits in a different order could lead to difficulties.

[10] This would not be the case if the two velocities had been evaluated at different time-arguments.

[11] Except if the random force is delta-correlated in time.

[12] A misconception of this author, corrected by Kraichnan (1995), is that the invariance of the four-fifths law to random Galilean transformation is related to the so-called localness of interactions (see Section 7.3).

For example, if $\ell \to 0$ before $\nu \to 0$, the third order structure function is expected to behave as ℓ^3; indeed, in the presence of viscosity, for small separations, velocity increments will be linear in $\boldsymbol{\ell}$ because the flow is expected to be smooth. Whether or not the $\nu \to 0$ and the $t \to \infty$ limits can be interchanged depends on the smoothness of the solution to the three-dimensional Euler equation (the Navier–Stokes equation with $\nu = 0$). If no singularities appear in finite time, then it is easy to show that, at any finite time, the energy dissipation tends to zero with the viscosity, and not to a finite positive limit as required for the derivation of the four-fifths law. At the moment the question of the regularity/singularity for three-dimensional Euler flow is open (see Sections 7.8 and 9.3).

Finally, we note that without the assumption of isotropy, a relation analogous to the four-fifths law can still be derived. With all the other assumptions unchanged, it may be shown from (6.8) that, in the limit $\nu \to 0$ and for small ℓ

$$-\frac{1}{4}\nabla_{\boldsymbol{\ell}} \cdot \langle |\delta \boldsymbol{v}(\boldsymbol{\ell})|^2 \delta \boldsymbol{v}(\boldsymbol{\ell})\rangle = \varepsilon. \tag{6.58}$$

This relation is essentially the same as (22.15) in Monin and Yaglom (1975).

An interesting question addressed by these authors is what happens if the flow is homogeneous at all scales while being isotropic only at small scales. Since only velocity increments appear in (6.58), when ℓ is small, this should become equivalent to the four-fifths law (6.53). However, (6.53) involves exclusively *longitudinal* velocity increments, while (6.58) involves both longitudinal and *transverse* velocity increments. Everything can be reexpressed in terms of third order moments of longitudinal velocity increments by use of (6.33). However, this relation has so far been proven only in the *globally isotropic case.*

A related question concerns the case in which the flow is isotropic but has only *homogeneous increments*, as assumed in the K41 theory (in both Kolmogorov's original formulation and our formulation in Section 6.1). In Section 4.5 we pointed out that random functions with homogeneous increments can sometimes be considered as limits of strictly homogeneous random functions. If a small-scale flow with homogeneous isotropic increments can be embedded in a large-scale homogeneous isotropic flow, then the four-fifths law remains valid. Observe that the embedding flow need not be a solution of the Navier–Stokes equation, as long as it is incompressible. Indeed, the derivation of the relations (6.32)–(6.34) involves only kinematic arguments.

At the moment we can only say that the best available experimental evidence (e.g., Figure 8.6, discussed in Section 8.3) supports the validity of the four-fifths law.

6.3 Main results of the Kolmogorov 1941 theory

6.3.1 The Kolmogorov–Obukhov law and the structure functions

We now return to the K41 theory, using the three basic hypotheses H1, H2 and H3 introduced in Section 6.1. First, we shall show that the four-fifths law implies

$$h = \frac{1}{3}. \tag{6.59}$$

Let us rewrite the four-fifths law (6.5) as

$$\langle (\delta v_{\|}(\ell))^3 \rangle = -\frac{4}{5}\varepsilon\ell, \tag{6.60}$$

where $\delta v_{\|}(\ell)$, the longitudinal velocity increment, is defined in (5.2). With the hypothesis H2, under rescaling of the increment ℓ by a factor λ, the l.h.s. of (6.60) changes by a factor λ^{3h}, while the r.h.s. changes by a factor λ. This immediately implies (6.59). Actually, to show just that $h = 1/3$, the assumption of *isotropy is not needed.* Indeed, from (6.17), which is valid irrespective of isotropy, and from the assumption of a scale-invariant velocity with exponent h, it is easily shown that $\Pi_K \propto K^{1-3h}$. This is independent of K only for $h = 1/3$. Here, we must stress that it would not be correct to infer $h = 1/3$ from the expression (2.52) for the energy flux. Indeed, this expression involves both the $\boldsymbol{v}_K^<$s and the $v_K^>$s. Using hypothesis H2 of Section 6.1, it is easily shown that $v_{K/\lambda}^> \stackrel{\text{law}}{=} \lambda^h v_K^>$. (Observe that $v_K^>$ involves only *small* scales.) There is, however, no simple transformation property for $\boldsymbol{v}_K^<$, so that it is not possible to conclude the argument. This observation shows how important it was to reexpress the energy flux solely in terms of velocity increments.

Let us now examine the consequences for the moments of the (longitudinal) velocity increments at inertial-range separations, assuming homogeneity and isotropy. We shall also assume that moments of arbitrary positive order $p > 0$ are finite (see Section 8.3). We define the (longitudinal) structure function of order p by

$$S_p(\ell) \equiv \langle (\delta v_{\|}(\ell))^p \rangle. \tag{6.61}$$

Note that the argument of the structure functions is here taken to be positive, since ℓ is the absolute value of the increment $\boldsymbol{\ell}$. An alternative definition, allowing for both positive and negative arguments x, is ($\boldsymbol{\ell}^0$ is an arbitrary unit vector)

$$S_p(x) \equiv \left\langle \left[(\boldsymbol{v}(\boldsymbol{r} + x\boldsymbol{\ell}^0) - \boldsymbol{v}(\boldsymbol{r})) \cdot \boldsymbol{\ell}^0 \right]^p \right\rangle, \tag{6.62}$$

which reduces to $S_p(\ell)$ for positive x. A consequence of isotropy is that the functions thus defined are even (odd) for even (odd) ps.[13]

From the self-similarity hypothesis H2 and (6.59) we infer that

$$S_p(\ell) \propto \ell^{p/3}. \tag{6.63}$$

Since $(\varepsilon\ell)^{p/3}$ has exactly the same dimensions as S_p, we have

$$S_p(\ell) = C_p \varepsilon^{p/3} \ell^{p/3}, \tag{6.64}$$

where the C_ps are dimensionless. The C_ps cannot depend on the Reynolds number, since the limit of infinite Reynolds number is already taken. For $p = 3$, it follows from (6.53) that $C_3 = -4/5$, which is clearly universal, i.e. independent of the particular flow under consideration. In the derivation given here nothing requires the C_ps for $p \neq 3$ to be universal. This universality was postulated in Kolmogorov's first 1941 paper, but was then questioned by Landau. We shall return to this matter at length in Section 6.4. Sreenivasan (1995) gives a compilation of values of the constant C_2 from a variety of experiments with R_λs ranging from about 50 to more than 10^4 and finds $C_2 = 2.0 \pm 0.4$. The constant C_2, denoted C by Kolmogorov, should in principle be called the Kolmogorov constant. Unfortunately, usage has reserved this name for the constant in front of the longitudinal energy spectrum, which is actually 4.02 times smaller.

We note that expression (6.64) for the structure functions involves only the energy dissipation rate ε, the scale ℓ and not the integral scale ℓ_0. It thus follows from the K41 theory that, if we take the limits $\nu \to 0$ and $\ell_0 \to \infty$ (order indifferent), while holding $\varepsilon > 0$ fixed, all the structure functions have finite limits. (Insofar as the constants C_p are finite.) There is an important *converse* to this statement: if the structure functions have finite nonvanishing limits when $\nu \to 0$ and $\ell_0 \to \infty$ while holding $\varepsilon > 0$ fixed, then these limits display K41 scaling. Indeed, for finite ℓ_0, dimensional analysis implies that the structure function of order p is given by the r.h.s. of (6.64) times a dimensionless function $\tilde{S}_p(\ell/\ell_0)$;

[13] Structure functions are sometimes defined with an absolute value of the velocity increment. This is justified for noninteger values of p, but should be avoided for odd ps. For example, Kolmogorov's four-fifths law concerns the third order structure function without an absolute value.

this function has a finite nonvanishing limit as $\ell_0 \to \infty$ or, equivalently, as $\ell \to 0$, thereby ensuring K41 scaling. Hence, deviations from K41, which will be discussed in Chapter 8, require that structure functions of order other than 3 have an *explicit dependence on the integral scale* at inertial-range separations.

We now return to consequences of K41. The fact that the second order structure function follows an $\ell^{2/3}$ law implies a $k^{-5/3}$ law for the energy spectrum. Indeed, from (4.60), (4.61) and (6.64), we obtain

$$E(k) \sim \varepsilon^{2/3} k^{-5/3}, \tag{6.65}$$

The experimental results presented in Section 5.1 support the K41 theory as far as the second order structure function (and thus the spectrum) is concerned. As we shall see later, the consistency between the K41 theory and experimental data on structure functions is questionable when $p > 3$.

6.3.2 *Effect of a finite viscosity: the dissipation range*

In Section 6.2.5, we showed that when the viscosity ν is small, there is an 'inertial range' in which direct energy injection and energy dissipation are both negligible. This inertial range was shown to extend *at least* down to scales comparable to the Taylor scale $\lambda = (5E/\Omega)^{1/2}$. We now use the improved results obtained in Section 6.3.1 (which involve additional unproved assumptions) to show that, in the K41 framework, the inertial range actually extends down to the 'Kolmogorov dissipation scale'

$$\eta \equiv \left(\frac{\nu^3}{\varepsilon}\right)^{1/4}. \tag{6.66}$$

We start from the energy-flux relation (6.45), and assume that $K \gg K_c$ so that $\mathscr{F}_K \simeq \varepsilon$. The dissipation term involves the cumulative enstrophy

$$\Omega_K = \frac{1}{2}\langle|\omega_K^<|^2\rangle = \int_0^K k^2 E(k)dk. \tag{6.67}$$

By substituting the inertial-range value (6.65) of the energy spectrum $E(k)$ into (6.67), we can find the wavenumber up to which the dissipation term $2\nu\Omega_K$ in (6.45) is negligible compared to the energy flux ε. This gives the following 'dissipation wavenumber' (order unity constants have been omitted)

$$K_d = \left(\frac{\nu^3}{\varepsilon}\right)^{-1/4}, \tag{6.68}$$

which is precisely the inverse of the 'Kolmogorov dissipation scale' η defined above. The range of scales comparable to or less than η is called the 'dissipation range'. In this range the energy input from nonlinear interactions and the energy drain from viscous dissipation are in exact balance. It is a misconception that nonlinear interactions may be ignored in the dissipation range.

Kolmogorov's (1941a) original derivation of the dissipation scale η was rather different:

Kolmogorov's first universality assumption.[14] *At very high, but not infinite Reynolds numbers, all the small-scale statistical properties are uniquely and universally determined by the scale ℓ, the mean energy dissipation rate ε and the viscosity ν (or, equivalently, by ℓ, ε and η).*

'Small-scale' is here understood as scales small compared to the integral scale, i.e. inertial-range and dissipation-range scales. By a simple dimensional argument, the first universality assumption implies the following universal form for the energy spectrum at large wavenumbers

$$E(k) = \varepsilon^{2/3} k^{-5/3} F(\eta k), \tag{6.69}$$

where $F(\cdot)$ is a universal dimensionless function of a dimensionless argument. By the second universality assumption of Kolmogorov (Section 6.1), $F(\cdot)$ tends to a finite positive limit (the Kolmogorov constant) for vanishing argument. The universality of the whole function $F(\cdot)$ has been questioned by Frisch and Morf (1981), using the same sort of argument that Landau developed for the Kolmogorov constant (see Section 6.4).

There were several early attempts to determine the functional form of $F(\cdot)$ at high wavenumbers. They will not be reviewed here (see, e.g., Monin and Yaglom 1975). The most interesting remark was made by von Neumann (1949). He observed that an analytic function has a Fourier transform which falls off exponentially at high wavenumbers. The logarithmic decrement is equal to the modulus δ of the imaginary part of the position of the singularity in complex space nearest to the real domain. Therefore, in von Neumann's view, exponential fall-off at high k was more likely than the rapid algebraic fall-off proposed by Heisenberg (1948). Actually, for a random homogeneous function, the situation is a bit more complicated: there is a probability distribution $P(\delta)$ and thus the form for the energy spectrum at high k is the Laplace transform of $P(\delta)$ near its minimum value $\delta_\star$ (Frisch and Morf 1981).

[14] Called by Kolmogorov the 'first hypothesis of similarity' and recast here in slightly different language.

An exponential fall-off (with possible algebraic prefactors) is obtained only if $\delta_\star > 0$, i.e. if there is a tubular region around the real domain in which (almost) all realizations are analytic. This will be referred to as *uniform analyticity*. The experimental results of Gagne (1987) suggest that this condition may be satisfied. We shall come back to the issue of complex singularities in connection with intermittency in the dissipation range (Section 8.2).

6.4 Kolmogorov and Landau: the lack of universality

An important issue in fully developed turbulence is Landau's objection to universality. In Sections 6.4.1–6.4.3 we shall examine the original formulation of Landau's slightly cryptic objection (Section 6.4.1), its modern reinterpretation (Section 6.4.2, which some readers may wish to examine first), and the evidence that Kolmogorov actually had a version of his theory which withstood Landau's objection (Section 6.4.3).

6.4.1 *The original formulation of Landau's objection*

In the first (Russian) edition of the book on fluid mechanics that Landau published with Lifshitz and which appeared in 1944, there was a footnote[15] which in later editions found its way to the main text. Hereafter is the full text of the remark. The English is taken from the most recent version of the book (Landau and Lifshitz 1987). The only change is the substitution of the notation of the present book for velocity increments, structure functions and the integral scale.

It might be thought that the possibility exists in principle of obtaining a universal formula, applicable to any turbulent flow, which should give $S_2(\ell)$ for all distances ℓ that are small compared with ℓ_0. In fact, however, there can be no such formula, as we see from the following argument. The instantaneous value of $\left(\delta v_\parallel(\ell)\right)^2$ might in principle be expressed as a universal function of the dissipation ε at the instant considered. When we average these expressions, however, an important part will be played by the manner of variation of ε over times of the order of the periods of the large eddies (with size $\sim \ell_0$), and this variation is different for different flows. The result of the averaging therefore cannot be universal.

It is possible that Landau made a similarly worded remark to Kolmogorov, shortly after the publication of the 1941 theory when they where both in the city of Kazan (on the Volga), a place to which many Moscow activities had been decentralized in the face of the approaching Nazi threat. This is, however, not clear, because the only remark on

[15] It will henceforth be referred to as the 'footnote remark'.

record which we found was made by Landau at the end of a seminar delivered in Kazan in January 1942. A résumé of this seminar and of Landau's remark were published (Kolmogorov 1942). According to the résumé, Kolmogorov first summarized his 1941 work and then proposed an application to turbulence modeling which is outside of the scope of the present book. At the end of the seminar Landau made the following remark.[16] The English is taken mostly from the translation by Spalding (1991).

L. Landau remarked that A.N. Kolmogorov was the first to provide correct understanding of the local structure of a turbulent flow. As to the equations of turbulent motion, it should be constantly born in mind, in Landau's opinion, that in a turbulent field the presence of curl of the velocity was confined to a limited region; qualitatively correct equations should lead to just such a distribution of eddies.

Both the 'footnote remark' and the 'Kazan remark' require some explanation, which will be given in the next section.

6.4.2 A modern reformulation of Landau's objection

We shall here give an interpretation of Landau's 'footnote remark' which is largely based on the one proposed by Kraichnan (1974). Landau tried to show that the C_ps in (6.64), for $p \neq 3$, cannot be *universal*, i.e. they must depend on the detailed geometry of the production of the turbulence.

The ε in (6.64) is a mean dissipation rate, the mean being taken over the attractor of the flow, i.e. the mean is a time-average. Let us now construct a superensemble, made of $N > 1$ experiments with different positive values of the mean dissipation rate, denoted ε_i $(i = 1, \cdots, N)$. The differences could be caused, for instance, by the flows having different integral scales. Let us tentatively assume that the C_ps are universal. We denote by $S_p^i(\ell)$ the structure functions for the ith flow. We have, by (6.64):

$$S_p^i(\ell) = C_p(\varepsilon_i)^{p/3}\ell^{p/3}. \tag{6.70}$$

Now, let us assume that it is legitimate to apply (6.64) to the superensemble (we shall come back to this). We define

$$S_p^{\text{super}}(\ell) = \frac{1}{N}\sum_i S_p^i(\ell) \quad \text{and} \quad \varepsilon^{\text{super}} = \frac{1}{N}\sum_i \varepsilon_i, \tag{6.71}$$

the superaveraged structure functions and dissipation rate, respectively.

[16] It will henceforth be referred to as the 'Kazan remark'.

From (6.64) and (6.70), we obtain:

$$\left(\frac{1}{N}\sum_i \varepsilon_i\right)^{p/3} = \frac{1}{N}\sum_i (\varepsilon_i)^{p/3}. \tag{6.72}$$

This relation is contradictory, except when $p = 3$.

The preceding argument depends crucially on the ability to consider the different flows as being part of a single superflow. This can be justified by considering a single flow in which the characteristic parameters change slowly in space on a scale much larger than the integral scale. Let us, for example, consider a wind tunnel in which a uniform flow of velocity V is incident on a grid made of parallel rods with a uniform mesh m. There are two types of rod, as sketched in Fig. 6.3; type A has diameter d_1 and type B has diameter $d_2 > d_1$. In assembling the grid, type A and type B are selected at random in such a way that the type is changed on average every M rods, where M is a large number (say, 1000). The turbulence downstream (say, 100 meshes) behind type-B rods has a larger integral scale than that behind type-A rods. Hence, the dissipation rate per unit mass ε_2 behind type-B rods is smaller than the dissipation rate ε_1 behind type-A rods. (For dimensional reasons, ε scales as V^3/d.) In this example, it is clear that the properties of the turbulent eddies at a given location can be significantly affected only by those rods behind which they are produced. However, all the parts of the flow are coupled (for example, by pressure effects), so that it is legitimate to treat the superensemble as a single flow.

We observe that Landau's 'footnote remark' was formulated in the temporal rather than in the spatial domain. This is precisely the picture we get if, in the above example, we endow the grid with a slow uniform motion parallel to itself; a fixed probe will then successively encounter the eddies associated to type-A and type-B rods.

There is another interesting implementation of Landau's 'footnote remark', which uses the 'Kazan remark'. The latter probably referred to the experimental fact that in the turbulent flow produced by a jet, vorticity seems to be confined to a roughly conical region with a very sharp outer boundary, beyond which the flow is found to be laminar, as suggested by Fig.1.14. This boundary has an instantaneous shape which is quite complex (possibly a fractal)[17] and, of course, it changes in time. A probe placed at some distance from the axis of the jet will

[17] The boundary is so complex that occasionally pockets of laminar fluid are found right on the axis of the jet.

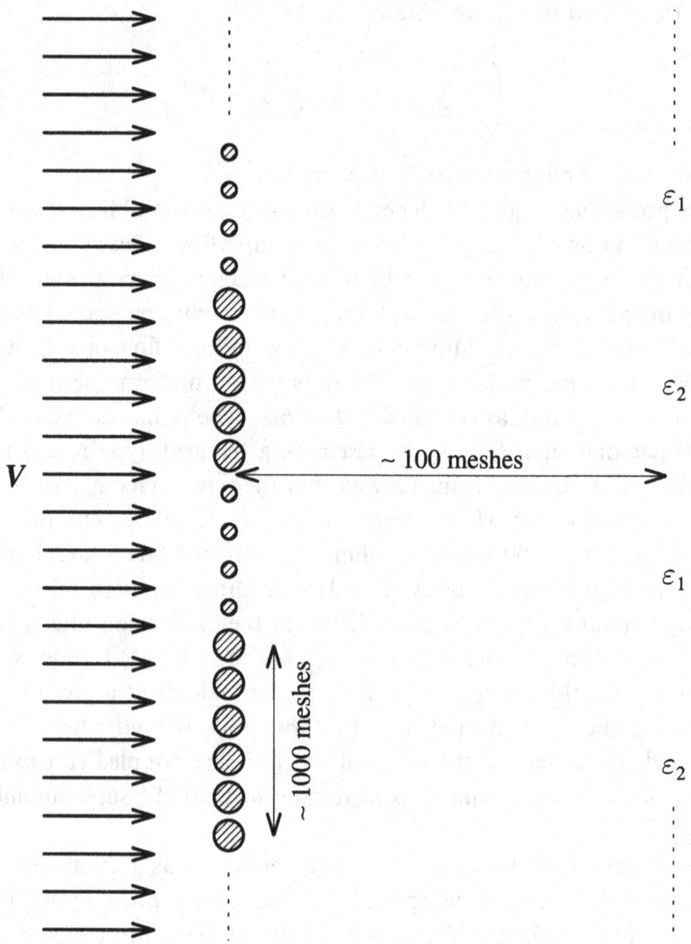

Fig. 6.3. An illustration of Landau's objection to the universality of the constants C_p in the structure functions, using a grid made of two sorts of rod.

thus find that the turbulence is 'on' only some fraction of the time, which decreases as the probe is moved away from the axis.[18] As a consequence, the (apparent) Kolmogorov constant will change with the distance to the axis (Kuznetsov, Praskovsky and Sabelnikov 1992).

Note that in all the above arguments, we (but not Landau) had to

[18] This phenomenon is known as 'external intermittency'; its relation to the intermittency discussed in Chapter 8 is not clear.

assume a *separation of scales* between the integral scale and the super-large scale on which the energy input is modulated. In other words, we had to assume that the production mechanism involves (at least) two scales. Otherwise, it is not clear that universality can be proven wrong.

Here, we wish to add a few words for the benefit of those readers familiar with renormalization group techniques, a subject discussed briefly in Section 9.6.4. Forster, Nelson and Stephen (1977) have investigated the problem of randomly forced turbulence when the forcing spectrum follows a power-law $F(k) = 2Dk^{3-\epsilon}$, with ϵ positive and small. The coefficient D then appears to the power 2/3 in the expression of the energy spectrum, so that Landau's nonuniversality argument can be carried over almost identically. Fournier and Frisch (1983a) have shown that, as long as very-long-range correlations are ruled out, the dimensionless constant in the energy spectrum (the analog of the Kolmogorov constant) can be calculated explicitly in terms of ϵ (for small ϵ) and is thus universal.

6.4.3 Kolmogorov and Landau reconciled?

Our presentation of the K41 theory in Chapter 6 uses scale-invariance rather than universality in deriving, for example, the two-thirds law. It is thus not inconsistent with Landau's 'footnote remark'. Actually, Kolmogorov himself was to some extent aware of the existence of an alternative formulation of his 1941 theory.

Let us indeed consider the third 1941 paper (Kolmogorov 1941c). After deriving the four-fifths law, Kolmogorov makes the following statement (adapted to our notation):

It is natural to assume that for large ℓ the ratio $S_3(\ell)/(S_2(\ell))^{3/2}$, i.e. the skewness of the distribution of probabilities for the difference $\delta v_{\|}(\ell)$, remains constant.

(In the context of the paper 'large' means at inertial-range scales.) In other words, *a particular form of scale-invariance* is postulated by Kolmogorov. Also, notice that he assumes that the skewness is 'constant' (independent of scale) rather than 'universal' (independent of the flow). From this assumption and (6.5) he then recovers the two-thirds law for $S_2(\ell)$ (his relation (9)) and observes that

...in Kolmogorov (1941a) the relation (9) was deduced from somewhat different considerations.

It seems therefore legitimate to refer to the scale-invariant version of the theory also as 'K41'. However, Kolmogorov, in his third 1941 did not try to give up universality, since the paper includes an attempt to estimate the value of the Kolmogorov constant from experimental data.[19] Kolmogorov did not try to modify the 1941 theory until the early 1960s, as discussed in Chapter 8.

6.5 Historical remarks on the Kolmogorov 1941 theory

The $k^{-5/3}$ law for the energy spectrum, which is an immediate consequence of the two-thirds law, is actually not explicitly written in Kolmogorov's first turbulence paper. It is found for the first time in Obukhov (1941a, b), who derived it from a closure argument: he guessed a simple dimensionally consistent expression of the energy flux $\Pi(k)$ in terms of the sole energy spectrum. According to Kolmogorov (1941c), Obukhov's derivation was 'independent' of his. Obukhov's argument cannot be used to predict expressions for structure functions of order higher than 2.

Kolmogorov's work remained unknown in the West until after the war when it received considerable exposure thanks to Batchelor (1946, 1947, 1953). Since the remarks made by Landau came to be known only later, the emphasis was on the *universality* aspects of Kolmogorov's theory.

The Kolmogorov theory was actually independently discovered several times by famous scientists. A case study of this has been made by Battimelli and Vulpiani (1982) from whose work most of the following lines are taken.

While detained at Farm Hall near Cambridge at the end of 1945, Heisenberg and von Weizsäcker developed a closure theory of fully developed turbulence quite similar to that of Obukhov, which also leads to the $k^{-5/3}$ law for the energy spectrum. These related theories were eventually published as companion papers (von Weizsäcker 1948; Heisenberg 1948).

Onsager arrived at the $k^{-5/3}$ law (actually a $k^{-11/3}$ law because he worked with the three-dimensional spectrum) by considering the energy cascade and requesting, like Kolmogorov, dependence only on the wavenumber and on the energy dissipation rate. He also stressed the universality of the factor in front of $\varepsilon^{2/3}k^{-11/3}$. He communicated his results in June 1945 in a letter to C.C. Lin (Onsager 1945a). He also communicated his results to T. von Kármán whose reaction was not enthusiastic (according to a letter the latter sent to C.C. Lin). A short

[19] Kolmogorov had a very strong interest in experimental aspects of turbulence and ascribed this in part to the influence of Ludvig Prandtl whom he may have met during his visit to Göttingen around 1930 (V.I. Arnold, private communciation).

abstract was published that same year (Onsager 1945b). A longer paper was published a few years later, after Onsager had become aware of the work of Kolmogorov and of the German physicists (Onsager 1949). Here, Onsager pointed out that an $\ell^{1/3}$ law for velocity increments means that, in some sense, the velocity is not smooth but only Hölder continuous of exponent one-third and brought up (probably for the first time in the subject) the issue of *singularities* to which we shall return in Sections 7.8 and 9.3.

7
Phenomenology of turbulence in the sense of Kolmogorov 1941

7.1 Introduction

In previous chapters we showed how it is possible to establish certain scaling laws for fully developed turbulence by starting from unproven but plausible hypotheses and then proceeding in a systematic fashion. By 'phenomenology' of fully developed turbulence one understands a kind of shorthand system whereby the same results can be recovered in a much simpler way, although, of course, at the price of less systematic arguments. Phenomenology of fully developed turbulence has some associated 'mental images', such as the 'Richardson cascade' (Section 7.3), which have played a very important role in the history of the subject. After recasting the K41 theory in phenomenological language and images, it also becomes possible to grasp intuitively some of the shortcomings which may be present. A considerable part of the existing work on turbulence rests on K41 phenomenology, particularly in applied areas such as the modeling of turbulent flow. Kolmogorov himself, with Ludwig Prandtl, was one of the pioneers of this important area of research, which is beyond the scope of this book (Kolmogorov 1942; see also Batchelor 1990, Spalding 1991 and Yaglom 1994). We shall give only some examples of what can be derived by phenomenology: counting degrees of freedom (Section 7.4), comparing macroscopic and microscopic length scales (Section 7.5), finding the probability distribution function of velocity gradients (Section 7.6) and finding the law of decay of the energy (Section 7.7).

Other examples of what phenomenology and simple scaling arguments can achieve may be found in Tennekes and Lumley (1972), Tennekes (1989) and Monin and Yaglom (1971, 1975). We finally give an example where the usual phenomenology can be very misleading: finite-time blow-up for ideal flow (Section 7.8).

7.2 Basic tools of phenomenology

Hereafter, the symbol '$\sim$' will mean 'equal within an order unity constant'. In estimates of orders of magnitude, no distinction will be made between a vector and its modulus; hence, vector notation will be mostly dropped. Order unity factors (e.g., $1/2$) will be dropped unless they appear repeated many times (e.g., in 2^{-n}).

The main ingredients of phenomenology are listed below.

- ℓ: the scale under consideration (typically taken between the integral scale ℓ_0 and the dissipation scale η).
- v_ℓ: the typical value of the velocity associated to scales $\sim \ell$. The correct definition is the r.m.s. value of the velocity subject to band-pass filtering, say of an octave around the wavenumber ℓ^{-1}. A working definition, for most instances, is to take

$$v_\ell \sim \sqrt{\langle \delta v_\parallel^2(\ell) \rangle}, \tag{7.1}$$

when $\ell \leq \ell_0$ and

$$v_\ell \sim \sqrt{\langle \boldsymbol{v}(\boldsymbol{r}+\boldsymbol{\ell}) \cdot \boldsymbol{v}(\boldsymbol{r}) \rangle}, \tag{7.2}$$

when $\ell > \ell_0$.

- v_0: the r.m.s. velocity fluctuation $\sim v_{\ell_0}$.
- t_ℓ the 'eddy turnover time'[1] associated with the scale ℓ:

$$t_\ell \sim \frac{\ell}{v_\ell}. \tag{7.3}$$

t_ℓ is the typical time for a structure of size $\sim \ell$ to undergo a significant distortion due to the *relative* motion[2] of its components, as indicated in Fig. 7.1.

In view of the conservation of volume (incompressibility), if pairs of points such as 1 and 2 in Fig. 7.1 diverge, other pairs of points such as 3 and 4 must come together. Thus, t_ℓ is also the typical time for the transfer of excitation (e.g., energy) from scales $\sim \ell$ to smaller ones. From this, we may estimate the energy flux from scales $\sim \ell$ to smaller scales, here denoted Π'_ℓ:[3]

$$\Pi'_\ell \sim \frac{v_\ell^2}{t_\ell} \sim \frac{v_\ell^3}{\ell}. \tag{7.4}$$

[1] Also called 'circulation time'.

[2] The *absolute* motion produces no distortion. Kolmogorov (1941a) pointed out in a footnote that *relative* velocities must be used.

[3] Π'_ℓ may be defined as the energy flux Π_K for $K \sim \ell^{-1}$.

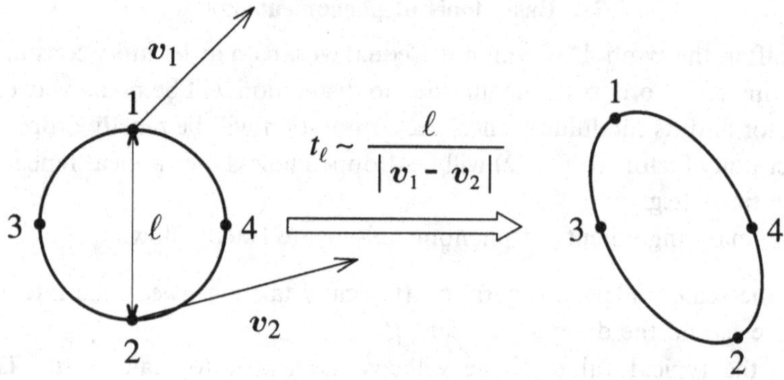

Fig. 7.1. Cross-section of a (roughly) spherical volume being squeezed into elliptical shape by fluid motion: points 1 and 2 separate, whilst points 3 and 4 get closer.

Indeed, the numerator is, within a factor 1/2, the amount of kinetic energy per unit mass associated with eddy motion on scales $\sim \ell$ and the denominator is the typical time for the transfer of such energy to smaller scales.

In the inertial range, where there is neither direct energy input nor direct energy dissipation, the energy flux should be independent of ℓ and equal to the (finite) mean energy dissipation rate ε:

$$\Pi'_\ell \sim \frac{v_\ell^3}{\ell} \sim \varepsilon. \tag{7.5}$$

Thus,

$$v_\ell \sim \varepsilon^{1/3}\ell^{1/3}, \tag{7.6}$$

which expresses in our phenomenological language that the velocity field is scale-invariant of exponent $h = 1/3$. By (7.3) and (7.6), the eddy turnover time is

$$t_\ell \sim \varepsilon^{-1/3}\ell^{2/3}. \tag{7.7}$$

One consequence is that when pairs of particles are released in a turbulent flow a distance ℓ apart, the rate of increase of the square of their distance, ℓ^2/t_ℓ, varies as $\ell^{4/3}$. The experimental discovery of this law by Richardson (1926) antedates the Kolmogorov 1941 theory.

Near the top of the inertial range, when $\ell \sim \ell_0$, (7.6) becomes

$$v_0 \sim \varepsilon^{1/3}\ell_0^{1/3}, \tag{7.8}$$

which may also be written as

$$\varepsilon \sim \frac{v_0^3}{\ell_0}. \tag{7.9}$$

This relation also follows from the four-fifths law (6.5) and has a wider applicability than the K41 theory: it only uses hypothesis H3 about the finiteness of the energy dissipation. Eq. (7.9) is actually used very frequently in the empirical modeling of turbulence.

The bottom of the inertial range, where viscosity becomes relevant, can be obtained phenomenologically as follows. The typical time for viscous diffusion to attenuate excitation on a scale $\sim \ell$ is[4]

$$t_\ell^{\text{diff}} \sim \frac{\ell^2}{\nu}. \tag{7.10}$$

Observe that t_ℓ^{diff} goes to zero with ℓ faster than t_ℓ, given by (7.7). It follows that, however small the viscosity, diffusion will become relevant below a scale obtained by equating the two times, *viz*

$$\eta \sim \left(\frac{\nu^3}{\varepsilon}\right)^{1/4}, \tag{7.11}$$

which is the Kolmogorov dissipation scale already introduced in Section 6.3.2.

This is the place to mention that in experiments, it is usually found that viscosity becomes relevant at scales (in the physical space) about 30 times[5] larger than η. This may be interpreted trivially by noting that phenomenology does not predict numerical constants. Less trivially, it suggests that nonlinear interactions are actually weaker than predicted by a turnover argument.

7.3 The Richardson cascade and the localness of interactions

It is useful to recast the phenomenology of the previous section in graphical form as shown on Fig. 7.2. The eddies of various sizes are represented as blobs stacked in decreasing sizes. The uppermost eddies have a scale $\sim \ell_0$. The successive generations of eddies have scales $\ell_n = \ell_0 r^n$, $(n = 0, 1, 2, ...)$ where $0 < r < 1$. The value $r = 1/2$ is the most common choice, but the exact value has no meaning. The smallest eddies

[4] This may be seen by noting that Fourier components with wavenumber $k \sim \ell^{-1}$ are attenuated proportionally to $e^{-\nu k^2 t}$.

[5] In Fourier space, the energy spectrum departs significantly from the $k^{-5/3}$ law beyond wavenumber $K_d/10 = 1/(10\eta)$; see Monin and Yaglom (1975).

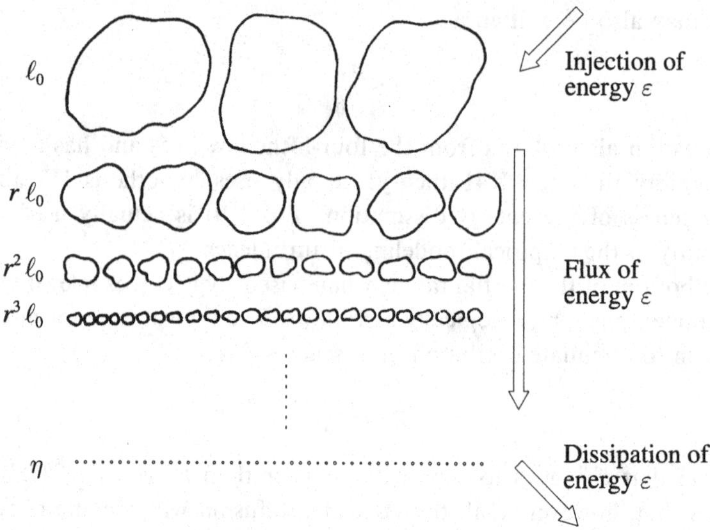

Fig. 7.2. The cascade according to the Kolmogorov 1941 theory. Notice that at each step the eddies are space-filling.

have scales $\sim \eta$, the Kolmogorov dissipation scale. The number of eddies per unit volume is assumed to grow with n as r^{-3n} to ensure that small eddies are as space-filling as large ones. Energy, introduced at the top at a rate ε (per unit mass), is 'cascading' down this hierarchy of eddies at the same rate ε and is eventually removed by dissipation at the bottom, still at the rate ε. The picture is, of course, not intended to be taken too literally: eddies could be much flatter than shown and the smaller ones are actually imbedded in the larger ones.[6] The main advantage of the cascade picture is that it brings out two basic assumptions of the K41 phenomenology.

The first assumption is *scale-invariance* within the inertial range. This would be violated if, for example, small eddies were less and less space-filling. Such an assumption leads to corrections to the K41 theory. This matter will be discussed in detail in Section 8.5.1. The second assumption is *localness*[7] of interactions. This means that, in the inertial range, the energy flux at scales $\sim \ell$ involves predominantly scales of comparable size, say from $r\ell$ to $r^{-1}\ell$.

[6] In the figure, where eddies are stacked on top of each other, their number grows only as r^{-n}.

[7] In turbulence, 'local' and 'localness' usually refer to scales, not to positions as in other areas of physics.

The traditional argument in favor of localness goes as follows. Consider an inertial-range eddy of scale ℓ. It will be swept along by the energy-containing eddies with scales $\sim \ell_0 \gg \ell$. This will not change the energy content of the eddy being swept. Note that a *uniform* sweep induces no distortion of fine-scale structures, because of the Galilean invariance of the Navier–Stokes equation. In homogeneous isotropic turbulence, the sweeping of inertial-range eddies by energy-containing eddies might be thought of as akin to a *random Galilean transformation* (Section 6.2.5). Actually, as observed by Kraichnan (1958, p. 42), velocity jumps of an appreciable fraction of the r.m.s. velocity can take place across very thin shear layers. It may therefore not always be legitimate to associate large velocities exclusively with large scales. Furthermore, Kraichnan (1959, p. 536) pointed out that there are slender vortex filaments extending 'throughout a substantial part of the turbulent domain' and that this leads to questioning some aspects of K41.[8]

Leaving the issue of sweeping, let us now consider distortion, which is controlled by shear, i.e. velocity gradients. The typical shear s_ℓ associated with scales $\sim \ell$ is

$$s_\ell \sim \frac{v_\ell}{\ell} \sim \varepsilon^{1/3}\ell^{-2/3}, \tag{7.12}$$

where v_ℓ is obtained from (7.6). The smallest shear is near the top of the inertial range $\ell \sim \ell_0$ and the largest near its bottom $\ell \sim \eta$. The shearing of an eddy of scale ℓ by an eddy of scale $\ell' \gg \ell$ is ineffective in producing distortions, because there is very little shear at such scales. The shearing of an eddy of scale ℓ by eddies of scale $\ell' \ll \ell$ is ineffective in producing distortions, because the shearing eddies are not acting coherently[9] over the scale ℓ. Hence, the predominant distortions come from scales $\ell' \sim \ell$.

Such phenomenological considerations about localness can be supplemented by a harder argument if we consider the exact expression (6.24) for the energy flux for homogeneous isotropic turbulence. When we substitute the four-fifths law (6.53) for the third order structure function, we find that the integral in (6.24) converges both at small ℓs (ultraviolet region) and at large ℓs (infrared region). It follows that the dominant contribution to the flux Π_K comes from $\ell \sim 1/K$. This result would

[8] At that time, Kraichnan had an alternative theory, the direct interaction approximation (DIA; see Section 9.5.3) which provided an incentive to look for shortcomings of K41; whatever the later fate of the DIA, his criticisms remain of considerable interest in view of the current work on vortex filaments (see Section 8.9).

[9] The effect on eddies of scale $\sim \ell$ may be represented by an 'eddy viscosity' (see Section 9.6); a phenomenological presentation of this may be found in Frisch and Orszag (1990).

remain true if, at inertial-range scales, the third order structure function $S_3(\ell)$ scaled as ℓ^{ζ_3} with $0 < \zeta_3 < 2$. Of course, ζ_3 can differ from 1 only if hypothesis H3 of Section 6.1 is violated.

To following historical and poetical note serves as concluding note about cascades. The modern concept of a cascade probably owes its origins to Lewis Fry Richardson (1922). He took inspiration from observations of clouds and from Jonathan Swift's verse:

So, nat'ralists observe, a flea
Hath smaller fleas that on him prey;
And these have smaller yet to bite 'em,
And so proceed *ad infinitum.*
Thus every poet, in his kind,
Is bit by him that comes behind.

The last two lines, which are not usually quoted, may also be relevant, if 'fluid dynamicist' is substituted for 'poet'.

In Kolmogorov's 1941 papers no explicit reference to Richardson is made, but in his 1962 paper Kolmogorov writes that the K41 hypotheses 'were based physically on Richardson's idea of the existence in the turbulent flow of vortices of all possible scales...'. Furthermore, Richardson's work is quoted in Obukhov's (1941a, b) papers, written under the direction of Kolmogorov. The graphical representation of the cascade shown in Fig. 7.2 is taken from Frisch, Sulem and Nelkin (1978), where it was used mostly to underline a possible shortcoming of the K41 theory to which we shall return in Section 8.5.1.

7.4 Reynolds numbers and degrees of freedom

The results of Section 7.2 may be recast in dimensionless form by using as reference length the integral scale ℓ_0, as reference velocity the r.m.s. velocity fluctuation v_0 and as reference time their ratio

$$t_0 \sim \frac{\ell_0}{v_0}, \tag{7.13}$$

the 'large eddy turnover time'. Using (7.6) and (7.9), we obtain the following expressions at inertial-range scales:

$$\frac{v_\ell}{v_0} \sim \left(\frac{\ell}{\ell_0}\right)^{1/3}, \quad \frac{t_\ell}{t_0} \sim \left(\frac{\ell}{\ell_0}\right)^{2/3}. \tag{7.14}$$

We now define the *integral-scale Reynolds number*

$$R \sim \frac{\ell_0 v_0}{\nu}. \tag{7.15}$$

This is the most frequently used Reynolds number in phenomenological work. Henceforth, 'Reynolds number' without further specification will always mean integral-scale Reynolds number. Observe that, when the Reynolds number is high, the Taylor-scale Reynolds number

$$R_\lambda \sim \frac{\lambda v_0}{\nu}, \tag{7.16}$$

defined in (5.7), is related to the integral-scale Reynolds number by

$$R_\lambda \sim R^{1/2}. \tag{7.17}$$

To show this, it suffices to use (6.55), (7.9) and $E \sim v_0^2$. Observe that only hypothesis H3 (finiteness of the energy dissipation rate for $R \to \infty$) is used in deriving the relation above.

We now give a very important relation concerning the viscous cutoff, the Kolmogorov dissipation scale η. From (7.11) and (7.9), we obtain

$$\frac{\ell_0}{\eta} \sim \left(\frac{\nu^3}{\ell_0^3 v_0^3}\right)^{-1/4} \sim R^{3/4}. \tag{7.18}$$

Hence, in the K41 theory, the inertial range spans a range of scales growing as the (3/4)th power of the Reynolds number. It follows that if we want to describe such a flow accurately in a numerical simulation on a uniform grid, the minimum number of grid points per (integral scale)^3 is

$$N \sim R^{9/4}. \tag{7.19}$$

One consequence of this is that the storage requirement of a (fully resolved) numerical simulation grows as $R^{9/4}$. Since the time step has usually to be taken proportional to the spatial mesh,[10] the total computational work needed to integrate the equations for a fixed number of large eddy turnover times, grows as R^3. This scaling shows that progress in achieving high Reynolds numbers in 'honest' (fully resolved) simulations, may be very slow.[11] Of course, this result is based on the assumed K41 scaling, but so far it has provided rather good empirical guidance for how much the Reynolds number can be scaled up when increasing the numerical resolution.

The figure $R^{9/4}$ is sometimes taken to measure the *number of degrees of freedom* of a turbulent flow. For this to be correct, we must assume: (i) that the K41 theory is right; (ii) that the motion at inertial-range

[10] In order to be able to follow the sweeping of the finest structure across one mesh.

[11] The state-of-the-art is now somewhere between $R \sim 10^3$ and $R \sim 10^4$.

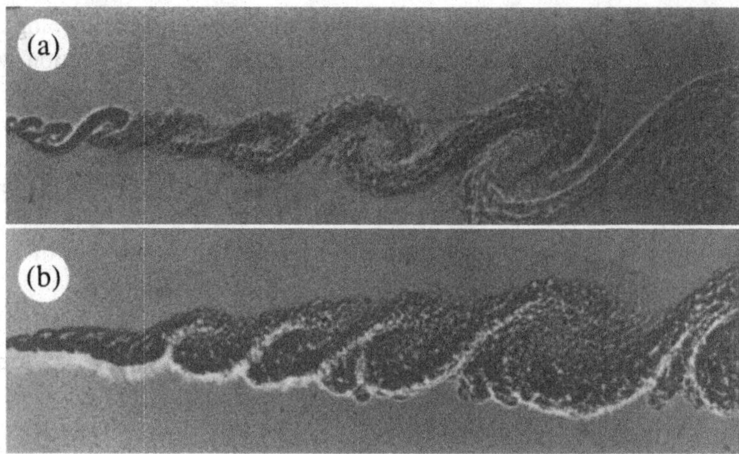

Fig. 7.3. Large-scale structure in a turbulent mixing layer (a). Coherent structure at higher Reynolds number (b) (Van Dyke 1982). Photographs J. Konrad (a) and M. Rebollo (b).

scales is almost totally disorganized. There is, however, considerable evidence that high-Reynolds-number turbulent flow is far from being totally disorganized. Experiments in which the turbulence is produced by interfacing two streams with different velocities (mixing layers) reveal 'coherent structures' on scales comparable to ℓ_0 (see, e.g., Fig. 7.3). There is also evidence that some order is present on smaller scales. Indeed, high-Reynolds-number simulations reveal the presence of vortex filaments in three-dimensional turbulent flow of the sort seen in Fig. 7.4, a matter we shall return to in Section 8.9. Here, we just point out that the number of degrees of freedom could be significantly smaller than $R^{9/4}$, thereby brightening the prospects for numerical simulations (but not on uniform grids).

From the viewpoint of dynamical systems, the number of degrees of freedom of a turbulent flow may be defined as the dimension of its attractor (see, e.g., Ruelle 1989). This definition is useful in studying the transition to turbulence when the dimensions are very low and can be accurately measured using, for example, the Grassberger and Procaccia (1983) scheme. No reliable method has been found to measure dimensions above, say 10 or 20. For fully developed turbulence, existing schemes are thus completely useless. Attempts have been made to obtain *upper bounds* for such dimensions.[12] This cannot be done rigorously, since

12 See, e.g., Constantin, Foias and Temam 1988.

Fig. 7.4. Intermittent vortex filaments in a three-dimensional turbulent fluid simulated on a computer (She, Jackson and Orszag 1991).

not enough is known about the uniqueness and regularity properties of the three-dimensional Navier–Stokes equation.[13] Even with unproven assumptions made about the viscous cutoff, such bounds do not improve on the K41 estimate $N \sim R^{9/4}$.

7.5 Microscopic and macroscopic degrees of freedom

A question frequently asked about fully developed turbulence goes as follows: since the dissipation scale decreases with the viscosity, could it not become so small that the hydrodynamic approximation breaks

[13] For two-dimensional flow there are known rigorous bounds, which are far above the number of grid points needed in a simulation.

down?[14] In this hypothetical situation, microscopic phenomena would become important in selecting the solution. This is certainly the case if the Navier–Stokes equation with finite but small viscosity generates genuine singularities, as has been speculated by Leray (1934) and never been disproved rigorously (see Sections 7.8 and 9.3.) It is also the case if no singularities are present but the dissipation scale is comparable to the molecular mean-free-path.

Actually, such possibilities are ruled out in the K41 framework. We shall now show that the ratio of the Kolmogorov dissipation scale η to the molecular mean-free-path λ_{mfp} is a growing rather than a decreasing function of the Reynolds number. In order to stay within the K41 framework of *incompressible turbulence*, we must assume that the turbulent flow has a small Mach number:

$$M = \frac{v_0}{c_s} \ll 1, \tag{7.20}$$

where c_s denotes the speed of sound. For most fluids, such as air or water, the speed of sound is comparable in order of magnitude to the thermal speed, the r.m.s. value of molecular velocities:

$$c_s \sim c_{\mathrm{th}}. \tag{7.21}$$

Furthermore, the (kinematic) viscosity, like any transport coefficient, is given approximately by

$$\nu \sim \lambda_{\mathrm{mfp}}\, c_{\mathrm{th}}. \tag{7.22}$$

It follows from (7.18), (7.20), (7.21) and (7.22) that (Corrsin 1959)

$$\frac{\eta}{\lambda_{\mathrm{mfp}}} \sim \frac{\eta}{\ell_0}\,\frac{\ell_0 v_0}{\lambda_{\mathrm{mfp}} c_{\mathrm{th}}}\,\frac{c_{\mathrm{th}}}{v_0} \sim M^{-1} R^{1/4}. \tag{7.23}$$

This ratio is very large because of the factor M^{-1}, and it grows indefinitely with R. One consequence of this is that, the higher the Reynolds number, the better the hydrodynamic approximation and the smaller the ratio of macroscopic to microscopic degrees of freedom. We stress once more that such conclusions hold only within the K41 framework.

[14] It is well known that such a breakdown happens for shock waves in a compressible fluid: their width can be comparable to the mean-free-path.

7.6 The distribution of velocity gradients

We show here that K41 phenomenology together with a simple argument à la Landau of the sort used in Section 6.4 can be used to relate the p.d.f. of the velocity and that of its gradient.

Let us assume that the turbulence is homogeneous and isotropic and describable by K41 phenomenology. The notation v_0 will now denote a *fluctuating* velocity characteristic of large eddies, rather than the r.m.s. velocity. Its p.d.f. is denoted $P_v(v_0)$. For simplicity, it is assumed that the integral scale ℓ_0 has no fluctuations. In view of (7.6), (7.9) and (7.11), the velocity increment over a distance ℓ and the dissipation scale are given by

$$v_\ell \sim v_0 \left(\frac{\ell}{\ell_0}\right)^{\frac{1}{3}}, \tag{7.24}$$

and

$$\eta \sim \nu^{\frac{3}{4}} \ell_0^{\frac{1}{4}} v_0^{-\frac{3}{4}}, \tag{7.25}$$

respectively. Hence, the velocity gradient (denoted s, for 'strain') is given by

$$s \sim \frac{v_\eta}{\eta} \sim v_0^{3/2} \nu^{-\frac{1}{2}} \ell_0^{-\frac{1}{2}}. \tag{7.26}$$

Eq. (7.26) defines a nonlinear change of variables from v_0 to s. It follows that the p.d.f. P_s of velocity gradients is given by (order unity constants are dropped)

$$P_s(s) \sim \nu^{\frac{1}{3}} \ell_0^{\frac{1}{3}} s^{-1/3} P_v\left(\nu^{\frac{1}{3}} \ell_0^{\frac{1}{3}} s^{\frac{2}{3}}\right). \tag{7.27}$$

Even within K41 phenomenology, this relation is of dubious validity for the *core* of the p.d.f.s, i.e. values of v or s comparable to their r.m.s. values. Indeed, arguments which are valid only within order unity constants are not suited for predicting precise functional forms. Thus, (7.27) may be applied only to the *tails* of the p.d.f.s and cannot be used to predict low order moments such as the skewness and the flatness of velocity derivatives. We shall come back to such coefficients in Section 8.5.6.

It is generally believed that the p.d.f. of the velocity is nonuniversal, since it depends on the detailed mechanism of production of the turbulent flow. Nonuniversality of the p.d.f. of the velocity translates, by (7.27), into nonuniversality of the tail of the p.d.f. of the velocity gradient. Actually, there are many instances where the p.d.f. of the velocity is found to be approximately Gaussian, for example in grid-generated turbulence in wind tunnels (Batchelor 1953) or for turbulence in the planetary

Fig. 7.5. The top three lines, inspected with a mirror, reveal Leonardo's knowledge about the decay of turbulence.

boundary layer (Van Atta and Park 1972). We then obtain for the p.d.f. of velocity gradients a modified exponential law involving a 4/3 power of the argument and a $s^{-\frac{1}{3}}$ prefactor.

The derivation in this section is taken from Frisch and She (1991). This paper also briefly discusses the modifications which are appropriate beyond the K41 theory. Further results are contained in Benzi, Biferale, Paladin, Vulpiani and Vergassola (1991). When the velocity has a Gaussian distribution, arguments of the kind presented here give for the p.d.f. of velocity gradients modified exponential distributions or more complex functional forms which are always decreasing much more slowly than Gaussians. Such behavior is consistent with data from experiments and numerical simulations (see, e.g., Vincent and Meneguzzi 1991).

7.7 The law of decay of the energy

It is an experimental fact that turbulence, once generated, decays quite slowly. This may actually have been the very first scientific observation ever made about turbulent flow. Indeed, Fig. 7.5, inspected with a mirror, will reveal the following notes made by Leonardo (Piumati 1894, fo. 74,v) around the year 1500, given with an English translation:

doue laturbolenza dellacqua sigenera
doue la turbolenza dellacq[a] simantiene plūgho
doue laturbolenza dellacqua siposa

where the turbulence of water is generated
where the turbulence of water maintains for long
where the turbulence of water comes to rest

In his second 1941 paper on turbulence Kolmogorov (1941b) made an attempt to predict the quantitative law of decay of turbulence. His argument has been found to be somewhat flawed. It is still of considerable interest because it can be extended in such a way that (obvious) flaws are eliminated. We shall once more use a nontraditional presentation, which, hopefully, is faithful to the spirit of Kolmogorov's work.

We shall say that the velocity has *infrared asymptotic self-similarity* with scaling exponent $h < 0$, if

$$v_\ell \sim C\ell^h, \qquad \text{for } \ell \to \infty, \tag{7.28}$$

where v_ℓ is defined by (7.2). We assume $h < 0$, since otherwise the velocity field is not homogeneous but has only homogeneous increments. The following may be shown.

Principle of permanence of large eddies. *If the turbulent flow is freely decaying (no external force) and initially possesses the property of infrared asymptotic self-similarity with a scaling exponent $-5/2 < h < 0$ and a constant C, then this property is preserved for all later times with the same h and C.*

This is a nontrivial result, to which we shall return below. Let us first examine its consequences. Eq. (7.28) implies that the velocity correlation function decreases as ℓ^{2h} for $\ell \to \infty$ and that the energy spectrum is proportional to k^{-1-2h} for $k \to 0$. Since $h < 0$, such a growing energy spectrum is unphysical at large wavenumbers (small scales). Let us therefore assume that there is a (time-dependent) integral scale ℓ_0, below which the turbulence is of the usual fully developed type. This requires that the (time-dependent) Reynolds number

$$R \sim \frac{\ell_0 v_0}{\nu} \gg 1. \tag{7.29}$$

The r.m.s. (time-dependent) velocity v_0 may be evaluated from (7.28) by requiring that $v_0 \sim v_{\ell_0}$. This leads to

$$v_0 \sim C\ell_0^h. \tag{7.30}$$

We now observe that, by (7.9), as long as $R \gg 1$, we can evaluate the rate of dissipation of the mean energy $E \sim v_0^2$ as follows:

$$\frac{d}{dt}v_0^2 \sim -\varepsilon \sim -\frac{v_0^3}{\ell_0}. \tag{7.31}$$

From (7.30) and (7.31) we obtain the following differential equation for the integral scale:

$$\frac{d}{dt}\left(C^2\ell_0^{2h}\right) \sim -\frac{C^3\ell_0^{3h}}{\ell_0}. \tag{7.32}$$

Provided that C is time-independent, (7.32) integrates to

$$\ell_0 \propto (t+a)^{\frac{1}{1-h}}, \qquad v_0 \propto (t+a)^{\frac{h}{1-h}}, \tag{7.33}$$

and hence

$$E \propto (t+a)^{\frac{2h}{1-h}}, \qquad R \propto (t+a)^{\frac{1+h}{1-h}}. \tag{7.34}$$

Observe that, h being negative, the r.m.s. velocity and the energy always decrease with time and the integral scale always increases.[15] When $h > -1$, the Reynolds number increases with time because the increase in the integral scale more than makes up for the decrease of the r.m.s. velocity. Hence, the flow has nontrivial asymptotics for large times. When $h < -1$ the above derivation ceases to be valid once the Reynolds number has dropped to values order unity.

In his derivation, Kolmogorov (1941b) also makes crucial use of (7.31), his eq. (22), a relation whose validity, as we have already observed, is broader than the K41 theory used by Kolmogorov to derive it. (He actually observes that 'The formula (22) may be established in different ways'.)

The main difference of substance between our derivation of (7.33)–(7.34) and Kolmogorov's is that he restricted his argument to the case $h = -5/2$, thereby obtaining a $t^{-10/7}$ law for the energy decay at large times. He made this particular choice because of a claim by Loitsyansky (1939), according to whom

$$\Lambda = \int_0^\infty b_{nn}(r,t) r^4 dr \tag{7.35}$$

should not depend on time. (Here, $b_{nn}(r,t)$ denotes the longitudinal velocity correlation function.) Batchelor and Proudman (1956; see also Proudman and Reid 1954) showed that Loitsyansky's derivation was invalidated by an effect involving long-range pressure correlations.

However, there are good reasons to believe that the 'principle of permanence of large eddies', as formulated at the beginning of this section, holds for $h > -5/2$. Expressed in Fourier-space language, this principle states that if initially the energy spectrum behaves as

$$E(k) \sim C' k^s, \quad s = -1 - 2h, \quad \text{for } k \to 0, \tag{7.36}$$

then the same property will hold at later times with the same h (or s) and C'. When $s = 2$, the coefficient of k^2 is proportional to the mean

[15] This increase in the integral scale does not involve a transfer of energy to larger scales (inverse cascade). It happens because smaller eddies die faster than larger ones.

square linear momentum and is thus conserved (Saffman 1967a, b). In the general case, the constancy of the coefficient of k^s when $s < 4$ (or $h > -5/2$) is a consequence of the energy transfer relation (6.37) and the observation that the transfer function $T(k)$ is proportional to k^4 for small k. There is no rigorous proof of this result which has been established so far only by diagrammatic or renormalization group methods (see Sections 9.5 and 9.6.4).

When $s = 4$ the coefficient of k^4 in the energy spectrum should become time-dependent. Finding the law of decay is then an open problem. It has so far been solved only within a closure framework, yielding a law of decay of the energy $\propto t^{-n}$ with n significantly smaller than Kolmogorov's value 10/7 (Lesieur and Schertzer 1978; Frisch, Lesieur and Schertzer 1980; Lesieur 1990).

7.8 Beyond phenomenology: finite-time blow-up of ideal flow

According to Richardson, energy introduced at the scale ℓ_0, cascades down to the scale η where it is dissipated. Consider the total time $T_\star$ which is the sum of the eddy turnover times associated with all the intermediate steps of the cascade. From (7.7), the eddy turnover time varies as $\ell^{2/3}$. If we let the viscosity ν, and thus η, tend to zero, $T_\star$ is the sum of an infinite *convergent* geometric series. Thus it takes a *finite time* for energy to cascade to infinitesimal scales. We also know that in the limit $\nu \to 0$, the enstrophy Ω goes to infinity as ν^{-1} (to ensure a finite energy dissipation).

From such observations, it is tempting to conjecture that ideal flow (the solution of the Euler equation), when initially regular,[16] will spontaneously develop a singularity in a finite time (finite-time blow-up).

This would be incorrect for at least two reasons. Firstly, the kind of phenomenology discussed in this chapter is meant only to describe the (statistically) steady state in which energy input and energy dissipation balance each other. The inviscid ($\nu = 0$) initial-value problem is not within its scope. Secondly, a basic assumption needed for K41 is that the symmetries of the Navier–Stokes or Euler equation are recovered in a statistical sense (hypothesis H1 of Chapter 6). This requires the flow to be *highly disorganized*. Kolmogorov himself was clearly aware of this,

[16] For example, by having only large-scale motion initially, so that the flow is very smooth, actually analytic.

since in a footnote to his first 1941 paper he wrote:

> ... *In virtue of the chaotic*[17] *mechanism of translation of motion from the pulsations of lower orders to the pulsations of higher orders, ... the fine pulsations of higher orders are subjected to approximately space-isotropy statistical régime* ...

Complex spatial structures have never been observed in numerical simulations of inviscid flow with smooth initial conditions. Note that inviscid flow has frozen-in vortex lines (Lamb 1932; Batchelor 1970), the topology of which cannot change since no viscous reconnection can take place.

There is yet another phenomenological argument, not requiring K41, which suggests finite-time blow-up. Consider the vorticity equation (2.15) for inviscid flow, rewritten as

$$D_t\boldsymbol{\omega} = \boldsymbol{\omega} \cdot \nabla \boldsymbol{v}, \tag{7.37}$$

where $D_t \equiv \partial_t + \boldsymbol{v} \cdot \nabla$ denotes the Lagrangian derivative. Observe that $\nabla \boldsymbol{v}$ has the same dimensions as $\boldsymbol{\omega}$ and can be related to it by an operator involving Poisson-type integrals. (For this use the fact that $\nabla^2 \boldsymbol{v} = -\nabla \wedge \boldsymbol{\omega}$.) It is then tempting to predict that the solutions of (7.37) will behave as the solution of the scalar nonlinear equation

$$D_t s = s^2, \tag{7.38}$$

which blows up in a time $1/s(0)$ when $s(0) > 0$. Actually, (7.38) is just the sort of equation one obtains in trying to find rigorous *upper bounds* to various norms when studying the well-posedness of the Euler problem (see Section 9.3). This is precisely why the well-posedness 'in the large' (i.e. for arbitrary $t > 0$) is an open problem in three dimensions.[18] This problem has been singled out by Saffman (1981) as 'one of the most challenging of the present time for both the mathematician and the numerical analyst'.

The evidence is that the solutions of the Euler equation behave in a way much tamer than predicted by (7.38). Since such evidence cannot be obtained by experimental means, one has to resort to numerical simulations. We briefly summarize here what is known from numerical simulations about inviscid three-dimensional flow with smooth initial conditions.

The most reliable method for investigating blow-up uses a combination of the spectral technique (see the footnote on page 70 of Section 5.2) and

[17] 'khaotitsheskogo' in the original

[18] In two dimensions, the absence of singularities for any $t > 0$ has been proven by Hölder (1933) and Wolibner (1933); see also Kato (1967) and Sulem and Sulem (1983).

the method of tracing of complex singularities (Sulem, Sulem and Frisch 1983). The latter monitors the imaginary part $\delta(t)$ of the complex-space singularity nearest to the real space as a function of real time t. This $\delta(t)$ is called the 'width of the analyticity strip'. The method takes advantage of two rigorous results: (i) that a hypothetical real singularity of the Euler equations occurring at time $t_\star$ is preceded by a positive $\delta(t)$ which shrinks to zero at $t = t_\star$ (Bardos and Benachour 1977); (ii) that the width $\delta(t)$ of the analyticity strip can be measured directly from the spatial Fourier transform of the solution which falls off (roughly) as $e^{-\delta(t)k}$. As long as $\delta(t)$ is sufficiently large (in practice about two meshes), the energy spectrum[19] has a rather conspicuous exponential tail; its logarithmic decrement is equal to $2\delta(t)$. (The factor 2 is present because the energy spectrum is proportional to the *square* of the Fourier amplitude.)

This procedure has been applied by Brachet *et al.* (1983) to the Taylor–Green vortex, introduced in Section 5.2. The simulation was done on a grid of 256^3 points. The maximum wavenumber was $256/3 \approx 85$. Fig. 7.6(a) shows the energy spectrum $E(k, t)$ at various times in linear–log coordinates, so that the exponential tails appear as straight lines. Fig. 7.6(b) shows the evolution of $\delta(t)$, also in linear–log coordinates: except for very short times and for as long as $\delta(t)$ can be reliably measured, it displays almost perfect exponential decrease. If this result can be safely extrapolated to later times, it follows that the Taylor–Green vortex will *never* develop a real singularity: there is no inviscid blow-up. When this result was obtained in 1981, it came as a rather big surprise.[20] Indeed, based on the kind of phenomenology described above and also on results from closure (see Section 9.5), there was a widespread belief that finite-time blow-up would take place.[21]

But is it safe to extrapolate the behavior of $\delta(t)$? About ten years later it became possible to extend the Brachet *et al.* (1983) calculation, using a grid of 864^3 points and also to study flows with random initial conditions without the somewhat special symmetries of the Taylor–Green vortex which helped in reducing computational work (Brachet,

[19] In numerical simulation of decaying turbulence, it is customary to define the energy spectrum as an angular average, without any ensemble averaging.

[20] Results by Morf, Orszag and Frisch (1980), using the Taylor series up to t^{44} for the enstrophy and analytic continuation by Padé approximants, did suggest finite-time blow-up; Brachet *et al.* (1983) extended the series up to t^{80} and could not support the previous results.

[21] I shared such a belief, but G.I. Taylor did not, as appears from a brief statement made to S.A. Orszag in 1969 which was communicated to me privately. As for A.N. Kolmogorov, I am not aware of anything he has said on this matter.

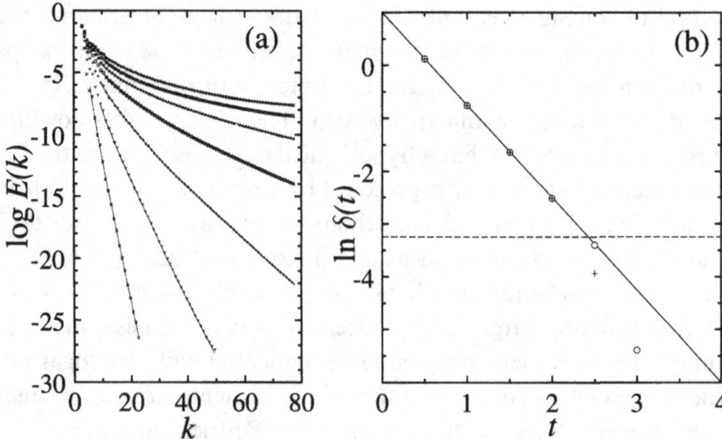

Fig. 7.6. Spectral simulation of the inviscid Taylor–Green vortex using 256^3 Fourier modes. Evolution of the energy spectrum in linear–log coordinates; from bottom to top: output from time $t = 0.5$ in increments of 0.5 (a). Time-dependence of the width of the analyticity strip $\delta(t)$ in linear–log coordinates; the circles and plus signs correspond to 256^3 and 128^3 Fourier modes, respectively; the dashed line gives the threshold of reliability (b) (Brachet *et al.* 1983).

Meneguzzi, Vincent, Politano and Sulem 1992). Again, exponential decrease of $\delta(t)$ was observed.

However formidable a 864^3 simulation may look, it can only explore a span of scales of about 300, because it uses a *uniform* grid. Pumir and Siggia (1990) developed a different approach using grid-refinement 'where needed' and were thereby able to explore a span of scales of up to 10^5. Still, no blow-up was observed (Pumir and Siggia 1990). Somewhat paradoxically, simulations by Grauer and Sideris (1991) and Pumir and Siggia (1992) of two-dimensional axisymmetric flow with a poloidal component of the velocity (an instance for which there is no regularity theorem) have given some evidence of finite-time blow-up. This is, however, a controversial issue (see, e.g., E and Shu 1994).

All these simulations have also given us a qualitative explanation for why ideal flow is much more regular then predicted by naive phenomenology: the exponential decrease of $\delta(t)$ corresponds to an exponential flattening of vorticity 'pancakes' of the sort shown in Fig. 7.7. The vorticity in such structures has a very fast dependence on the spatial coordinate transverse to the pancake, so that the flow is to leading order one-dimensional. If the flow were exactly one-dimensional, the nonlinearity would vanish (as a consequence of the incompressibility condition).

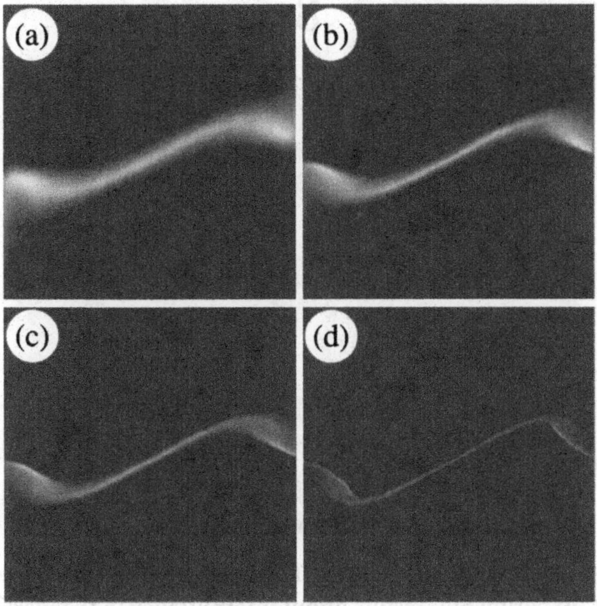

Fig. 7.7. Vorticity pancakes observed in an inviscid simulation of the Taylor–Green vortex. Output at time $t = 3.0$ (a), $t = 3.25$ (b), $t = 3.50$ (c) and $t = 4.0$ (d) (Brachet, Meneguzzi, Vincent, Politano and Sulem 1992).

This *depletion of nonlinearity* explains why the growth of the vorticity is much slower than predicted by (7.38) which ignores this phenomenon. We shall come back to the issue of depletion in Sections 9.3, 9.5 and 9.7.

The present section was meant mostly to underline a severe shortcoming of phenomenology. We can nevertheless conclude it with a more optimistic note. Standard phenomenology was here found to *overestimate* the strength of nonlinearity for ideal flow. Maybe it also overestimates it for viscous flow. This would be good news: as shown at the end of Section 7.2, standard K41 phenomenology predicts that *viscous flow* never blows up, something for which there is still no proof in three dimensions. Leray's (1934) result, the best available for more than half a century, cannot rule out the presence of singularities at a set of instants which is very 'small'. We shall return to such matters for both viscous and inviscid flow in Section 9.3.

8
Intermittency

8.1 Introduction

This chapter is organized as follows. The basic concepts are introduced in Section 8.2. Experimental results about intermittency in the inertial range, based on velocity measurements, are presented in Section 8.3. Exact results, independent of any phenomenology, are presented in Section 8.4. Two broad classes of phenomenological models of intermittency are then discussed. In the first class (Section 8.5), intermittency is studied via velocity increments. It comprises the β-model (Section 8.5.1), the bifractal model (Section 8.5.2) and the multifractal model (Sections 8.5.3 and 8.5.4). Implications of the multifractal model for the dissipation range and for the skewness and flatness of velocity derivatives are presented in Sections 8.5.5 and 8.5.6, respectively. In the second class (Section 8.6), intermittency is studied via the fluctuation of the dissipation; inertial-range quantities are related to such fluctuations by a bridging ansatz, originally introduced by Obukhov and Kolmogorov (Section 8.6.2). Random cascade models are presented in Section 8.6.3; their multifractal behavior is shown to be a direct consequence of the probabilistic theory of 'large deviations', which is presented in an elementary fashion in Section 8.6.4. The lognormal model and its shortcomings are discussed in Section 8.6.5. Shell models, a class of deterministic nonlinear models which can display intermittency, are presented in Section 8.7.

The order chosen here for the presentation of the entire material on the theory of intermittency is pedagogical, not historical. Most of the latter aspects are discussed in Section 8.8. Recent trends in intermittency research are presented in Section 8.9.

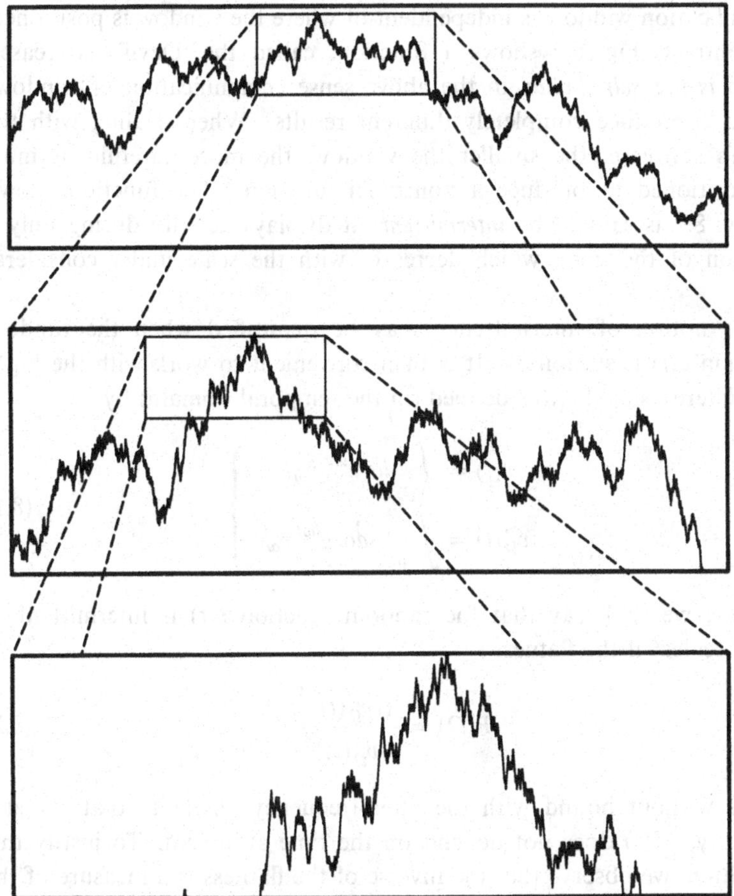

Fig. 8.1. A portion of the graph of the Brownian motion curve, enlarged twice, illustrating its self-similarity.

8.2 Self-similar and intermittent random functions

A central assumption of the K41 theory is the *self-similarity* of the random velocity field at inertial-range scales. As we shall see this symmetry may well be broken. The meaning of the concept of self-similarity as applied to a random function is illustrated in Fig. 8.1 in which a sample of a self-similar random function $v(t)$, here the Brownian motion function, is shown with two successive enlargements. It must be stressed that the 'general aspect' (actually, the statistical properties) within the magnification window is independent of where the window is positioned. In contrast, Fig. 8.2 shows a function, called the 'Devil's staircase'[1] which *is not self-similar* in the above sense: magnification of windows 1 and 2 produce completely different results. When dealing with the Devil's staircase, the smaller the window, the more carefully it must be positioned to produce a nontrivial function. The function shown in Fig. 8.2 is said to be *intermittent*: it displays activity during only a fraction of the time, which decreases with the scale under consideration.

The notion of intermittency may be quantified when the random function $v(t)$ is stationary. It is then convenient to work with the high-pass filtered signal $v^{>}_{\Omega}(t)$, defined (in the temporal domain) by

$$\left.\begin{aligned} v(t) &= \int_{\mathbb{R}^3} d\omega\, e^{i\omega t}\, \hat{v}_{\omega}, \\ v^{>}_{\Omega}(t) &= \int_{|\omega|>\Omega} d\omega\, e^{i\omega t}\, \hat{v}_{\omega}. \end{aligned}\right\} \tag{8.1}$$

Here, we shall say that the random function $v(t)$ is intermittent at small scales[2] if the flatness

$$F(\Omega) = \frac{\langle (v^{>}_{\Omega}(t))^4 \rangle}{\langle (v^{>}_{\Omega}(t))^2 \rangle^2} \tag{8.2}$$

grows without bound with the filter frequency Ω. Note that, by stationarity, $F(\Omega)$ does not depend on the time argument. To justify this definition, we observe that the inverse of the flatness is a measure of the fraction of the time the (high-pass) signal is 'on'. Consider the signals in Fig. 8.3. Part (a) shows a stationary signal $v(t)$ and part (b) shows the same signal 'chopped-off' a fraction $1 - \gamma$ of the total time. More precisely, $v_{\gamma}(t)$ is obtained from $v(t)$ by setting it equal to zero except during 'on intervals' which are randomly selected in such a way that the time the signal is on represents a fraction γ of the total time. Assuming that all relevant moments exist, it follows that

$$\langle v_{\gamma}^2 \rangle = \gamma \langle v^2 \rangle, \qquad \langle v_{\gamma}^4 \rangle = \gamma \langle v^4 \rangle. \tag{8.3}$$

[1] The Devil's staircase gives the fraction of the mass of the Cantor set in the interval $[0, t]$. It is constructed recursively. One starts with a uniform distribution in the interval $[0, 1]$ of unit total mass, removes the middle third and redistributes the removed mass evenly among the remaining intervals. The process is repeated indefinitely.

[2] There are other forms of intermittency, not discussed here, which appear in relation to transition to turbulence (see, e.g., Pomeau and Manneville 1980).

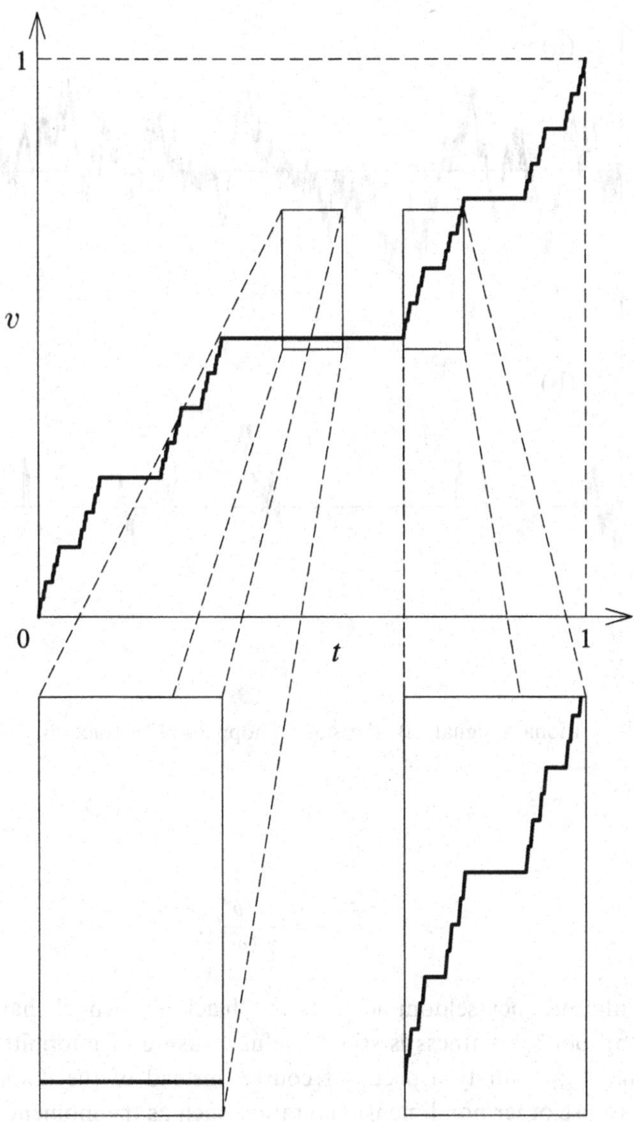

Fig. 8.2. The Devil's staircase: an intermittent function.

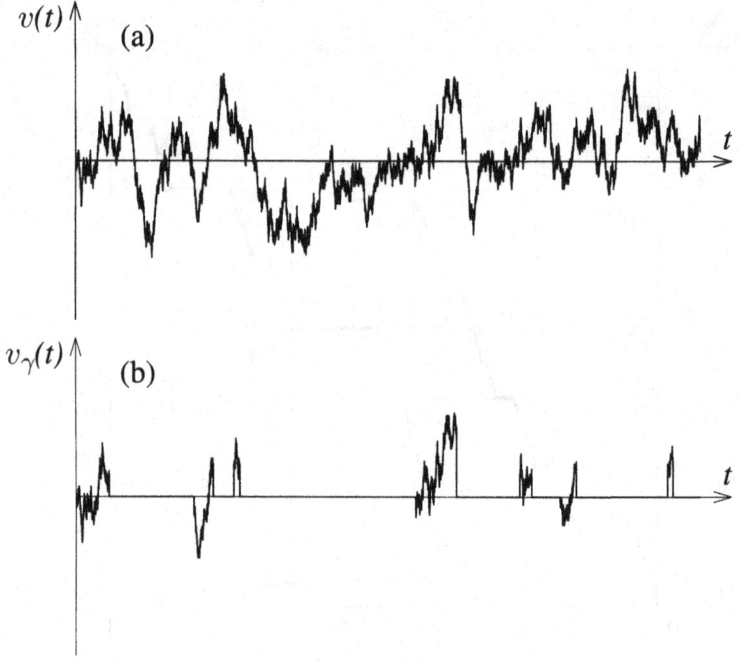

Fig. 8.3. A stationary signal (a); the same 'chopped-off' a fraction $1-\gamma$ of the time (b).

Hence,

$$F_\gamma = \frac{\langle v_\gamma^4 \rangle}{\langle v_\gamma^2 \rangle^2} = \frac{1}{\gamma} \frac{\langle v^4 \rangle}{\langle v^2 \rangle^2}. \tag{8.4}$$

True intermittency seldom achieves the 'black-and-white' character of Fig. 8.3(b), but the flatness is still a useful measure of intermittency for signals having a bursty aspect. Of course, instead of the flatness, it is possible to use other nondimensional ratios, such as the moment of order 6 divided by the cube of the moment of order 2. Odd order moments are, however, inappropriate, since they may vanish for reasons of symmetry or accidentally.

Observe that with our definition, neither Gaussian nor self-similar signals are intermittent, because their flatness $F(\Omega)$ is independent of Ω. In the Gaussian case, this is because the Gaussian property is conserved by any linear operation such as filtering and because a Gaussian random

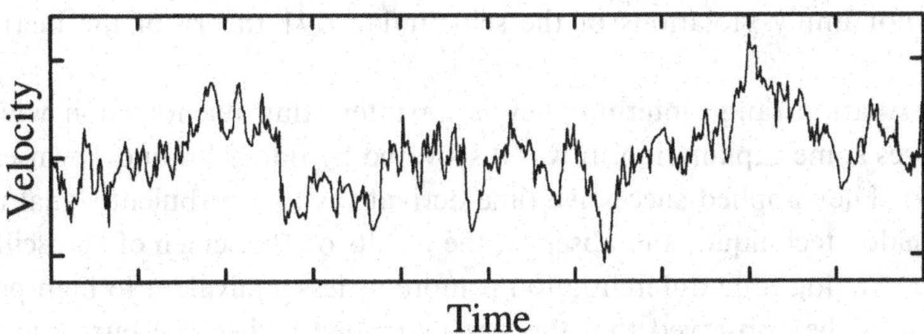

Fig. 8.4. Velocity signal from a jet with $R_\lambda \simeq 700$ (Gagne 1980).

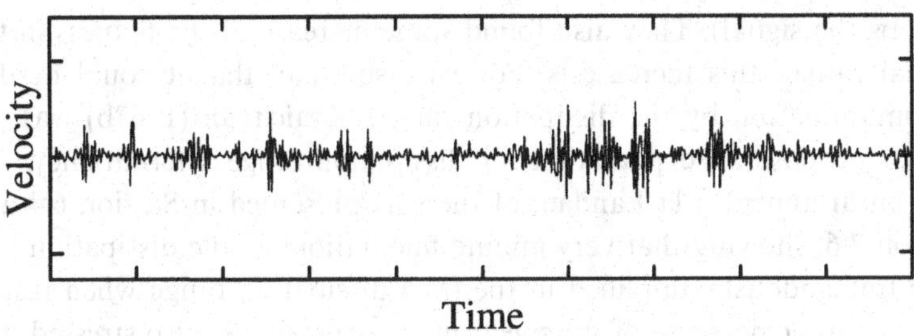

Fig. 8.5. Same signal as in Fig. 8.4, subject to high-pass filtering, showing intermittent bursts (Gagne 1980).

variable has a flatness of 3 (an immediate consequence of (4.7)). Let us now assume that $v(t)$ has self-similar increments with scaling exponent h. It is then easily shown that, for any $\lambda > 0$,

$$v^{>}_{\lambda\Omega} \overset{\text{law}}{=} \lambda^{-h} v^{>}_{\Omega}. \tag{8.5}$$

As a consequence, when Ω is changed into $\lambda\Omega$ in (8.2), both the numerator and the denominator are multiplied by λ^{-4h}, leaving the flatness unchanged.

Is turbulence self-similar or is it intermittent? Visual inspection of Fig. 8.4, a sample of a turbulent signal from a jet (Gagne 1980), suggests that it is more like the self-similar example of Fig. 8.1. If, however, the same signal is subject to high-pass filtering and the filter frequency Ω is chosen high enough, intermittent features appear as shown in Fig. 8.5. This sort of intermittency becomes very conspicuous only when the scale associated with Ω is comparable to or smaller than the Kolmogorov dissipation scale. It is thus a characteristic of the dissipation range and does not imply violations of the self-similar K41 theory of the inertial range.

Dissipation-range intermittency is an interesting phenomenon which deserves some explanation. It was discovered by Batchelor and Townsend (1949). They applied successive time-derivatives to a turbulent signal using analog techniques and observed the results on the screen of an oscilloscope. Analog time-differentiation is more or less equivalent to high-pass filtering. They observed that the signals tended to become bursty when the order of differentiation was increased. Kuo and Corrsin (1971) used band-pass filtering and found that the flatness increases drastically with mid-band frequency in the dissipation range (until experimental noise swamps the signal). They also found some increase of the flatness in the inertial range; this increase is, however, so small that it could be due to contamination by the dissipation range. Kraichnan (1967b) was the first to propose an explanation for dissipation-range intermittency. He used an argument à la Landau, of the sort presented in Section 6.4 and Section 7.6, showing that very minute fluctuations in the dissipation rate ε are tremendously amplified in the (far) dissipation range when faster-than-algebraic decrease of the spectrum is assumed. He also stressed that dissipation-range intermittency does not imply inconsistency with K41.

The first systematic explanation of dissipation-range intermittency was given by Frisch and Morf (1981), stimulated by the aforementioned data of Gagne. They considered a wide class of dynamical systems governed by ordinary or partial differential equations with solutions which are analytic in the time variable.[3] A standard result for Fourier transforms of analytic functions is that their high-frequency asymptotic behavior is dominated by contributions from those singularities in the complex time domain which are close to the real line.[4] From this, Frisch and Morf (1981) showed that each such singularity produces a burst in the high-pass filtered solution. The center of the burst coincides with the real part of the singularity and minus the logarithm of the amplitude is proportional to the imaginary part multiplied by the filter frequency Ω. Thus, when $\Omega \to \infty$, only those singularities with very small imaginary parts produce conspicuous bursts. Since such events are rare, this gives an increasing intermittency.

[3] Analyticity is easily proved for ordinary differential equations such as the Lorenz model and is a consequence of the conjectured boundedness of the velocity for the three-dimensional Navier–Stokes equation with positive viscosity.

[4] This is basically how von Neumann (1949) predicted an exponential fall-off for the energy spectrum at high wavenumbers; see Section 6.3.2.

8.3 Experimental results on intermittency

There is little doubt about the existence of intermittency in the dissipation range. A subtler question concerns the existence of intermittency in the inertial range, since the latter would invalidate the K41 theory. Historically, the first attempts to detect such intermittency involved measurements of fluctuations of the (local) energy dissipation. Such quantities involve simultaneously dissipation-range scales and inertial-range scales. Their interpretation is thus very delicate; we shall come back to the matter after the appropriate concepts have been introduced (see Section 8.6).

The K41 theory predicts that the structure function of order p scales with an exponent $p/3$ over inertial-range separations, something which can be tested experimentally. However, in his 1941 papers Kolmogorov made *explicit* predictions only for second and third order structure functions. This may be one of the reasons measurements of structure functions of order $p \geq 4$ were not attempted until the early 1970s (Van Atta and Chen 1970). Measurements of *high-order* structure functions are actually rather difficult. Indeed, by definition, they require accurate measurements of high-order moments of velocity increments. These involve the tail of the corresponding probability distribution functions, i.e. very rare events. Hence, it is necessary to process very long records of the turbulent signal. Early measurements of structure functions were limited by recording capabilities. As shown by Anselmet, Gagne, Hopfinger and Antonia (1984) this can lead to vastly overestimating the discrepancy between the measured exponents and their K41 prediction. They give, nevertheless, reasonably convincing evidence that significant discrepancies remain. Their experiments were performed in the laboratory at Taylor-scale based Reynolds numbers R_λ of up to 852. The experiment was repeated by Gagne and Hopfinger in the S1 wind tunnel at $R_\lambda = 2700$ (Gagne 1987). Although this gave a much wider inertial range, it mostly confirmed the previous results by Anselmet, Gagne, Hopfinger and Antonia (1984).

Fig. 8.6 shows the results for the (longitudinal) structure functions of order 2, 3 and 6, obtained by Gagne and Hopfinger. It is adapted from Gagne (1987). The second order structure function for the same data was presented in Fig. 5.1. The presentation of the data in Fig. 8.6 uses compensating power-law factors to identify exponents of structure functions. The results suggest that structure functions follow power-laws in the inertial range:

$$S_p(\ell) = \left\langle \left(\delta v_\parallel(\ell)\right)^p \right\rangle \propto \ell^{\zeta_p}, \tag{8.6}$$

where the ζ_ps are called the exponents of structure functions.

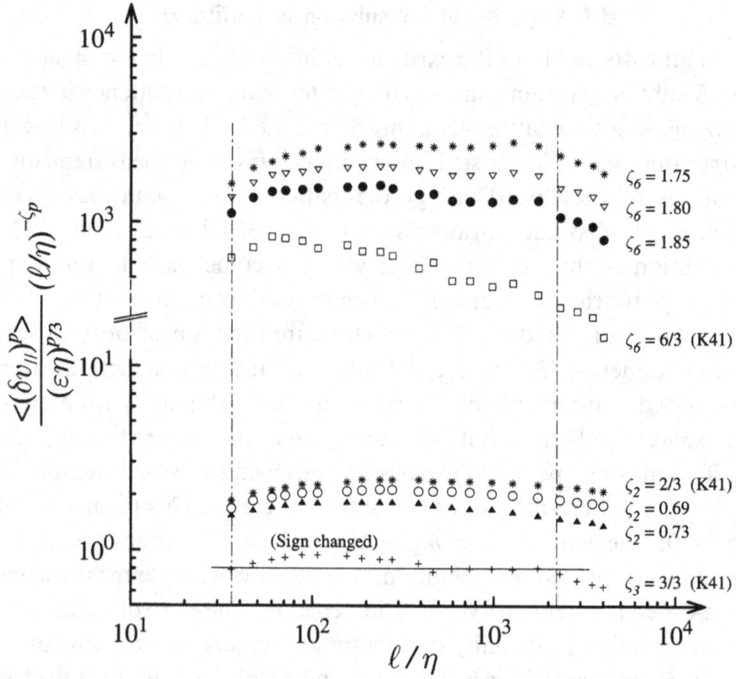

Fig. 8.6. Structure functions of order 2, 3 and 6 in the time domain, compensated by 'guessed' power-law factors for the S1 data (Gagne 1987).

We mention here that it is not *a priori* evident that structure functions of high order exist. This depends on the shape of the tails of the p.d.f.s of velocity increments: if it is algebraic, moments beyond a certain order are infinite. Actually the experimental evidence is that the tails decrease in a roughly exponential manner (Van Atta and Chen 1970; Gagne, Hopfinger and Frisch 1990; Gagne and Castaing 1991). Of course, there could be an algebraic tail lost in the noise beyond the exponential tail. Intermittency models assuming such 'hyperbolic' tails can be constructed and studied (see, e.g., Schertzer and Lovejoy 1984, 1985; Schmitt, Lavallée, Schertzer and Lovejoy 1992).

Let us now examine the results about structure functions (for more detail, see Anselmet, Gagne, Hopfinger and Antonia 1984 and Gagne 1987). The third order structure function (represented with the sign reversed to make it positive) is in good agreement with the four-fifths law: for ℓ/η from 500 to 2000 the data fall almost exactly on the value $4/5$. For lower values of ℓ/η, around the Taylor scale, the values are about 20% higher.

Deviations from the four-fifths law are intriguing because this law should apply irrespective of the validity of K41. Actually, there are many possible causes of deviations. They include: lack of asymptoticity (e.g., contamination by the dissipation range), lack of homogeneity and/or isotropy, violations of the Taylor hypothesis, violation of the hypothesis H3 of the finiteness of the energy dissipation when $\nu \to 0$, inaccurate determination of the mean energy dissipation rate ε, and poor quality of the data.

Anselmet, Gagne, Hopfinger and Antonia (1984) proposed using the range of scales over which the third order structure function is reasonably close to 4/5 as an operational definition of the inertial range. This is shown by the two vertical dash-dotted lines in Fig. 8.6, which delineate the inertial range.

We have already discussed the second order structure function in Section 5.1. With the 'compensated' representation of Fig. 8.6, we see that it is difficult to make a choice between the three values of ζ_2 tried (including the K41 value 2/3). As we have already stressed in Section 5.1, the Fourier representation in terms of the energy spectrum gives a wider inertial range and thereby a more accurate definition of the exponent ζ_2. According to the measurements performed in 1994 by Y. Gagne and M. Marchand (private communication) at the S1 wind tunnel, ζ_2 is too close to the K41 value 2/3 to detect any discrepancy. Measurements of exponents can also be done by *wavelet transform* methods (Muzy, Bacry and Arnéodo 1993). For the calculation of structure functions, this amounts to using a filter emphasizing the Fourier components of scale ℓ of the velocity, instead of just the increment over a distance ℓ. Perhaps the shape of the filter could be optimized to improve the quality of the scaling.

The data for the sixth order structure function shown in Fig. 8.6 suggest that the ζ_6 is closer to 1.8 than to its K41 value of 6/3. The value $\zeta_6 = 1.8$ is also proposed in Anselmet, Gagne, Hopfinger and Antonia (1984). A consequence of $\zeta_2 \geq 2/3$ and $\zeta_6 < 6/3$ is that the (hyper-)flatness

$$F_6(\ell) = \frac{S_6(\ell)}{(S_2(\ell))^3} \tag{8.7}$$

grows as a power-law when $\ell \to 0$ (while staying within the inertial range). Thus intermittency, measured by $F_6(\ell)$, becomes arbitrarily strong at very high Reynolds numbers and small inertial-range scales. It would be harder to reach such a conclusion using the flatness $F_4(\ell)$ since the results of Anselmet, Gagne, Hopfinger and Antonia (1984) do not show any measurable discrepancy of ζ_4 from its K41 value of 4/3. Actually, there is considerable interest in the sixth order structure function because the discrepancy from K41,

$$\mu \equiv 2 - \zeta_6, \tag{8.8}$$

can be interpreted in various models of intermittency as the codimension (3 minus the dimension) of dissipative structures (see below). Even though the discrepancy from the K41 value for ζ_6 appears to be quite strong, there is no absolute guarantee that this is genuine. Anselmet, Gagne, Hopfinger and Antonia (1984) have checked that the size of their sample is sufficient to ensure convergence of the statistics, but there could be systematic errors of the sort already mentioned in connection with the third order structure function. An additional uncertainty in determining the exponent stems from the undulating character of the sixth order structure function (shared by those of higher order). Novikov (1969) pointed out the possibility of log-periodic corrections to scaling, i.e. power-law behavior with correction terms involving functions which are periodic in the logarithm of the scale (or the wavenumber). Smith, Fournier and Spiegel (1986) showed that log-periodic corrections to scaling arise in fractal models with *lacunarity*, i.e. having a preferred ratio of scales (the simplest example being the triadic cantor set). According to Benzi and coworkers the undulations depend strongly on the global geometry of the flow and are particularly affected by recirculation (R. Benzi, private communication). The data obtained by Maurer, Tabeling and Zocchi (1994) in their helium gas experiment (discussed briefly in Section 5.1) for the structure functions of order 2, 3, 4 and 6, shown in Fig. 8.7, are nearer a pure power-law than those from the S1 wind tunnel. The helium experiment has, however, a fairly large turbulence intensity (the r.m.s. velocity fluctuations are about 25% of the mean flow). Hence, the spatial scale $\ell = U\tau$ associated with a given time-lag τ (by the Taylor hypothesis, using the instantaneous velocity U) undergoes large modulations which, after averaging, may smooth out the undulations.

A new processing technique has been proposed by Benzi, Ciliberto, Baudet *et al.* (1993), Benzi, Ciliberto, Tripiccione *et al.* (1993; see also Benzi, Ciliberto, Baudet and Ruiz Chavarria 1995), which may greatly improve the accuracy of exponents and which overcomes in particular the difficulties stemming from the undulations. Instead of plotting $S_p(\ell)$ vs ℓ, they plot $S_p(\ell)$ vs $S_{p'}(\ell)$ in log–log coordinates. This gives much straighter graphs.[5] One reason is that the undulations, whatever their

[5] Special instances of this procedure may be found in Anselmet, Gagne, Hopfinger and Antonia (1984, Section 5.2).

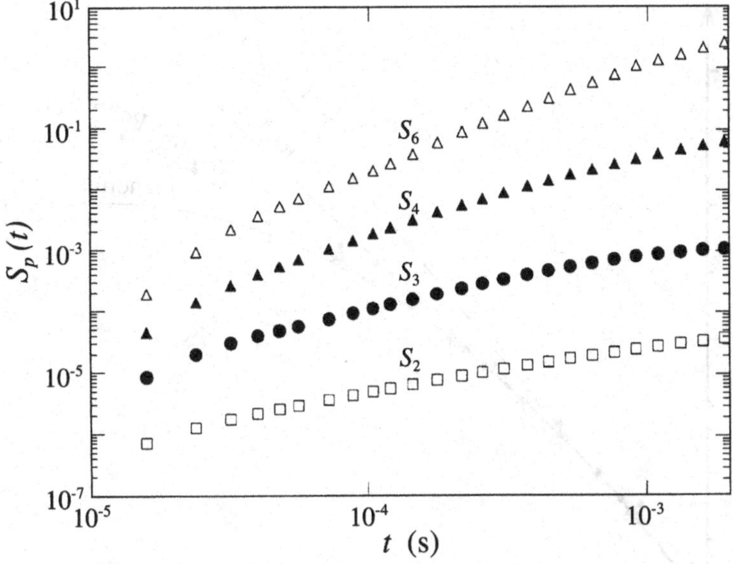

Fig. 8.7. Structure functions of order 2, 3, 4 and 6 (as labeled) in the time domain in log–log coordinates for a low temperature helium gas flow between counter-rotating cylinders with $R_\lambda = 1200$. The units are $\text{cm}^3\,\text{s}^{-3}$ for order 3 and arbitrary for the other orders (Maurer, Tabeling and Zocchi 1994).

origin, appear to be correlated among the different orders of structure functions (as seen from Fig. 8.6).[6] Another reason, stressed by Benzi, Ciliberto, Baudet *et al.* (1993), is that the functional forms of the structure functions at the beginning of the dissipative fall-off are the same, down to about five times the Kolmogorov scale. They refer to this as *extended self-similarity* (ESS). If the exponent for $S_{p'}(\ell)$ is known (say, for $p' = 3$), the exponent for $S_p(\ell)$ follows. In this way, using various high-Reynolds-number data (including those from the S1 wind tunnel), ESS gives the following numbers (Benzi, Ciliberto, Baudet and Ruiz Chavarria 1995): $\zeta_2 = 0.70$, $\zeta_3 = 1.00$ (by assumption), $\zeta_4 = 1.28$, $\zeta_5 = 1.53$, $\zeta_6 = 1.77$, $\zeta_7 = 2.01$, $\zeta_8 = 2.23$. The corresponding data points are plotted as black triangles in Fig. 8.8.

An alternative way to improve the accuracy on scaling exponents would be to fit the structure functions to power-laws with log-periodic corrections as was done by Sornette and Sammis (1995) to improve the prediction of the time of occurrence of large earthquakes.

[6] This correlation is particularly strong when considering structure functions of neighboring orders (A. Noullez, private communication).

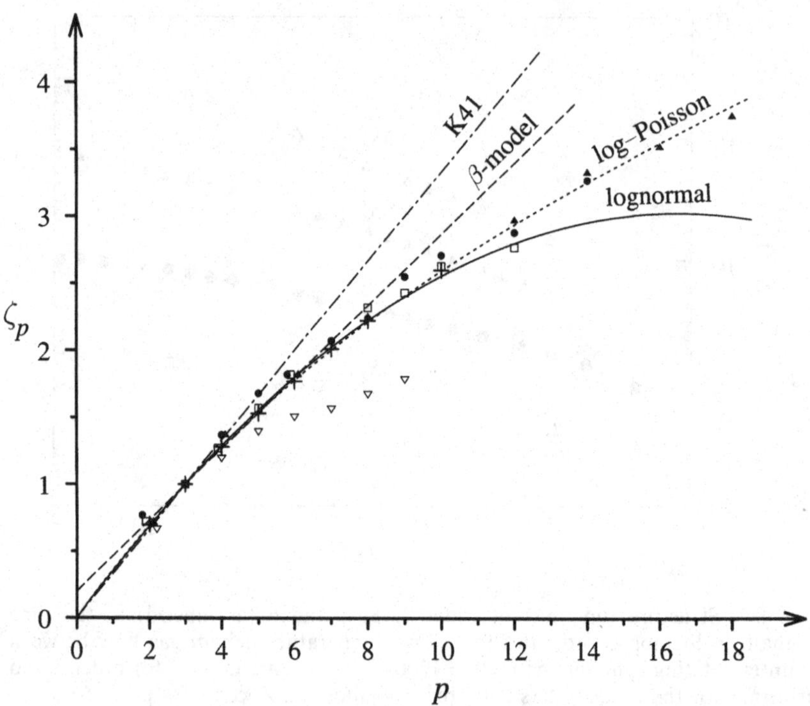

Fig. 8.8. Exponent ζ_p of structure functions in the time domain of order p vs p. Inverted white triangles: data from Van Atta and Park (1972); black circles, white squares and black triangles: data from Anselmet, Gagne, Hopfinger and Antonia (1984) with $R_\lambda = 515$, 536, 852, respectively; + signs: S1 data processed by 'ESS' (see p. 131). Straight chain line: $\zeta_p = p/3$ (K41); dashed line: β-model (eq. (8.31)) with $D = 2.8$; solid line: lognormal model (eq. (8.122)) with $\mu = 0.2$; dotted line: log–Poisson model (eq. (8.141)).

Anselmet, Gagne, Hopfinger and Antonia (1984) have summarized all their results in a graph of ζ_p vs p which is reproduced in Fig. 8.8. The straight chain line through the origin of slope $1/3$ is the K41 prediction. The straight dashed line and the curved continuous line correspond to various models of intermittency which will be discussed later. The inverted white triangles, correspond to early data of Van Atta and Park (1972), which suffered from unconverged statistics (beyond $p = 4$). The same problem also appears for the data of Anselmet, Gagne, Hopfinger and Antonia beyond a value of p somewhere between 10 and 12. In addition, as already stressed, there may be other causes of errors. Even with such *caveat*s the data of Fig. 8.8 have played an important role in turbulence theory because they have directly motivated the introduction of the *multifractal model* to be discussed in Section 8.5.3 (see also Section 8.8).

This entire section can be summarized as: It is plausible but not certain that there are intermittency corrections to the K41 theory of the inertial range.

8.4 Exact results on intermittency

In this section we examine some 'exact' constraints which have to be satisfied by the exponents ζ_p of structure functions. By 'exact' we understand constraints which can be derived from probabilistic inequalities or the basic physics of incompressible flow without recourse to additional assumptions beyond the very existence of such exponents. One instance, already discussed at length, is Kolmogorov's four-fifths law which implies $\zeta_3 = 1$. Our additional exact results concern structure functions of even order.

The only assumption needed here is that structure functions of even order have power-law behavior for large Reynolds numbers at inertial-range scales. Specifically, we assume the following.

S1 In the limit $R = \ell_0 v_0/\nu \to \infty$, the structure functions of even order $2p > 0$ possess the exponents ζ_{2p}, i.e. for $\ell \to 0$, one has to leading order:

$$\frac{\langle(\delta v_\parallel(\ell))^{2p}\rangle}{v_0^{2p}} \simeq A_{2p}\left(\frac{\ell}{\ell_0}\right)^{\zeta_{2p}}, \tag{8.9}$$

where A_{2p} is a positive numerical constant (not necessarily universal).

S2 For large finite R, the scaling (8.9) still holds, as intermediate asymptotics, over a range of scales (the inertial range) increasing with R at least as a power-law:

$$1 \gg \frac{\ell}{\ell_0} \gg R^{-\alpha}, \qquad \alpha > 0. \tag{8.10}$$

We now establish three propositions.

P1 For any three positive integers $p_1 \leq p_2 \leq p_3$, we have the convexity inequality:

$$(p_3 - p_1)\zeta_{2p_2} \geq (p_3 - p_2)\zeta_{2p_1} + (p_2 - p_1)\zeta_{2p_3}. \tag{8.11}$$

P2 Under assumption S1, if there exist two consecutive even numbers $2p$ and $2p+2$ such that

$$\zeta_{2p} > \zeta_{2p+2}, \tag{8.12}$$

then the velocity of the flow (measured in the reference frame of the mean flow) cannot be bounded.

P3 Under assumption S2 and under the assumptions made in P2, if the Mach number based on v_0 is held fixed, and the Reynolds number is increased indefinitely,[7] then the maximum Mach number of the flow also increases indefinitely.

Proof. It follows from the Hölder inequality for moments of random variables (Feller 1968b) that

$$\langle(\delta v_\parallel)^{2p_2}\rangle^{2p_3-2p_1} \leq \langle(\delta v_\parallel)^{2p_1}\rangle^{2p_3-2p_2} \langle(\delta v_\parallel)^{2p_3}\rangle^{2p_2-2p_1}. \tag{8.13}$$

Substituting expression (8.9) for the structure functions, and letting $\ell \to 0$, we obtain (8.11) which expresses that the graph of ζ_{2p} vs p is concave; this proves P1. Let us now denote by $U_{\max}$ the maximum velocity, taken over space and time. We have at any instant of time

$$|\delta v_\parallel(\ell)| \leq 2U_{\max}. \tag{8.14}$$

With ensemble averages reinterpreted as time averages, it follows from (8.14) that

$$\langle(\delta v_\parallel(\ell))^{2p+2}\rangle \leq 4U_{\max}^2\langle(\delta v_\parallel(\ell))^{2p}\rangle. \tag{8.15}$$

Assuming $\ell \ll \ell_0$ and using (8.9), we obtain

$$\frac{U_{\max}^2}{v_0^2} \geq \frac{1}{4}\frac{A_{2p+2}}{A_{2p}}\left(\frac{\ell}{\ell_0}\right)^{-(\zeta_{2p}-\zeta_{2p+2})}. \tag{8.16}$$

Using (8.12) and letting $\ell \to 0$, we find that $U_{\max} = \infty$. This proves proposition P2.

We now define

$$M_0 = \frac{v_0}{c_s}, \qquad M_{\max} = \frac{U_{\max}}{c_s}, \tag{8.17}$$

which are respectively the Mach number based on the r.m.s. velocity and on the maximum velocity (in the frame of the mean flow). We select a scale ℓ:

$$\frac{\ell}{\ell_0} = R^{-\alpha/2}, \tag{8.18}$$

[7] For example, by considering a sequence of grid-generated turbulent flows with ever-increasing mesh, all using the same fluid and the same flow velocity.

which, by (8.10), is within the inertial range. Substituting (8.18) into (8.16) and using (8.17), we obtain:

$$M_{\max}^2 \geq \frac{1}{4}\frac{A_{2p+2}}{A_{2p}} M_0^2 R^{(\zeta_{2p}-\zeta_{2p+2})\alpha/2}. \tag{8.19}$$

Proposition P3 follows readily. *QED.*

A Mach number, measured in the reference frame of the mean flow, which becomes arbitrarily large, violates a basic assumption needed in obtaining the *incompressible* Navier–Stokes equation. It does not, however, violate the basic physics since it is conceivable that, at extremely high Reynolds numbers, supersonic velocity would appear. Anyway, (8.12) is not consistent with a uniform (in Reynolds number) validity of the incompressible Navier–Stokes equation.

We may summarize the findings of this section as follows: If structure functions of even order follow power-laws with exponents ζ_{2p} and if the incompressible approximation does not break down at high Reynolds numbers, then the graph of ζ_{2p} vs p is concave and nondecreasing.[8]

8.5 Intermittency models based on the velocity

8.5.1 The β-model

There is a very simple way to modify the phenomenological model introduced in Section 7.3, so as to incorporate a form of intermittency. Fig. 8.9, which is to be compared with Fig. 7.2, shows the idea of the *β-model*: at each stage of the Richardson cascade, the number of 'daughters' of a given 'mother-eddy' is chosen such that the fraction of volume occupied is decreased by a factor β $(0 < \beta < 1)$. The factor β is an adjustable parameter of the model. Otherwise, nothing is changed in the presentation of the cascade made in Section 7.3. In the β-model, the fraction p_ℓ of the space which is 'active', i.e. within a daughter-eddy of size $\ell = r^n \ell_0$ decreases as a power of ℓ. Indeed,

$$p_\ell = \beta^n = \beta^{\frac{\ln(\ell/\ell_0)}{\ln r}} = \left(\frac{\ell}{\ell_0}\right)^{3-D}, \tag{8.20}$$

where

$$3 - D \equiv \frac{\ln \beta}{\ln r}. \tag{8.21}$$

[8] Additional exact results, giving inequalities for the exponents ζ_p, may be found in Constantin and Fefferman (1994) and are summarized in Constantin (1994).

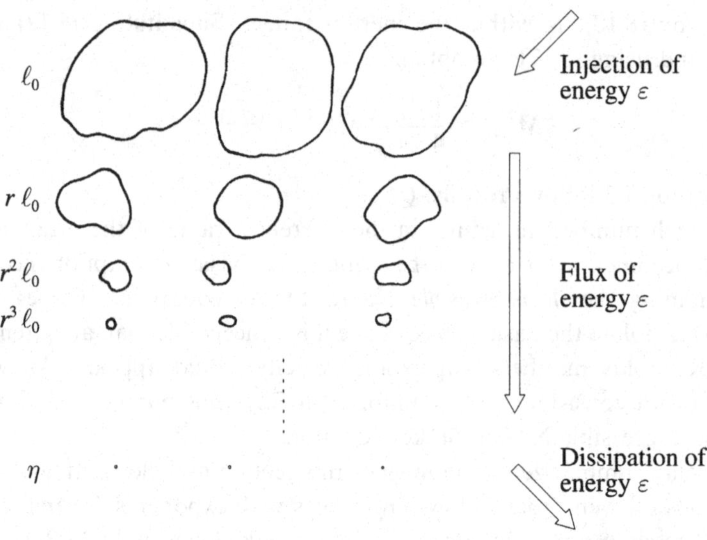

Fig. 8.9. The cascade according to the β-model. Notice that with each step the eddies become less and less space-filling.

The notation chosen for D is justified by the observation that it can be interpreted as a *fractal dimension* (Mandelbrot 1977; Falconer 1990).

This statement is made intuitive by Fig. 8.10 which shows within a unit cube three objects, a point, a curve and a surface having respectively the dimension D (in the ordinary sense of manifolds) of zero, one and two. We now ask: What is the probability p_ℓ that a ball of (small) radius ℓ, the center of which is chosen in the cube with a random uniform distribution, will intersect such an object? The answer follows immediately from the geometric construction shown in Fig. 8.10. For the surface, the center of the intersecting ball has to be within a sandwich of thickness 2ℓ; for the curve, it has to be within a sausage of radius ℓ and for the point it has to be within a ball of radius ℓ. Thus, in all cases

$$p_\ell \propto \ell^{3-D}, \qquad \ell \to 0. \tag{8.22}$$

This probability would scale the same way if little cubes of size ℓ were used instead of balls. If the embedding space is d-dimensional instead of three-dimensional, (8.22) becomes

$$p_\ell \propto \ell^{d-D}, \qquad \ell \to 0. \tag{8.23}$$

The quantity $d - D$ is called the *codimension*.

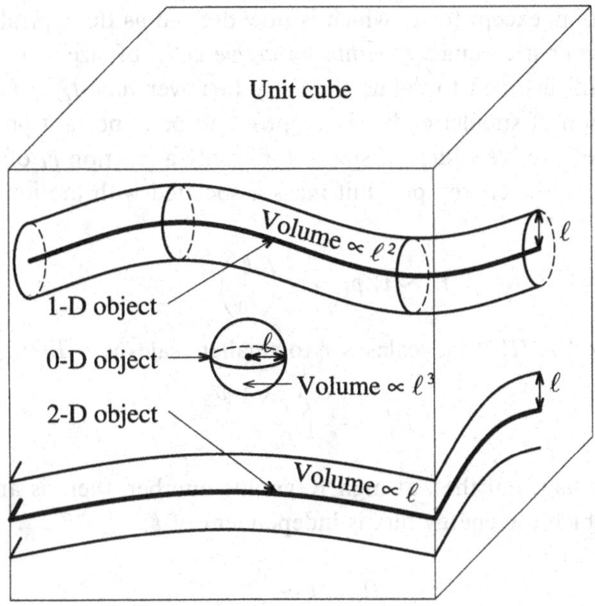

Fig. 8.10. The probability that a sphere of radius ℓ encounters an object of dimension D behaves as ℓ^{3-D} as $\ell \to 0$.

For more exotic objects, such as Cantor sets, (8.23) can be used as a definition of a dimension, which need not be an integer. This dimension is closely related to the *Kolmogorov capacity dimension* also called *covering dimension.*[9]

For nonnegative D, i.e. when $\beta \geq r^3$, there is on average at least one daughter per mother-eddy. D is then the dimension of the set $\mathcal{S}$ on which the cascade accumulates after indefinite iteration. When $\beta < r^3$, the dimension D is *negative*, the average number of daughters is less than unity, so that the cascade terminates almost surely after a finite number of steps and $\mathcal{S}$ is empty. However, there is a finite probability p_ℓ, given by (8.22), of observing an eddy of any finite size ℓ. Hence, negative dimensions may be viewed here as controlling the rarefication of sequences of sets converging to an empty set (Mandelbrot 1990, 1991).

Returning now to the β-model, we derive its scaling laws by adapting the standard K41 phenomenology used in Section 7.2. The notation is as

[9] For a rigorous but still elementary presentation of the Kolmogorov capacity dimension, see Ruelle (1989). This dimension is not to be confused with the *capacitary dimension* which, by a theorem of Frostman, is equal to the Hausdorff dimension; for details, see Kahane and Salem (1994, Chapter III) or the thesis of Frostman (1935).

in that section, except for v_ℓ which is now defined as the typical velocity difference over a distance ℓ *within an active eddy* of size $\sim \ell$. Observe that it is still justified to define the eddy turnover time t_ℓ as ℓ/v_ℓ, since the formation of smaller eddies is supposed to be important only within active eddies. Active eddies of size $\sim \ell$ fill only a fraction p_ℓ of the total volume; thus, the energy per unit mass associated with motion on scale $\sim \ell$ is

$$E_\ell \sim v_\ell^2 p_\ell = v_\ell^2 \left(\frac{\ell}{\ell_0}\right)^{3-D}. \tag{8.24}$$

The energy flux Π'_ℓ from scales $\sim \ell$ to smaller scales is $\sim E_\ell/t_\ell$. Thus,

$$\Pi'_\ell \sim \frac{v_\ell^3}{\ell}\left(\frac{\ell}{\ell_0}\right)^{3-D}. \tag{8.25}$$

We assume as usual that, at high Reynolds number, there is an inertial range in which the energy flux is independent of ℓ:

$$\Pi'_\ell \sim \varepsilon \sim \frac{v_0^3}{\ell_0}. \tag{8.26}$$

The relation $\varepsilon \sim v_0^3/\ell_0$ has already been derived in Section 7.2 where it was stressed that it does not use K41; alternatively, it may be seen as a consequence of (8.25) evaluated for $\ell \sim \ell_0$.

From (8.25) and (8.26) we obtain

$$v_\ell \sim v_0 \left(\frac{\ell}{\ell_0}\right)^{\frac{1}{3}-\frac{3-D}{3}}, \tag{8.27}$$

and

$$t_\ell \sim \frac{\ell}{v_\ell} \sim \frac{\ell_0}{v_0}\left(\frac{\ell}{\ell_0}\right)^{\frac{2}{3}+\frac{3-D}{3}}. \tag{8.28}$$

Eq. (8.27) may be viewed as the statement that the velocity field has the *scaling exponent*

$$h = \frac{1}{3} - \frac{3-D}{3} \tag{8.29}$$

on the set $\mathscr{S}$ of (fractal) dimension[10] D on which the cascade accumulates. This reformulation will be useful for the generalization to multifractals (Section 8.5.3).

[10] Actually, the Kolmogorov capacity dimension should be used. Fung and Vassilicos (1991) pointed out that spiral structures, which are quite common in turbulent flow, can have a capacity dimension with a nontrivial fractal value, while possessing an integer Hausdorff dimension.

We turn now to the structure functions. At a phenomenological level, it is difficult to distinguish between *longitudinal* structure functions (defined in (6.61)) and those involving other components. We shall here avoid such distinctions and simply denote the structure function of order p by $\langle \delta v_\ell^p \rangle$. There are two contributions to this quantity: a factor v_ℓ^p coming from active eddies and an 'intermittency factor' $p_\ell = (\ell/\ell_0)^{3-D}$ which gives the fraction of the volume filled by active eddies of scale ℓ. Using (8.27), we thus obtain

$$S_p(\ell) = \langle \delta v_\ell^p \rangle \sim v_0^p \left(\frac{\ell}{\ell_0}\right)^{\zeta_p}, \tag{8.30}$$

with

$$\zeta_p = \frac{p}{3} + (3 - D)\left(1 - \frac{p}{3}\right). \tag{8.31}$$

It is seen that for the β-model the exponent ζ_p is a linear-plus-constant function of the order p. Note that for $p = 6$ the discrepancy from the K41 value is equal to the codimension $3 - D$.

Specializing to $p = 2$, we find that the second order structure function has the exponent $2/3 + (3 - D)/3$; hence, the energy spectrum in the inertial range satisfies

$$E(k) \propto k^{-\left(\frac{5}{3} + \frac{3-D}{3}\right)}, \tag{8.32}$$

which is *steeper* than the Kolmogorov–Obukhov $k^{-5/3}$ spectrum. For $p = 3$, we obtain $\zeta_3 = 1$, as required by Kolmogorov's four-fifths law.

The viscous cutoff for the β-model is obtained, just as in Section 7.2, by equating the eddy turnover time t_ℓ given by (8.28) and the viscous diffusion time. This gives the following dissipation scale

$$\eta \sim \ell_0 R^{-\frac{3}{1+D}} \sim \ell_0 R^{-\frac{1}{1+h}}, \tag{8.33}$$

where $R = \ell_0 v_0 / \nu$ as usual.

The β-model has a rather long history which is better presented in the perspective of other models (see Section 8.8).

The K41 theory is recovered when $D = 3$; indeed, this assumption suppresses the intermittency. When $D < 2$, the exponent h becomes negative, so that v_ℓ increases when the scale decreases. As a consequence the dissipation scale η, given by (8.33), can become smaller than the mean-free-path; we shall return to this question in Section 8.5.5. When $D < 0$, the spectrum becomes steeper than $k^{-8/3}$. Here, it is of interest to point out a rigorous result by Sulem and Frisch (1975). They considered solutions of the Navier–Stokes equation with a finite total energy (an assumption consistent with spatial periodicity but not with homogeneity), in the limit of vanishing viscosity, and showed that, if the spectrum is steeper than $k^{-8/3}$, then the energy flux Π_K tends to zero for $K \to \infty$. Finally, we note that the viscous cutoff, given by (8.33), ceases to exist when the dimension $D \leq -1$.

As far as scaling is concerned, there is a single adjustable parameter in the β-model, the dimension D associated with the intermittent cascade. How does it compare to the experimental data presented in Section 8.3? The straight dashed line in Fig. 8.8 corresponds to the β-model for $D = 2.8$. As stated by Anselmet, Gagne, Hopfinger and Antonia (1984), the β-model fits the data rather well for values of the order p up to 8 but not at all beyond $p = 12$. We have, however, emphasized in Section 8.3 that the present limitations on the experimental determination of exponents do not allow us to rule out K41 completely, let alone the β-model. Still the data of Anselmet, Gagne, Hopfinger and Antonia, taken without further questioning, suggest that the graph of ζ_p vs p is *not a straight line*. This has led to all sorts of generalizations of the β-model. As a pedagogical intermediate step between the β-model and the full multifractal model of Section 8.5.3, we shall now discuss the 'bifractal model'.

8.5.2 The bifractal model

As we have seen in Section 8.5.1, the β-model is equivalent to the statement that the velocity field has a scaling exponent h on a set $\mathscr{S}$ of fractal dimension D, such that h and D are related by (8.29). A natural extension is to assume *bifractality*: there are now two sets $\mathscr{S}_1$ and $\mathscr{S}_2$, both imbedded in the physical space $\mathbb{R}^3$. Near $\mathscr{S}_1$ the velocity has scaling exponent h_1 and near $\mathscr{S}_2$ it has scaling exponent h_2. Specifically, we assume that

$$\frac{\delta v_\ell(r)}{v_0} \sim \begin{cases} \left(\dfrac{\ell}{\ell_0}\right)^{h_1}, & r \in \mathscr{S}_1, \ \dim \mathscr{S}_1 = D_1 \\ \left(\dfrac{\ell}{\ell_0}\right)^{h_2}, & r \in \mathscr{S}_2, \ \dim \mathscr{S}_2 = D_2. \end{cases} \tag{8.34}$$

Here, $\delta v_\ell(r)$ denotes the velocity increment between the point r and another point a distance ℓ away. As in Chapter 7, all vector notation has been eliminated, since this kind of phenomenology is not able to distinguish between components. The above scaling laws are meant to hold at inertial-range scales. Thus, after the spatial increment ℓ and the velocity increment $\delta v_\ell(r)$ have been divided by ℓ_0 and v_0, all remaining constants are order unity.

With this assumption we can calculate the structure function of order p (at inertial-range separations). The scaling exponent h_1 gives a contribution $(\ell/\ell_0)^{ph_1}$ which must be multiplied by the probability $(\ell/\ell_0)^{3-D_1}$ of being within a distance ℓ of the set $\mathscr{S}_1$, and similarly for the other exponent. We thus obtain

$$\frac{\langle \delta v_\ell^p \rangle}{v_0^p} = \mu_1 \left(\frac{\ell}{\ell_0}\right)^{ph_1} \left(\frac{\ell}{\ell_0}\right)^{3-D_1} + \mu_2 \left(\frac{\ell}{\ell_0}\right)^{ph_2} \left(\frac{\ell}{\ell_0}\right)^{3-D_2}, \tag{8.35}$$

where μ_1 and μ_2 are order unity constants.

Thus, all the structure functions comprise the superposition of two power-laws. In the inertial range, when $\ell \ll \ell_0$, the power-law with the smallest exponent will dominate. We thus obtain

$$\langle \delta v_\ell^p \rangle \propto \ell^{\zeta_p}, \qquad \zeta_p = \min\left(ph_1 + 3 - D_1,\ ph_2 + 3 - D_2\right). \tag{8.36}$$

It is seen that, depending on the value of the exponent p, the first or second type of singularity dominates. This is reminiscent of what Berry (1982) calls a 'battle of catastrophes'. By this he understands the competition of two or more singularities which determines the asymptotic behavior of certain integrals appearing in problems such as the study of the twinkling of light in a random medium (Berry 1977), the distribution of circulation times of fluid particles along closed random streamlines, etc.[11]

As an illustration of bifractality, let us take a mixture of K41 turbulence and β-model turbulence: $D_1 = 3$, $h_1 = 1/3$, $0 < D_2 < 3$ and $h_2 = 1/3 - (3 - D_2)/3$. We obtain

$$\zeta_p = \begin{cases} p/3 & 0 \le p \le 3 \\ p/3 + (3 - D_2)(1 - p/3) & p \ge 3. \end{cases} \tag{8.37}$$

Observe that the parameters have been chosen in such a way that $\zeta_3 = 1$, to be consistent with the four-fifths law. Note that ζ_p can also be defined for noninteger positive p; in that case it is necessary to take the absolute value of δv_ℓ in the definition of the structure function. The graph of ζ_p given by (8.37) is shown in Fig. 8.11. It has a kink at $p = 3$. This is known as a 'phase transition': as far as structure functions are concerned this model behaves exactly as in the K41 theory for $p \le 3$ and displays intermittency only for larger values of p.

[11] This work is quoted here, not just because of its beauty, but also because it could provide and alternative to the 'multifractal' interpretation of intermittency, to be discussed in Section 8.5.3.

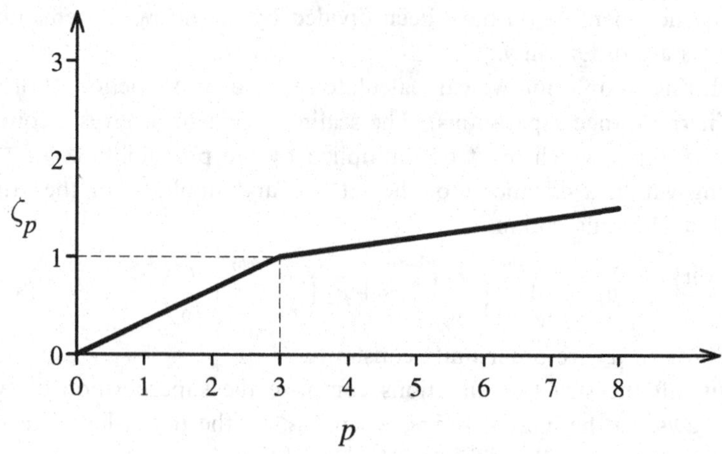

Fig. 8.11. Exponent ζ_p for the 'bifractal model'. Notice the change of slope at $p = 3$.

A more natural model exists which is exactly bifractal, namely Burgers' equation:

$$\partial_t v + v\partial_x v = v\partial^2_{xx}v. \tag{8.38}$$

We shall not here open the Pandora's box of Burgers' equation, how it does (and often does not) relate to the turbulence problem. For this, see the books by Burgers (1974), Gurbatov, Malakhov and Saichev (1991) and Barabási and Stanley (1995), the articles by Kardar, Parisi and Zhang (1986), She, Aurell and Frisch (1992), Sinai (1992) and Vergassola, Dubrulle, Frisch and Noullez (1994) and references therein.

The solutions of Burgers' equation with smooth initial data and/or smooth forcing,[12] when considered in the limit of vanishing viscosity, develop after some time the kinds of structures sketched in Fig. 8.12. There are isolated shocks ($D = 0$ and $h = 0$) connected by smooth ramps ($D = 1$ and $h = 1$). As observed by Aurell, Frisch, Lutsko and Vergassola (1992) this implies bifractality: $\zeta_p = 1$ for $p \geq 1$ and $\zeta_p = p$ for $0 \leq p \leq 1$. In other words, at inertial-range separations, structure functions of fractional positive order less than one are dominated by the smooth ramps, while those of order greater than one are dominated by the shocks.

[12] The smoothness assumption is important: different results may be obtained when the force has a power-law spectrum, as assumed in Forster, Nelson and Stephen (1977) or Kardar, Parisi and Zhang (1986).

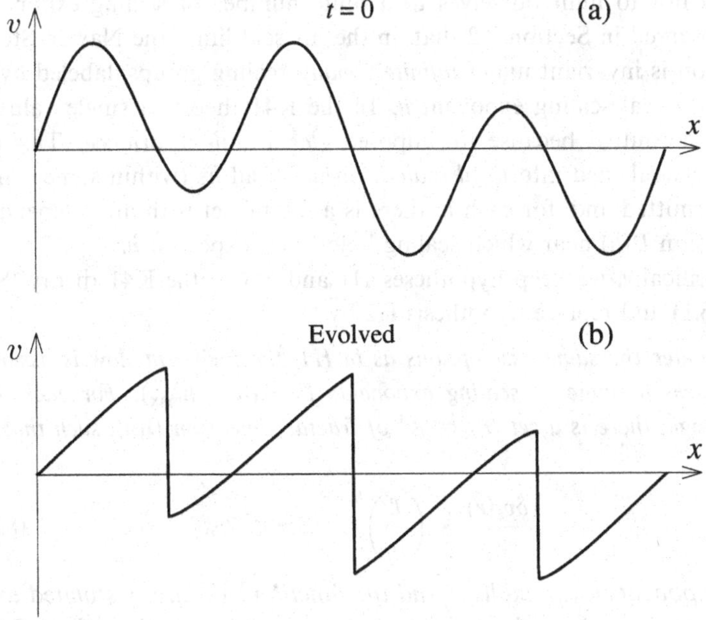

Fig. 8.12. Solution of Burgers' equation in the limit of zero viscosity: initial time (a); after the formation of shocks (b).

The fact that the β-model scaling (with $D = 0$) for structure functions of integer order $p \geq 0$ *is exact* for solutions of Burgers' equation is useful when testing statistical theories of turbulence, particularly those based on diagrammatic methods: these methods often use formal machinery which is applicable both to the Navier–Stokes equation and to Burgers' equation (see Section 9.5).

8.5.3 The multifractal model

The data points of Anselmet, Gagne, Hopfinger and Antonia (1984) for the function ζ_p, shown in Fig. 8.8, appear to be on a *curved* line, rather than a straight line as in the β-model or a broken straight line as in the bifractal model. This has led Parisi and Frisch (1985) to develop the *multifractal model.* Historical aspects are again postponed to Section 8.8.

Based on our experience with bifractality it is rather clear that a curved function ζ_p will be obtained if we assume *a continuous infinity* of scaling exponents, rather than just two. Actually, there is a more basic reason not to limit ourselves to a finite number of scaling exponents. We observed in Section 2.2 that, in the inviscid limit, the Navier–Stokes equation is invariant under *infinitely many* scaling groups, labeled by an arbitrary real scaling exponent h. In the K41 theory, a single value of h was permitted because we imposed *global scale-invariance*. This can now be weakened into *local scale-invariance*: all hs (within some range) are permitted and, for each h, there is a fractal set with an h-dependent dimension $D(h)$ near which scaling holds with exponent h.

Specifically, we keep hypotheses H1 and H3 of the K41 theory (Section 6.1) and replace hypothesis H2 by

$\mathbf{H_{mf}}$ *Under the same assumptions as in H1, the turbulent flow is assumed to possess a range of scaling exponents $I = (h_{\min}, h_{\max})$. For each h in this range, there is a set $\mathcal{S}_h \subset \mathbb{R}^3$ of fractal dimension $D(h)$, such that, as $\ell \to 0$,*

$$\frac{\delta v_\ell(\boldsymbol{r})}{v_0} \sim \left(\frac{\ell}{\ell_0}\right)^h, \qquad \boldsymbol{r} \in \mathcal{S}_h. \tag{8.39}$$

The exponents $h_{\min}$ and $h_{\max}$ and the function $D(h)$ are postulated to be universal, i.e. independent of the mechanism of production of the flow.

From this *multifractal* assumption, proceeding as in the preceding section on bifractals, we derive the expression for the structure function of order p:

$$\frac{S_p(\ell)}{v_0^p} \equiv \frac{\langle \delta v_\ell^p \rangle}{v_0^p} \sim \int_I d\mu(h) \left(\frac{\ell}{\ell_0}\right)^{ph+3-D(h)}. \tag{8.40}$$

Here, the measure $d\mu(h)$ gives the weight of the different exponents. (As we shall see, its expression is irrelevant.) The factor $(\ell/\ell_0)^{ph}$ is the contribution from (8.39) and the factor $(\ell/\ell_0)^{3-D(h)}$ is the probability of being within a distance $\sim \ell$ of the set $\mathcal{S}_h$ of dimension $D(h)$. Note that the sum in (8.35) has now become an integral over the range I of scaling exponents. In the limit $\ell \to 0$ the power-law with the smallest exponent dominates and we obtain by a steepest descent argument

$$\lim_{\ell \to 0} \frac{\ln S_p(\ell)}{\ln \ell} = \zeta_p, \tag{8.41}$$

where

$$\zeta_p = \inf_h \, [ph + 3 - D(h)]. \tag{8.42}$$

The weights $d\mu(h)$ have indeed disappeared from the asymptotic expressions of the structure functions. More loosely, ignoring logarithmic corrections etc, (8.41) may be written as:

$$\frac{S_p(\ell)}{v_0^p} \sim \left(\frac{\ell}{\ell_0}\right)^{\zeta_p}, \qquad \ell \to 0. \tag{8.43}$$

By Kolmogorov's four-fifths law, the exponent for the third order structure function must be unity:

$$\zeta_3 = \inf_h \left[3h + 3 - D(h)\right] = 1. \tag{8.44}$$

In the relation (8.42) between the dimensions $D(h)$ and the exponents of structure functions ζ_p we recognize a *Legendre transformation* (Sewell 1987). There is a simple associated geometrical construction illustrated in Fig. 8.13(a): the quantity $3 - \zeta_p$ is the maximum signed vertical distance between the graph of $D(h)$ and the line through the origin with slope p. If $D(h)$ has a decreasing derivative, i.e. it is *concave*, then for a given value of p the maximum is attained at the unique value $h_\star(p)$ such that

$$D'\left(h_\star(p)\right) = p, \tag{8.45}$$

and ζ_p is given by

$$\zeta_p = p h_\star(p) + 3 - D\left(h_\star(p)\right). \tag{8.46}$$

From there it follows that (8.42) can be inverted as

$$D(h) = \inf_p \left(ph + 3 - \zeta_p\right). \tag{8.47}$$

Indeed, from (8.45) and (8.46) we obtain

$$\frac{d\zeta_p}{dp} = h_\star(p) + \left[p - D'\left(h_\star(p)\right)\right] \frac{dh_\star(p)}{dp} = h_\star(p). \tag{8.48}$$

It is readily seen that (8.46) and (8.48) imply (8.47). Note that, even if $D(h)$ is not concave, its Legendre transform ζ_p defined by (8.42) will be concave; however, the inversion formula returns then not $D(h)$ but its *concave hull*, i.e. the lowest concave graph lying above the graph of $D(h)$.

The inversion formula (8.47) can, in principle, be used to extract the function $D(h)$ from experimental or numerical data about the exponents ζ_p. The corresponding geometrical construction is shown in Fig. 8.13(b). In practice this is not a very well-conditioned operation for data such as those shown in Fig. 8.8, because the slope of the graph is poorly determined, except for small values of p where it is very close to $1/3$ with $D(h)$ very close to 3; these are, of course, the K41 values.

An important consequence of the inversion formula (8.48) is that the scaling exponent h is the slope of the graph of ζ_p for the value of p

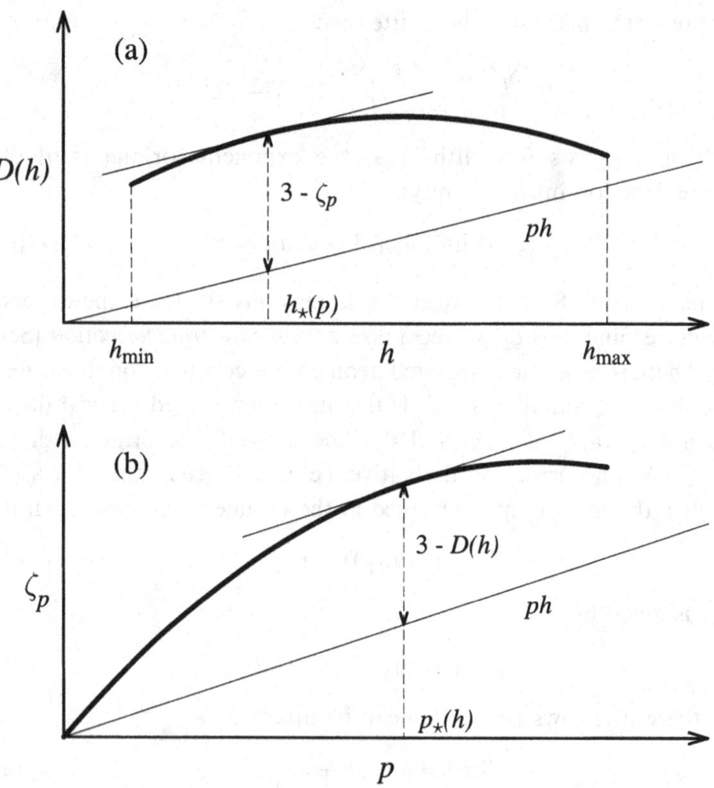

Fig. 8.13. Geometrical construction of the Legendre transform (a) and the inverse Legendre transform (b).

which minimizes $ph + 3 - \zeta_p$. Hence, the range I of possible values of p is the range of slopes of the graph of ζ_p. In view of proposition P3 of Section 8.4, the finiteness of the Mach number requires ζ_p to be a nondecreasing function of p. Hence, *negative scaling exponents h are ruled out in the multifractal model* for incompressible flow.

8.5.4 A probabilistic reformulation of the multifractal model

The phenomenological presentation of the multifractal model, given in the preceding section, has two major shortcomings. First, it assumes the existence of *singularities* and, second, it does not distinguish between positive and negative velocity increments. As we have seen in Section 7.8 it is still debatable if singularities can appear after a finite time at zero viscosity (Euler flow). By Occam's razor,[13] it is better to dispense with them. As for the confusion between positive and negative velocity increments, it leads to a similar confusion between $\langle \delta v_\ell^p \rangle$ and $\langle |\delta v_\ell|^p \rangle$. For odd values of p this is undesirable, particularly when $p = 3$ for which we have the Kolmogorov four-fifths law, which involves no absolute value.

Both shortcomings can be overcome if we use the p.d.f. $P_\ell^{\mathrm{inc}}(\delta v_\parallel)$ of (longitudinal) velocity increments over a distance ℓ, rather than individual realizations of the Eulerian velocity field. (The limits $t \to \infty$ and $\nu \to 0$ are assumed to have been taken.) With the assumptions of homogeneity, stationarity and isotropy, this p.d.f. does indeed depend only on ℓ. The structure functions are then given by

$$S_p(\ell) = \int_{\mathbb{R}} \delta v_\parallel^p \, P_\ell^{\mathrm{inc}}(\delta v_\parallel) \, d\delta v_\parallel. \tag{8.49}$$

We shall here ignore the (unlikely) possibility that the distribution of velocity increments would not possess a density with respect to the Lebesgue measure.

We now observe that, starting from the multifractal assumption H_{mf} of the preceding section, the probability of having $\delta v_\ell \propto \ell^h$ (ignore the sign for a moment) should be proportional to $\ell^{3-D(h)}$. This leads us to the probabilistic reformulation of the multifractal assumption, in which we distinguish positive and negative velocity increments

$\mathrm{H}_{\mathrm{pmf}}$ *Under the same assumptions as in H1, there is a universal function $D(h)$ which maps real scaling exponents h to scaling dimensions $D \le 3$ (including negative values and the value $-\infty$), such that for any h, the probability of velocity increments satisfies*

$$\lim_{\ell \to 0} \frac{\ln \bar{P}_\ell^{\mathrm{inc}}\left(\pm \ell^h\right)}{\ln \ell} = 3 - D(h). \tag{8.50}$$

Here, $\bar{P}_\ell^{\mathrm{inc}}(\delta v_\parallel)$ denotes the *cumulative* probability, obtained by integrating the p.d.f. $P_\ell^{\mathrm{inc}}(\delta v_\parallel)$, say from $\delta v_\parallel$ to $2\delta v_\parallel$.[14]

The value $D = -\infty$ takes care of nonexistent scaling exponents. Note that no difference is obtained in the limit $\ell \to 0$ if, instead of the argument ℓ^h of $\bar{P}$, one uses $v_0(\ell/\ell_0)^h$, which has the dimension of a velocity. Note also that we assumed the same limit for positive and negative velocity increments, although in principle they could be different.

[13] From Willem of Occam or Ockham (1290–approx. 1349). Occam's razor, aimed at unnecessary theological subtleties, ruled out any 'plurality of reasons' founded neither in experience nor in Scripture.

[14] In most instances of genuine multifractality, the same result is obtained with the p.d.f. P and the cumulative probability $\bar{P}$.

Our probabilistic definition of multifractality is closely related to what is called the 'abstract large deviations principle' in which the function $D(h)$ is known as the 'rate or Cramér function' (see, e.g., Varadhan 1984). We shall come back to large deviations theory in Section 8.6.4.

The 'traditional' definition of multifractality implies the probabilistic one, but the converse is not true. In order to be able to consider p.d.f.s, we must work with the (assumed) invariant measure of the Navier–Stokes dynamical system (Section 3.4). The quantity $P_\ell^{\text{inc}}(\delta v_\|)$ is the limit for $\nu \to 0$ of the p.d.f. of increments in the viscous case. The fact that we assume this limit to exist and to have the singular behavior predicted by (8.50) tells us nothing about the behavior for $\nu \to 0$ of individual realizations: they need not be singular, let alone fractal. Actually, if the Euler equation happens not to have any finite-time singularities (Section 7.8), then for any finite time the limit, as $\nu \to 0$, of individual realizations cannot be singular; hence, $D(h)$ cannot be interpreted as the dimension of such an empty set. As to the limit $t \to \infty$, there is no reason to expect that it exists at the level of *individual realizations*, whether for $\nu > 0$ or $\nu = 0$. Only the probability distributions have limits. Hence, in the absence of finite-time singularities for the Euler equation, $D(h)$ cannot be interpreted as the dimension of a singular set. To sum up: *(multi)fractality does not require finite-time singularities.*

The consequences of the probabilistic multifractal hypothesis for the structure functions can be derived mostly as in the preceding section, but it is useful to distinguish between positive and negative velocity increments. Let us define the positive-increment (resp. negative-increment) structure functions $S_p^+(\ell)$ (resp. $S_p^-(\ell)$) as the contribution to the integral (8.49) coming from positive (resp. negative) velocity increments. Clearly,

$$S_p(\ell) = S_p^+(\ell) + S_p^-(\ell). \tag{8.51}$$

We then obtain, for $\ell \to 0$

$$S_p^\pm(\ell) \propto \ell^{\zeta_p}, \tag{8.52}$$

where ζ_p is still given by (8.42). From (8.51) and (8.52) it follows that for even orders $2p$ the structure function $S_{2p}(\ell)$ scales with the exponent ζ_{2p}. For odd orders $2p+1$, there is the possibility of cancellation, so that the actual exponent ζ_{2p+1} could be larger than the expression given by (8.42). This relation can, however, be used to predict the exponents for structure functions of noninteger positive orders,[15] provided the absolute

[15] There are problems for negative ps to which we shall return in Section 8.6.2.

values of the velocity increments are used in the definitions. Hence, ζ_p may be considered as a function on the real half-line $p \geq 0$.

The probabilistic reformulation of multifractality provides in principle a way of measuring $D(h)$ directly in terms of the p.d.f. of velocity increments, without performing an inverse Legendre transform. This requires, however, a careful measurement of the p.d.f. over a considerable range of scales ℓ. So far, only limited data are available (see, e.g., Gagne, Hopfinger and Frisch 1990 and references therein). Such measurements could also reveal that different functions $D(h)$ are needed for positive and negative velocity increments.

Finally, we observe that the probabilistic reformulation of multifractality can be made in terms of p.d.f.s of *wavelet transforms* of the velocity field. This allows the explicit construction of random functions with multifractal scaling (Benzi, Biferale *et al.* 1993).

8.5.5 The intermediate dissipation range and multifractal universality

The multifractal model presented in the previous sections is a possible interpretation of experimentally observed discrepancies from K41 at inertial-range scales, where viscosity has no direct influence. This model has also implications for the behavior at smaller scales, where viscosity matters. It leads to a prediction about the shape of the energy spectrum at large wavenumbers, beyond the inertial range, (this section) and to predictions for the skewness and flatness of velocity derivatives (Section 8.5.6). In particular, a new form of 'multifractal' universality is predicted, which is inconsistent with Kolmogorov's *first universality assumption* (Section 6.3.2) and which can, in principle, be tested experimentally. The key idea is that, when focusing on smaller and smaller scales, the inertial-range contributions of the various scaling exponents h are successively turned off, thereby producing an *intermediate dissipation range* with quasi-algebraic fall-off of the energy spectrum.

We first observe that the phenomenological argument given in Section 7.2 for deriving the expression (7.11) of the Kolmogorov dissipation scale η can be extended to multifractal turbulence. Indeed, following Paladin and Vulpiani (1987a), we can use the *local* scaling relation (8.39) to construct an eddy turnover time $t_\ell = \ell/\delta v_\ell$ which depends on the scaling exponent h. Then, equating this eddy turnover time to the viscous diffusion time ℓ^2/ν, we obtain an h-dependent dissipation scale:

$$\frac{\eta(h)}{\ell_0} \sim R^{-\frac{1}{1+h}}, \qquad R \sim \frac{\ell_0 v_0}{\nu}. \tag{8.53}$$

For $h = 1/3$ (the K41 value) the Kolmogorov dissipation scale (7.11) is recovered. We observe that (8.53) has already been obtained (eq. (8.33)) in Section 8.5.1 for the β-model.

The main difference with K41 and the β-model is that in the multifractal case the scaling exponent h has a whole range of values $I = (h_{\min}, h_{\max})$. Hence, there is a *range of dissipation scales* extending from $\eta_{\min} \sim \ell_0 R^{-1/(1+h_{\min})}$ to $\eta_{\max} \sim \ell_0 R^{-1/(1+h_{\max})}$. Paladin and Vulpiani (1987a) observed that with multifractal scaling the K41 estimate (7.19) of the number of degrees of freedom must be modified. We have seen in Section 8.5.3 that h has to be nonnegative to be compatible with the constraint that ζ_p be nondecreasing (Section 8.4). It can also be easily checked that the result of Section 7.5, namely that the dissipation scale is much larger than the molecular mean-free-path, could otherwise be upset.[16]

Let us now, following Frisch and Vergassola (1991), turn to the effect of the h-dependence of the dissipation scale for the structure functions. First, (8.39) must be modified: for a given scaling exponent h, the power-law behavior is valid only at scales $\ell \gg \eta(h)$. At values less than this, smooth behavior must be assumed. For example, we could take, instead of (8.39),

$$\frac{\delta v_\ell(r)}{v_0} \sim \left(\frac{\ell}{\ell_0}\right)^h f\left(\frac{\ell}{\eta(h)}\right), \qquad r \in \mathscr{S}_h. \tag{8.54}$$

where $f(x) \to 1$ for $x \to +\infty$ and $x^h f(x)$ tends to zero smoothly for $x \to 0+$. Actually, the precise functional form does not matter for the subsequent derivations and we shall just take a step function: $f(x) = 1$ for $x \geq 1$ and $f(x) = 0$ for $0 \leq x < 1$. With the effect of viscosity thus included, the expressions (8.40) of the structure functions become

$$\frac{S_p(\ell)}{v_0^p} \equiv \frac{\langle \delta v_\ell^p \rangle}{v_0^p} \sim \int_{\eta(h)<\ell} d\mu(h) \left(\frac{\ell}{\ell_0}\right)^{ph+3-D(h)}. \tag{8.55}$$

Let $h_\star(p)$ be the scaling exponent which minimizes $ph + 3 - D(h)$, defined implicitly by (8.45). There is an associated dissipation scale $\eta\,(h_\star(p))$. We must now distinguish two asymptotic regions.

(i) **Inertial range**: $\ell_0 \gg \ell \gg \eta\,(h_\star(p))$. Viscous effects are negligible and the previously established power-law (8.43) remains valid. Note that the dissipation scale now depends on the order of the structure

[16] The negative scaling exponents reported by Bacry, Arnéodo, Frisch, Gagne and Hopfinger (1990) were later shown to be an artifact of the processing technique (Vergassola, Benzi, Biferale and Pisarenko 1993).

function. Since the graph of ζ_p is concave, $h_\star(p)$ is a decreasing function of p. Hence, the dissipation scale $\eta\,(h_\star(p))$ *decreases* with p. For $p = 2$, we know that $h_\star$ is very close to $1/3$ and hence the dissipation scale is very close to the K41 value $\eta \sim \ell_0 R^{-3/4}$. Experimental attempts to check that smaller dissipation scales are obtained for higher order structure functions have been made, for example, by van de Water, van der Vorst and van de Wetering (1991). We shall return below to the practical difficulties involved in probing scales smaller than the Kolmogorov dissipation scale.

(ii) **Intermediate dissipation range**: $\eta\,(h_\star(p)) \gg \ell \gg \eta\,(h_{\min})$. In this range we have the following situation. The influence of the scaling exponent $h_\star(p)$ which gave the leading order contribution in the inertial range is suppressed by viscous dissipation, but smaller scaling exponents h such that $\eta(h) \le \ell$ are still felt. Hence, we expect to obtain for the structure function of order p a *quasi-algebraic* behavior, i.e. a power-law with a scale-dependent exponent $ph(\ell) + 3 - D(h(\ell))$, where $h(\ell)$ is the largest 'acceptable' scaling exponent, i.e. such that the associated dissipation scale is equal to ℓ. This heuristic argument thus suggests the following relation in the intermediate dissipation range:

$$\left.\begin{aligned} &\frac{S_p(\ell)}{v_0^p} \sim \left(\frac{\ell}{\ell_0}\right)^{ph(\ell)+3-D(h(\ell))}, \quad \eta\,(h_\star(p)) \gg \ell \gg \eta\,(h_{\min}), \\ &\eta(h(\ell)) \equiv \ell_0 R^{-\frac{1}{1+h(\ell)}} \sim \ell. \end{aligned}\right\} \tag{8.56}$$

A more systematic derivation requires some precautions since we have *two* expansion parameters: the scale ℓ (which tends to zero) and the Reynolds number R (which tends to infinity). The trick is to let $R \to \infty$ and simultaneously $\ell \to 0$ in such a way that ℓ_0/ℓ behaves as a power of R. By application of steepest descent (Bender and Orszag 1978) to (8.55) we then obtain the following result: Let $R \to \infty$ and $\ell \to 0$ while

$$\lim \frac{\ln(\ell_0/\ell)}{\ln R} = \theta, \qquad 0 < \theta < (1 + h_{\min})^{-1}, \tag{8.57}$$

then

$$\lim \frac{\ln S_p(\ell)}{\ln R} = -\left[ph(p,\theta) + 3 - D(h(p,\theta))\right]\theta, \tag{8.58}$$

where

$$h(p,\theta) = \begin{cases} h_\star(p) & \text{for } 0 < \theta < (1 + h_\star(p))^{-1} \\ \theta^{-1} - 1 & \text{for } (1 + h_\star(p))^{-1} \le \theta < (1 + h_{\min})^{-1}. \end{cases} \tag{8.59}$$

As written here, (8.57)–(8.59) encompass the expression of the structure functions in both the inertial range (upper line of (8.59)) and the intermediate dissipation range (lower line).

It is of particular interest to express the result for $p = 2$ in terms of the energy spectrum $E(k)$ rather than the structure function. Noting that a term $\propto \ell^p$ in the structure function gives a term $\propto k^{-1-p}$ in the energy spectrum, we obtain the following result: Let $R \to \infty$ and $k \to \infty$ while

$$\lim \frac{\ln(k\ell_0)}{\ln R} = \theta, \qquad 0 < \theta < (1 + h_{\min})^{-1}, \tag{8.60}$$

then

$$\lim \frac{\ln E(k)}{\ln R} = F(\theta), \tag{8.61}$$

where

$$F(\theta) \equiv \begin{cases} -\frac{5}{3}\theta, & \text{for } 0 < \theta < \frac{3}{4}, \\ -2 - 2\theta + \theta D(\theta^{-1} - 1), & \text{for } \frac{3}{4} \le \theta < (1 + h_{\min})^{-1}. \end{cases} \tag{8.62}$$

Here, we have for simplicity used $h_\star(2) \approx 1/3$, so that the inertial-range spectrum follows the $-5/3$ law. In the 'intermediate dissipation range', at wavenumbers ranging from $\ell_0^{-1}R^{3/4}$ to $\ell_0^{-1}R^{1/(1+h_{\min})}$, the energy spectrum decreases quasi-algebraically, i.e. as a power-law the exponent of which changes logarithmically with the wavenumber in a way controlled by the function $D(h)$. If the minimum scaling exponent $h_{\min} > 0$, there exists another range of wavenumbers beyond the intermediate dissipation range at which it is still legitimate to use hydrodynamics (since the scales remain large compared to the mean-free-path). The multifractal model does not predict anything about this 'far dissipation range'. If uniform analyticity holds, the spectrum will fall off exponentially for the reason pointed out at the end of Section 6.3.2.

Eq. (8.62) is just the $-5/3$ law in the inertial range. But, as soon as scales below the Kolmogorov dissipation scale are included, the prediction of the multifractal model differs significantly from (6.69) which expresses Kolmogorov's first universality assumption. A consequence of K41 is that $\ln E(k)$ should be a universal function of $\ln k$ (in the inertial range and the intermediate dissipation range). In contrast, with the multifractal assumption, and assuming that $D(h)$ is a *universal* function, we find that $\ln E(k)/\ln R$ is a universal function of $\ln k/\ln R$. This is called *multifractal universality*. Our result can also be expressed in more operational terms for experimental purposes. Suppose we have a collection of experimental energy spectra corresponding to different Reynolds numbers (all sufficiently high to be in the domain of applicability of our

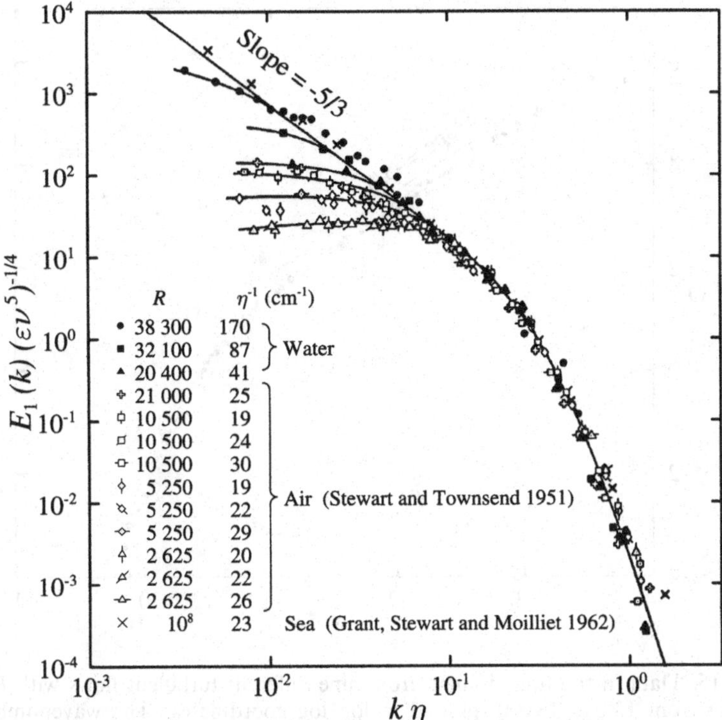

Fig. 8.14. Normalized longitudinal velocity spectrum in the time domain according to different authors (Gibson and Schwarz 1963).

asymptotic laws). According to K41, when plotted in log–log coordinates, all the spectra in the inertial and intermediate dissipation ranges can be collapsed onto a single curve after suitable shifts are performed.[17] According to the multifractal scaling (8.62), the same will hold, but only after the $\ln k$ and the $\ln E(k)$ coordinates have been 'renormalized' by a factor $1/\ln R$. Without this renormalization, the log–log plotted energy spectra would appear to be less and less curved in the (intermediate) dissipation range as the Reynolds number is increased.[18]

In principle it should be possible to distinguish experimentally between universality à la Kolmogorov and multifractal universality. Fig. 8.14, taken from Gibson and Schwarz (1963),[19] shows the kind of single-curve

[17] The shifts are needed because the different data have different r.m.s. velocities and integral scales.

[18] Almost identical remarks can be made about structure functions.

[19] Also reproduced as Fig. 75 in Monin and Yaglom (1975, p. 486).

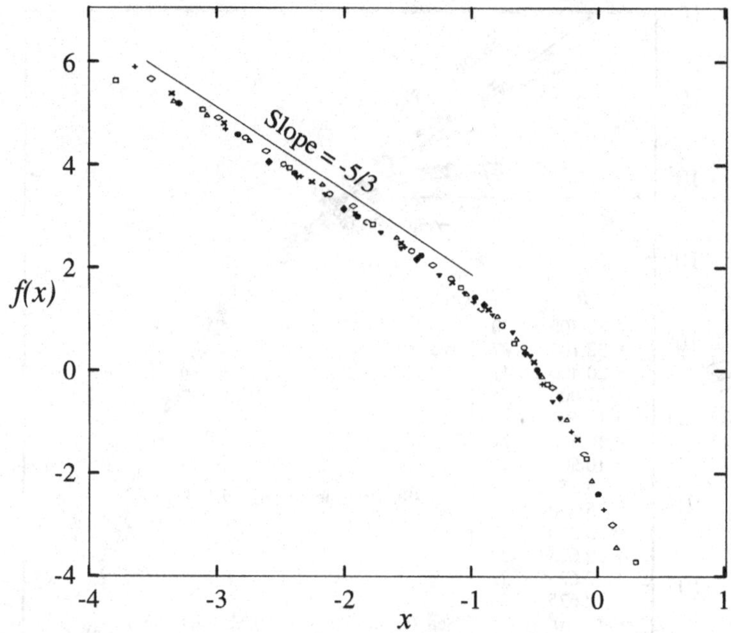

Fig. 8.15. Data in the time domain from nine different turbulent flows with R_λs ranging from 130 to 13 000, plotted in log–log coordinates. The wavenumber (horizontal) and the energy spectrum (vertical) have been divided by $\ln(R_\lambda/R_\star)$ with $R_\star = 75$ and the resulting curves have been shifted to give the best possible superposition (Gagne and Castaing 1991).

collapse obtained using the Kolmogorov first universality assumption, whereas Fig. 8.15, taken from Gagne and Castaing (1991), is a similar attempt using the assumed multifractal universality. Both include a large set of experimental data over a very substantial range of Reynolds numbers: fourteen experiments with Reynolds numbers ranging from two thousand to one hundred million for Gibson and Schwarz; nine experiments with R_λ ranging from 130 to 13000 for Gagne and Castaing. Which one gives a better collapse? At first, multifractal universality looks better. This is emphasized by Gagne and Castaing who also point out that the multifractal model is just one of the possible interpretations; they actually offer an alternative interpretation.[20] It must be stressed, however, that the very nice collapse of the data when plotted in the multifractal

[20] Still another interpretation of the same data set, in terms of a combination of a $k^{-5/3}$ law and a k^{-1} law, is given by She and Jackson (1993).

way happens to a large extent because in the variable $\theta = \ln(k\ell_0)/\ln R$ all the inertial ranges have the same span (roughly from $\theta = 0$ to $\theta = 3/4$).

In order to truly discriminate, it will be necessary to have reliable measurements at scales much smaller than the Kolmogorov dissipation scale η, so as to have a substantial (intermediate) dissipation range. Doing this with hot-wire probes is difficult. With current technology, hot-wire probes have usually a diameter of the order of 1 μm and a length of a fraction of a millimeter. Thus, signals are smoothed over a distance of a fraction of a millimeter. The Kolmogorov microscale η does not change very much with the Reynolds number: if the Reynolds number is raised by increasing solely the integral scale (so as not to affect the Mach number), the dissipation scale grows as the fourth root of the Reynolds number. For most experiments η is between 0.1 and 1 mm (see, e.g., Table 1 of Meneveau and Sreenivasan 1991). Novel nonintrusive optical techniques, such as RELIEF (Miles, Lempert, Zhang and Zhang 1991; Noullez, Wallace, Lempert, Miles and Frisch 1996), mentioned in Section 5.1, seem to have the capability to resolve scales down to 25 μm or less.

8.5.6 The skewness and the flatness of velocity derivatives according to the multifractal model

We begin with a few elementary facts about nondimensionalized moments of velocity derivatives.

As we have seen in Section 6.4.3, in his second derivation of the 1941 theory, Kolmogorov (1941c) used the skewness of velocity increments $S_3(\ell)/(S_2(\ell))^{3/2}$ which was postulated to be ℓ-independent as long as ℓ is in the inertial range. If we ignore the restriction that ℓ should be in the inertial range and let $\ell \to 0$, we obtain (here, $u = v_1$ and $x = x_1$):

$$S_3 = \lim_{\ell \to 0} \frac{S_3(\ell)}{[S_2(\ell)]^{3/2}} = \frac{\langle(\partial_x u)^3\rangle}{\langle(\partial_x u)^2\rangle^{3/2}}, \tag{8.63}$$

which is the definition of the (velocity-derivative) skewness for homogeneous and isotropic turbulence.[21] We shall also be interested in nondimensionalized moments of higher orders:

$$S_n = \frac{\langle(\partial_x u)^n\rangle}{\langle(\partial_x u)^2\rangle^{n/2}}. \tag{8.64}$$

Within the K41 framework, the finiteness of the skewness in the limit

[21] To avoid confusion of notation with the structure function, observe that the skewness has no ℓ-argument.

$\nu \to 0$ can be derived from Kolmogorov's first universality assumption (6.69). It would, however, not be correct to deduce the finiteness from the sole ℓ-independence of the skewness of velocity increments at inertial-range separations. Indeed, this would require the limits $\nu \to 0$ and $\ell \to 0$ to be interchanged.

When intermittency corrections to K41 are taken into account, the skewness $S_3(\ell)/(S_2(\ell))^{3/2}$ behaves as $\ell^{1-3\zeta_2/2}$ and becomes unbounded for $\ell \to 0$ (unless $\zeta_2 = 2/3$). However, there is good experimental and evidence that the skewness increases only very slowly with the Reynolds number, if at all. The measured values for $-S_3$ are typically in the range 0.3–0.5 for laboratory experiments and computer simulations (Monin and Yaglom 1975; Vincent and Meneguzzi 1991), rising to about 0.7 for high-Reynolds-number atmospheric data (Wyngaard and Tennekes 1970).

It is a classical result that the skewness is a measure of vortex stretching. This may be shown as follows for homogeneous isotropic turbulence. We observe that the third order moment of the velocity derivative and the physical-space flux (defined in (6.7)) are related by

$$\langle(\partial_x u)^3\rangle = -\frac{2}{35}\nabla_\ell{}^2\varepsilon(\ell)\Big|_{\ell=0}. \tag{8.65}$$

This follows from (6.36) and the observation that $\langle(\partial_x u)^3\rangle$ is equal to $(1/6)\partial^3 S_3(\ell)/\partial\ell^3$, evaluated at $\ell = 0$. In view of (6.41), we can also express the third order moment in terms of the transfer function:

$$\langle(\partial_x u)^3\rangle = -\frac{2}{35}\int_0^\infty k^2 T(k)dk. \tag{8.66}$$

It may be shown that $\langle(\partial_x u)^2\rangle = (2/15)\Omega$, where $\Omega = \int_0^\infty k^2 E(k)dk$ is the enstrophy. Hence, the skewness is given by (Batchelor 1953)

$$S_3 = -\left(\frac{135}{98}\right)^{1/2} \frac{\int_0^\infty k^2 T(k)dk}{\left(\int_0^\infty k^2 E(k)dk\right)^{3/2}}. \tag{8.67}$$

By (6.37), the quantity $\int_0^\infty k^2 T(k)dk$ is the time-rate-of-change of the enstrophy due to nonlinear interactions. Since enstrophy is (one-half) mean square vorticity and since vortex stretching is constantly taking place in three dimensions at high Reynolds numbers, one expects this quantity to be positive and thus the skewness to be negative (Batchelor and Townsend 1947).

We now turn to the consequences of the multifractal model. Our arguments will be essentially an adaptation of work by Nelkin (1990).

The main result is the following. *For high Reynolds numbers the non-dimensionalized moments of velocity derivatives S_n behave as power-laws of the Reynolds number R*

$$S_n \sim R^{\xi_n}, \qquad \xi_n = p(n) - 3n/2, \tag{8.68}$$

where p(n) is the unique solution of

$$\zeta_p = 2n - p. \tag{8.69}$$

Here, the ζ_ps are the usual exponents for the structure functions.

For the proof, we shall work in units where the r.m.s. velocity and the integral scale are order unity, so that the Reynolds number $R = O(1/\nu)$. Let us first calculate $\langle(\partial_x u)^n\rangle$. Proceeding as in the previous section, we find that the contribution of the scaling exponent h to $\partial_x u$ is

$$\partial_x u \sim \ell^{h-1}\big|_{\ell=\eta(h)}. \tag{8.70}$$

Note that the derivative comes mostly from the neighborhood of the h-dependent cutoff $\eta(h) = \nu^{1/(1+h)}$, given by (8.53). We now take the nth power of $\partial_x u$ and average; this introduces a codimension factor as usual. Finally, we integrate over all scaling exponents to obtain

$$\langle(\partial_x u)^n\rangle \sim \int_I d\mu(h)\, \ell^{n(h-1)+3-D(h)}\big|_{\ell=\nu^{1/(1+h)}}. \tag{8.71}$$

For $\nu \to 0$, the usual steepest descent argument gives (with possible logarithmic corrections omitted):

$$\langle(\partial_x u)^n\rangle \propto \nu^{\rho_n}, \tag{8.72}$$

where

$$\rho_n = \inf_h \frac{n(h-1)+3-D(h)}{1+h}. \tag{8.73}$$

The minimum is found by setting equal to zero the derivative of the r.h.s. with respect to h. Hence,

$$\rho_n = \frac{n(h_\star - 1) + 3 - D(h_\star)}{1+h_\star}, \tag{8.74}$$

where $h_\star$ is the solution of

$$\left[n - D'(h_\star)\right](1+h_\star) - n(h_\star - 1) - 3 + D(h_\star) = 0. \tag{8.75}$$

To solve (8.75) we use a trick. We introduce an exponent p related to $D'(h_\star)$ by (8.45) and we express $D(h_\star)$ in terms of ζ_p, using (8.46). Eq. (8.75) then reduces to (8.69). For any $n > 0$, (8.69) has a unique solution $p(n)$, because $\zeta_p + p$ increases monotonically and ranges from zero to infinity. Reexpressing ρ_n, given by (8.74), in terms of n, $p(n)$, $h_\star$ and $\zeta_{p(n)}$, we find that the (explicit) h-dependence cancels out and we are left with

$$\rho_n = n - p(n). \tag{8.76}$$

For the special case $n = 2$, eq. (8.69) has the obvious solution $p = 3$, which is a consequence of $\zeta_3 = 1$. Hence, $\rho_2 = -1$, so that

$$\langle(\partial_x u)^2\rangle \propto \nu^{-1}. \tag{8.77}$$

This is equivalent to the statement that the energy dissipation, which is proportional to $\nu\langle(\partial_x u)^2\rangle$, goes to a finite limit as $\nu \to 0$. The relation (8.77), obtained by Frisch and Vergassola (1991), shows the consistency of the method used to calculate moments of velocity derivatives. Using (8.76) and (8.77) in (8.64), we finally obtain (8.68). *QED.*

Eqs. (8.68) and (8.69) provide a very simple practical method for estimating exponents of nondimensionalized moments of derivatives in terms of the ζ_ps. One just has to find the intersection of the graph of ζ_p with the line $\zeta_p = 2n - p$. Since, for not too large p's, the ζ_p's are very close to their K41 values $p/3$, the intersections will be close to $p = 3n/2$, the small discrepancy being precisely ξ_n.

Let us now consider the special case of the skewness ($n = 3$). The number p is now close to 4.5. Based on the data of Anselmet, Gagne, Hopfinger and Antonia (1984), the exponents ζ_4 and ζ_5 are too close to their K41 values for the discrepancy to be measurable. Hence, ξ_3 is also too small to be measurable. If we use the figures of Benzi, Ciliberto, Baudet and Ruiz Chavarria (1995), based on their extended self-similarity analysis of the Gagne–Hopfinger S1 wind-tunnel data, we find $\xi_3 = 0.07$, which is quite small. Thus, multifractal intermittency is consistent with a skewness almost independent of the Reynolds number. For the flatness ($n = 4$), we obtain $\xi_4 = 0.15$, based on Anselmet, Gagne, Hopfinger and Antonia (1984) and $\xi_4 = 0.18$, based on Benzi, Ciliberto, Baudet and Ruiz Chavarria (1995).

Experimental values, quoted or measured by Wyngaard and Tennekes (1970), give a flatness rising from $S_4 \approx 6.$ for $R_\lambda = 200$ to a scatter of values in the range 20–40 when R_λ is around 3000. Since $R_\lambda \propto \nu^{-1/2}$, this is consistent with ξ_4 in the range 0.22–0.35, values somewhat higher than suggested by the analysis above. More recent experiments, using the same helium gas facility as described on p. 65, give a flatness of around 12 ± 2 for R_λ between 200 and 3000 (Tabeling, Zocchi, Belin, Maurer and Willaime 1995). If confirmed, this implies that ξ_4 is extremely small or vanishing.

8.6 Intermittency models based on the dissipation

A central quantity in the K41 theory is the mean energy dissipation. Landau's objection to one form of the Kolmogorov theory, as discussed in Section 6.4, centered around (large-scale) *fluctuations* of the dissipation. It is thus not surprising that much experimental effort went and still goes into the study of such fluctuations (at scales large and small). In Section 8.6.1 we shall show how multifractality can be defined and measured in terms of the fluctuations of the local dissipation rather than in terms of velocity increments. The relation between the two types of multifractality will be discussed in Section 8.6.2. Multiplicative random models of the dissipation leading to multifractality will be presented in Sections 8.6.3–8.6.5 together with some probabilistic background on 'large deviations theory'. Most historical comments will be postponed to Section 8.8.

8.6.1 Multifractal dissipation

The key quantity needed to define multifractality in terms of the dissipation is the *local space average* of the energy dissipation over a ball of radius ℓ centered at the point $\boldsymbol{r}$, first considered by Obukhov (1962) and Kolmogorov (1962):

$$\varepsilon_\ell(\boldsymbol{r}) = \frac{1}{(4/3)\pi\ell^3} \int_{|\boldsymbol{r}'-\boldsymbol{r}|<\ell} d^3r' \frac{1}{2}\nu \sum_{ij} \left[\partial_j v_i(\boldsymbol{r}') + \partial_i v_j(\boldsymbol{r}')\right]^2 . \tag{8.78}$$

Let us denote by $P_\ell^{\text{diss}}(\varepsilon)$ the p.d.f. of ε_ℓ. (The limits $t \to \infty$ and $\nu \to 0$ are assumed to have been taken.) By the assumption of homogeneity, $P_\ell^{\text{diss}}(\varepsilon)$ depends only on the radius ℓ of the ball and not on its position. The definition of multifractality is now essentially the same as the one given for velocity increments, in the probabilistic formulation of Section 8.5.4.

Definition. *The dissipation is said to be* multifractal *if there is a function $F(\alpha)$ which maps real scaling exponents α to scaling dimensions $F \le 3$ (including negative values and the value $-\infty$), such that for any α*

$$\lim_{\ell\to 0} \frac{\ln \bar{P}_\ell^{\text{diss}}\left(\ell^{\alpha-1}\right)}{\ln \ell} = 3 - F(\alpha), \tag{8.79}$$

where $\bar{P}_\ell^{\text{diss}}(\varepsilon)$ is the cumulative probability of ε_ℓ.

Note that, ε_ℓ being a positive quantity, there is no need to distinguish positive and negative arguments as was the case for velocity increments. The choice of the exponent $\alpha - 1$ rather than α is for historical reasons.

The more traditional definition of multifractality, analogous to that given in Section 8.5.3 for the velocity, which has the drawback that it assumes — possibly nonexistent — singularities for individual realizations, is the following:

$$\left.\begin{array}{l} \dfrac{\varepsilon_\ell(\boldsymbol{r})}{v_0^3/\ell_0} \sim \left(\dfrac{\ell}{\ell_0}\right)^{\alpha-1} \quad \text{as } \ell \to 0, \\ \text{for} \quad \boldsymbol{r} \in \mathscr{D}_\alpha \subset \mathbb{R}^3; \quad \dim \mathscr{D}_\alpha = F(\alpha). \end{array}\right\} \tag{8.80}$$

Note that $\varepsilon_\ell(\boldsymbol{r})$ and ℓ have been respectively divided by v_0^3/ℓ_0 (the order of magnitude of the mean dissipation) and ℓ_0 (the integral scale) to obtain dimensionless quantities. Expressed in words, the singularity definition of multifractality states that the dissipation, considered as a measure, has singularities of exponent $\alpha - 1$ on sets of dimension $F(\alpha)$.

By essentially the same arguments as in Section 8.5.3, we infer from one or other definition that, if the dissipation is multifractal, moments of ε_ℓ follow power-laws at small ℓ:

$$\langle \varepsilon_\ell^q \rangle \sim \left(\frac{v_0^3}{\ell_0}\right)^q \left(\frac{\ell}{\ell_0}\right)^{\tau_q}, \qquad \tau_q = \min_\alpha \left[q(\alpha - 1) + 3 - F(\alpha)\right]. \tag{8.81}$$

The order q of the moment need not be an integer, since $\varepsilon_\ell > 0$. It may even be negative, if $P_\ell^{\mathrm{diss}}(\varepsilon)$ goes to zero sufficiently rapidly with ε. Note that τ_q and $F(\alpha)$ are again related by a Legendre transformation, so that τ_q is a concave function of q. Observe that our definition of τ_q is not the 'standard' one found in the literature, for example in Meneveau and Sreenivasan (1991):

$$\tau_q^{\mathrm{our}} = \tau_q^{\mathrm{standard}} + 3(1 - q). \tag{8.82}$$

(Had we worked in d dimensions, there would be a $d(1 - q)$.) We find it more convenient to define τ_q as the exponent for moments of the local space average of the dissipation, although the other choice may be more convenient for studying multifractals in dynamical systems (Halsey, Jensen, Kadanoff, Procaccia and Shraiman 1986).

In the literature one also often finds the results stated in terms of the Renyi dimensions (Grassberger 1983; Hentschel and Procaccia 1983; see also Paladin and Vulpiani 1987b):

$$D_q = \frac{\tau_q}{q - 1} + 3. \tag{8.83}$$

Grassberger and Procaccia (1984) showed that D_0 is the fractal dimension of the support of the measure (here the dissipation), D_1 the information dimension and D_2 the correlation dimension. Unfortunately, because of the factor $q-1$ in the denominator of (8.83), D_q is not a concave function of q and it is usually better to work with τ_q.

Meneveau and Sreenivasan (1987, 1991) have obtained considerable evidence that the dissipation of fully developed turbulence is multifractal. It is very difficult to measure the local dissipation directly in a turbulent fluid, so that various modifications are needed. One-dimensional space averages of the dissipation are used as 'representatives' of the three-dimensional averages. Specifically, one considers a one-dimensional line **L** with coordinate x and, instead of (8.78), one uses

$$\varepsilon_\ell(x) \equiv \frac{1}{2\ell}\int_{|x'-x|<\ell} dx' \frac{1}{2}\nu \sum_{ij}\left[\partial_j v_i(x') + \partial_i v_j(x')\right]^2 . \tag{8.84}$$

Similarly, instead of (8.80), it is assumed that, for small ν, the local dissipation $\varepsilon_\ell(x)$ behaves as follows:

$$\frac{\varepsilon_\ell(x)}{v_0^3/\ell_0} \sim \left(\frac{\ell}{\ell_0}\right)^{\alpha-1} \quad \text{as } \ell \to 0 \quad \text{for } x \in \mathscr{D}'_\alpha; \qquad \dim \mathscr{D}'_\alpha = f(\alpha), \tag{8.85}$$

where $\mathscr{D}'_\alpha = \mathscr{D}_\alpha \cap \mathbf{L}$ and

$$f(\alpha) \equiv F(\alpha) - 2. \tag{8.86}$$

Note that the 'dimension' $f(\alpha)$ has decreased by two units because one-dimensional cuts are taken. In the literature $f(\alpha)$ is often called the multifractal (singularity) spectrum. For reasons which will become clear in Section 8.6.4, we shall call it the (dissipation) Cramér function. Additional modifications are needed to make the determination of $f(\alpha)$ practical, as listed below.

(i) Instead of the full squared rate-of-strain tensor in (8.84), one uses $(\partial u/\partial x)^2$, the square of the streamwise derivative of the streamwise velocity component.

(ii) When measurements are made in the time domain, the Taylor hypothesis is used to substitute time-derivatives for space-derivatives.

(iii) Time-derivatives are approximated by time-differences over the sampling time of the probe-recorded signal.

The third item is particularly delicate, since hot-wire probes may be barely capable of resolving the very small scales needed to measure

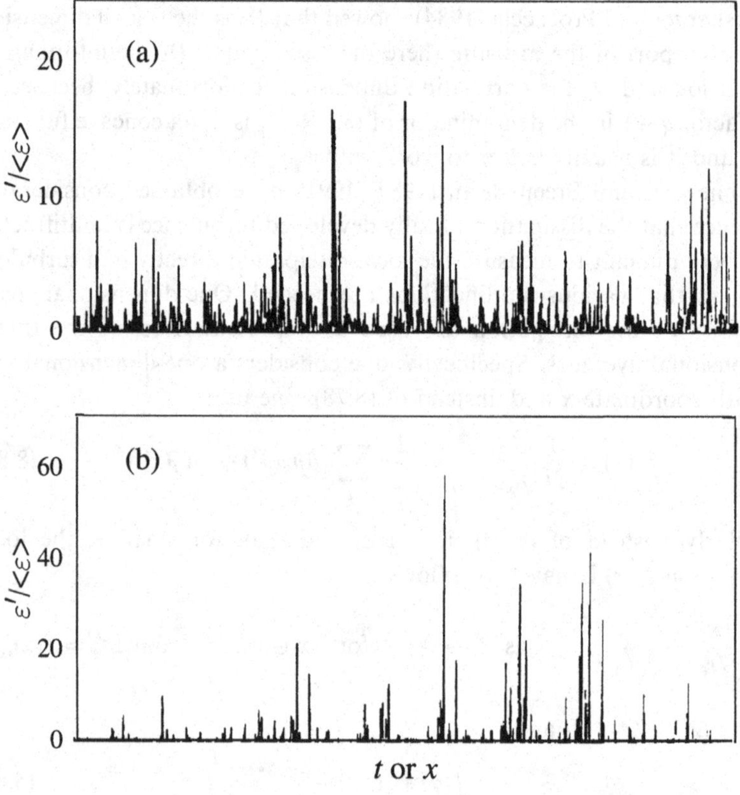

Fig. 8.16. Typical signals of a 'representative' of the local dissipation: (a) was obtained in a laboratory boundary layer at moderate Reynolds number, and (b) in the atmospheric surface layer at high Reynolds number (Meneveau and Sreenivasan 1991).

derivatives. In Aurell, Frisch, Lutsko and Vergassola (1992) the reader will find a critical discussion of this and other possible causes of spurious multifractality. [22]

Let us now turn to the experimental results of Meneveau and Sreenivasan (1991) which, in spite of the aforementioned difficulties, provide good evidence for multifractality of the dissipation. Fig. 8.16 shows typical signals (for two flows) of the representative of the local dissipation, namely $(\partial u_1/\partial t)^2$ normalized by its mean. The highly intermittent aspect of such signals is striking. Fig. 8.17 shows the multifractal spectrum

[22] Spurious multifractality from bad data processing can arise in many areas outside of turbulence; a case study for seismic data may be found in Eneva (1994).

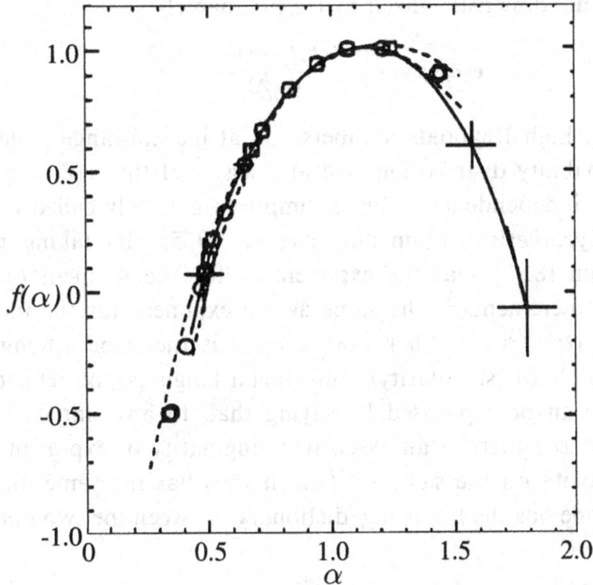

Fig. 8.17. Cramér function $f(\alpha)$ for the dissipation in the time domain measured in the atmospheric surface layer at high Reynolds number (Meneveau and Sreenivasan 1991).

computed by performing an inverse Legendre transformation on the function τ_q. The latter is obtained by identifying power-laws in moments of ε_ℓ, plotted vs ℓ in log–log coordinates. The fact that the function $f(\alpha)$ achieves a maximum, around unity, near $\alpha = 1$ is indicative that there is a nonsingular ($\alpha = 1$) background of space-filling ($f(\alpha) = 1$) dissipation, as in the K41 theory.

Relating multifractality of the dissipation and multifractality of the velocity is the subject of the next section.

8.6.2 Bridging multifractality based on the velocity and multifractality based on the dissipation

Kolmogorov (1962) had to relate the statistical properties of fluctuations of velocity increments to those of the space-averaged dissipation. For this, following a suggestion of Obukhov (1962), he was essentially relying on his 1941 result stating that velocity increments over a distance ℓ scale as $(\varepsilon\ell)^{1/3}$. The same formula may be used with ε_ℓ instead of ε. Specifically, in our notation, Kolmogorov assumed that the velocity increment, nondimensionalized by $(\ell\varepsilon_\ell)^{1/3}$, namely

$$\tilde{\boldsymbol{v}}(\boldsymbol{\ell},\tau) = \frac{\boldsymbol{v}(\boldsymbol{r}+\boldsymbol{\ell},t+\tau)-\boldsymbol{v}(\boldsymbol{r},t)}{(\ell\varepsilon_\ell)^{1/3}}, \tag{8.87}$$

has, at very high Reynolds numbers and at inertial-range scales, a universal probability distribution. He also assumed that $\tilde{\boldsymbol{v}}(\boldsymbol{\ell},\tau)$ and ε_ℓ are statistically independent.[23] This assumption is usually called the *refined similarity hypothesis* (Monin and Yaglom 1975). By taking $\tau = 0$, it follows from (8.87) that the exponent ζ_p for the moment of order p of velocity increments is the same as the exponent for the moment of order p of $(\ell\varepsilon_\ell)^{1/3}$. In other words: δv_ℓ has the same scaling properties as $(\ell\varepsilon_\ell)^{1/3}$. In (singularity) multifractal language the refined scaling hypothesis can be rephrased by saying that, to any singularity of exponent α of $\ell\varepsilon_\ell$, there is an associated singularity of exponent $h = \alpha/3$ for the velocity on the same set (which thus has the same dimension). Therefore one has the following 'dictionary' between the two multifractal formalisms:

$$h = \frac{\alpha}{3}, \qquad D(h) = F(\alpha) = f(\alpha) + 2, \qquad \zeta_p = \frac{p}{3} + \tau_{p/3}. \tag{8.88}$$

Meneveau and Sreenivasan (1991) have found that Kolmogorov's relation (8.88) between ζ_p and τ_q is consistent with the current quality of experimental data, at least over the range where both can be measured. This last restriction is not just a matter of accuracy. As observed by Bacry, Arnéodo, Frisch, Gagne and Hopfinger (1990), moments of ε_ℓ can stay finite for negative orders, but moments of velocity increments of order ≤ -1 are infinite, because the p.d.f. of velocity increments is finite and nonvanishing at zero velocity increment (Gagne, Hopfinger and Frisch 1990). Muzy, Bacry and Arnéodo (1993) observed that this difficulty is not present if, instead of velocity increments, one uses their wavelet-transform modulus-maxima method. Stolovitzky, Kailasnath and Sreenivasan (1992) pointed out that Kolmogorov's formulation (8.87) does not preclude the divergence of negative-order structure functions, since the random variables $\tilde{\boldsymbol{v}}$ is not expected to have finite moments for negative orders. Frisch (1991) observed that $\ell^3\varepsilon_\ell$, being a space integral, is an additive quantity (when one-dimensional rather than three-dimensional space averages are used the additive quantity is $\ell\varepsilon_\ell$); similarly, $\delta v_\parallel(\boldsymbol{r},\ell)$

[23] Actually, Kolmogorov assumed that the nondimensionalized velocity increment, *conditioned* upon a given value of ϵ_ℓ, is independent of ϵ_ℓ, as long as the local fluctuating Reynolds number $R_\ell = \ell(\ell\epsilon_\ell)^{1/3}/\nu$ is much larger than unity; for further discussion of this, see Stolovitzky and Sreenivasan (1994).

is an additive quantity (if A, B and C are three consecutive points on a line, the longitudinal velocity difference between points A and C is the sum of the difference between A and B and the difference between B and C); however, the *cube* of an additive quantity is not additive. Furthermore, a similar 'dictionary' can be set up for Burgers' equation. It may then be shown that inconsistencies already arise for moments of positive fractional orders less than unity (Aurell, Frisch, Lutsko and Vergassola 1992).

Actually, the only case where (8.88) can be established in a systematic way is for $p = 3$. It is then just a consequence of Kolmogorov's four-fifths law (6.5). Further criticism of the refined similarity hypothesis, presented by Hosokawa and Yamamoto (1992), has been found questionable by Praskovsky (1992).

Before the shortcomings of the lognormal model were widely realized (Section 8.6.5), there was considerable interest in the exponent

$$\mu = -\tau_2 = 2 - \zeta_6, \tag{8.89}$$

which, by (8.83), is equal to $3 - D_2$, where D_2 is the correlation dimension of the dissipation. (In the lognormal model, μ is the only free parameter.) The value of μ has been the subject of some debate. Early values, $\mu \approx 0.5$ were revised downward to $\mu \approx 0.2$ when it was realized by Anselmet, Gagne, Hopfinger and Antonia (1984) that the mean dissipation should not be subtracted before calculating the dissipation correlation function. The value of 0.2 is also consistent with $\zeta_6 \approx 1.8$ measured by Anselmet, Gagne, Hopfinger and Antonia (1984).

We finally mention that the results derived in Sections 8.5.5 and 8.5.6 for the intermediate dissipation range, the skewness and the flatness can, in principle, be derived within the multifractal formalism based on the dissipation. This is, however, not natural since in the derivations a key role is played by the eddy turnover time $\ell/\delta v_\ell$, a quantity not readily available in the formalism based on the dissipation. Hence, it is tempting to use a *single dissipation scale* (the Kolmogorov scale $\eta \sim \nu^{3/4}$), instead of making it dependent on the scaling exponent (h or α). This is why the results reported in Keller and Yaglom (1970) and Monin and Yaglom (1975, Section 25.4) are not entirely consistent with ours.

8.6.3 Random cascade models

One of the simplest methods — and historically the first — for obtaining multifractal dissipation measures is to construct multiplicative random cascade models of the sort introduced by the Russian school and studied by Mandelbrot. (We shall return to historical aspects in Section 8.8.) These are constructed somewhat in the same spirit as the β-model. They can be formulated either in terms of velocity fluctuations or in terms of fluctuations of the locally space-averaged dissipation. We shall use the latter.

We begin with a cube of side ℓ_0 in which we take the dissipation to be uniform. The value per unit volume is ε, a nonrandom positive quantity. This defines the generation $n = 0$. To obtain generation $n = 1$, we subdivide the cube into eight equal cubes of side $\ell_1 = \ell_0/2$. (The factor of 2 is assumed only for simplicity.) In each smaller cube we multiply the dissipation by independent realizations of a random variable W, which is subject to the following constraints:

$$W \geq 0, \quad \langle W \rangle = 1, \quad \langle W^q \rangle < \infty, \; \forall q > 0. \tag{8.90}$$

At the nth generation, there are 2^{3n} cubes of side

$$\ell = \ell_0 2^{-n}. \tag{8.91}$$

In each of them the dissipation (per unit volume) is uniform and assumes a value of the form

$$\varepsilon_\ell = \varepsilon W_1 W_2 \ldots W_n, \tag{8.92}$$

where the W_is are independently and identically distributed. The process is repeated indefinitely.

By (8.90), the (ensemble) average of any of the ε_ℓs is still equal to ε but ε_ℓ is usually not related to the sum of the $\varepsilon_{\ell/2}$s over the eight subcubes of side $\ell/2$ within a given cube of side ℓ. In other words, the cascade is *nonconservative*. Furthermore, the multiplication of many random variables leads to very large fluctuations.

It is a straightforward matter to calculate moments of ε_ℓ. From (8.91) and (8.92), we obtain for any positive integer q

$$\langle \varepsilon_\ell^q \rangle = \varepsilon^q \left(\frac{\ell}{\ell_0} \right)^{\tau_q}, \tag{8.93}$$

where

$$\tau_q = -\log_2 \langle W^q \rangle. \tag{8.94}$$

By the hypotheses (8.90), all the moments of the dissipation are finite.[24] From the bridging relation (8.88) we obtain the following expression for the exponents of structure functions:

$$\zeta_p = \frac{p}{3} - \log_2 \langle W^{p/3} \rangle. \tag{8.95}$$

A particularly simple result is obtained with the *black and white*[25] choice of Novikov and Stewart (1964). Here, only two values are permitted for the random variable W,

$$W = \begin{cases} 1/\beta & \text{with probability } \beta, \\ 0 & \text{with probability } 1-\beta. \end{cases} \tag{8.96}$$

From (8.94) and (8.95), we obtain for the Novikov–Stewart model

$$\tau_q = -(1-q)\log_2 \beta, \tag{8.97}$$

$$\zeta_p = \frac{p}{3} - \left(1 - \frac{p}{3}\right)\log_2 \beta, \tag{8.98}$$

which is exactly the result (8.31) for the (unifractal) β-model of Section 8.5.1.

Benzi, Paladin, Parisi and Vulpiani (1984) have proposed a *random β-model* in which the factor β is randomly and independently selected at each step of the cascade, with the same law. Just as the ordinary β-model is equivalent to the Novikov–Stewart model, as far as scaling is concerned, making an *arbitrary* random choice of β is equivalent to the general random cascade model. Benzi, Paladin, Parisi and Vulpiani (1984) also proposed a restricted choice with a single free parameter x between zero and 1, for which the p.d.f. of β is $P(\beta) = x\,\delta(\beta - 0.5) + (1-x)\,\delta(\beta - 1)$. A good fit to the data of Anselmet, Gagne, Hopfinger and Antonia (1984) is obtained for $x = 0.125$.

The graph of τ_q is a straight line in the Novikov–Stewart model; however, as soon as W takes more than one nonvanishing value, (8.94) produces a nontrivial concave graph, a consequence of the Schwarz inequality (Feller 1968b, Section V.8). Hence, the random cascade models usually have *multifractal* scaling properties. In order to gain a deeper

[24] There is alternative definition of ε_ℓ: subdivide the cube of side $\ell = 2^{-n}\ell_0$ into 2^{3m} subcubes of side $2^{-m}\ell$, sum the previously defined dissipations over these 2^m subcubes and divide by 2^{3m} (because our dissipations are per unit volume) and, finally, let $m \to \infty$. The limit exists but may have divergent moments beyond a certain order q_c (Mandelbrot 1974b; Kahane and Peyrière 1976; Collet and Koukiou 1992). There are interesting mathematical problems arising in this approach but the divergence of moments is an artifact of the nonconservative character of the cascade.

[25] A term suggested by Mandelbrot (1974b).

understanding of why such scaling holds, we digress in the next section into the theory of large deviations.

8.6.4 Large deviations and multifractality

Large-deviations theory, as already mentioned in Section 4.4, is concerned with the very low probability events where the normalized sums or integrals of random variables or functions differ from their mean values by an amount much larger than the standard deviation. The theory originated with the work of Cramér (1938). For a general exposition we refer the reader to Varadhan (1984) and Ellis (1985). For the application to random cascade models, we shall need only the case of independent variables, which is now briefly reviewed, following the exposition of Lanford (1973). A presentation of multifractality in terms of large deviations may also be found in Evertsz and Mandelbrot (1992).

Let m_i ($i = 1, 2, \ldots$) be identically distributed and independent random variables.[26] We know that under suitable conditions (Feller 1968b),

$$S_n = \frac{1}{n}\sum_{i=1}^{n} m_i \to \langle m \rangle \quad \text{for } n \to \infty, \tag{8.99}$$

the convergence being almost sure by the strong law of large numbers. The theory of large deviations gives estimates of the probability that the partial averages S_n are, for large n, close to an arbitrary value x, which may differ from $\langle m \rangle$ by $O(1)$.

Let

$$h(n; a, b) \equiv \ln \mathrm{Prob}\{a < S_n < b\}. \tag{8.100}$$

The function $h(n; a, b)$ is nonpositive and *superadditive*, i.e.

$$h(n + n'; a, b) \geq h(n; a, b) + h(n'; a, b). \tag{8.101}$$

This follows immediately from the observation that, by the assumed independence of the m_is,

$$\begin{aligned} &\mathrm{Prob}\left\{ a < \frac{1}{n+n'}\sum_{1}^{n+n'} m_i < b \right\} \geq \\ &\mathrm{Prob}\left\{ a < \frac{1}{n}\sum_{1}^{n} m_i < b \right\} \cdot \mathrm{Prob}\left\{ a < \frac{1}{n'}\sum_{n+1}^{n+n'} m_i < b \right\}. \end{aligned} \tag{8.102}$$

[26] In the application to random cascade models, $m = -\log_2 W$.

A simple consequence of superadditivity is that the limit

$$s(a,b) \equiv \lim_{n\to\infty} \frac{h(n;a,b)}{n} = \sup_n \frac{h(n;a,b)}{n} \tag{8.103}$$

exists. (Hint: Pick two integers n and n_0, perform the Euclidian division $n = qn_0 + r$, apply superadditivity and then let n and n_0 successively tend to infinity.) The value $-\infty$ is allowed for the limit. We then define

$$s(x) \equiv \sup_{a<x<b} s(a,b). \tag{8.104}$$

The function $s(x)$ is negative or zero (by construction) and is easily shown to be concave.

The large-deviations theorem, loosely stated, is that for large n (if you are a mathematician, skip the next line)

$$\mathrm{Prob}\left\{\frac{1}{n}\sum_{i=1}^{n} m_i \approx x\right\} \sim e^{ns(x)}. \tag{8.105}$$

In statistical mechanics, large-deviations theory is used to prove rigorous results about the approach to the thermodynamic limit. In that context the function $s(x)$ can be identified with the *entropy*; in the literature on large deviations, the negative of $s(x)$ is often called the 'rate function'. Mandelbrot (1991) has proposed calling $s(x)$ the *Cramér function*. We shall adopt his terminology.

A simple example of large deviations, taken from Lanford (1973), is the coin-tossing problem. The variable m takes only two values, say 0 and 1, with equal probabilities. The probability of S_n can be explicitly calculated from the binomial formula. Use of Stirling's formula on factorials then gives the following expression for the Cramér function:

$$s(x) = \begin{cases} -x\ln x - (1-x)\ln(1-x) - \ln 2 & \text{for } 0 \le x \le 1 \\ -\infty & \text{otherwise.} \end{cases} \tag{8.106}$$

Note that the function $s(x)$ defined by (8.106) behaves parabolically near its maximum $x = 1/2$, $s = 0$. This is quite general and equivalent to the statement that those deviations which are $O(n^{-1/2})$ have essentially a Gaussian distribution. Larger deviations, which are $O(1)$, cannot be correctly described as Gaussian.

In the general case, the Cramér function $s(x)$ can be expressed in terms of the logarithm of the characteristic function of the random variable m. Let us assume that the p.d.f. of the random variable m decreases faster than exponentially at large arguments; this ensures the existence of

$$Z(\beta) \equiv \langle e^{-\beta m} \rangle, \tag{8.107}$$

for any real β. The function $Z(\beta)$ is the characteristic function, as defined in (4.6), evaluated for an imaginary argument $z = i\beta$.

The functions $s(x)$ and $\ln Z(\beta)$ are Legendre transforms of each other:

$$\ln Z(\beta) = \sup_x [s(x) - \beta x], \tag{8.108}$$

$$s(x) = \inf_\beta [\ln Z(\beta) + \beta x]. \tag{8.109}$$

A rigorous but still elementary proof of (8.108) and (8.109) may be found in Lanford (1973). Here, we give a simplified proof based on (8.105). We observe that

$$Z^n(\beta) = \left\langle e^{-\beta(m_1+\ldots+m_n)} \right\rangle. \tag{8.110}$$

By the large-deviations theorem, for large n, the sum $m_1 + \ldots + m_n$ is close to nx with a probability $\sim e^{ns(x)}$. Hence, the contribution to $Z^n(\beta)$ coming from sums close to nx is $\sim e^{\{n[-\beta x+s(x)]\}}$. When integrating over all possible xs the dominant contribution will come from that x which maximizes $-\beta x + s(x)$. Thus,

$$Z^n(\beta) \sim \exp\left\{ n \sup_x [s(x) - \beta x] \right\}. \tag{8.111}$$

Taking logarithms, we obtain (8.108). As for (8.109), it is just the inverse Legendre transform relation (cf. Section 8.5.3). This concludes our digression on large deviations.

We return to the random cascade model of Section 8.6.3. The dissipation measure ε_∞, which is the limit obtained by the construction defined at the beginning of this section, is *multifractal* in the sense of Section 8.6.1. Indeed, setting

$$W_i \equiv 2^{-m_i}, \tag{8.112}$$

we obtain from (8.91), (8.92) and the large-deviations theorem

$$\frac{\varepsilon_\ell}{\varepsilon} = 2^{-(m_1+\ldots+m_n)} \approx 2^{-nx} \approx \left(\frac{\ell}{\ell_0}\right)^x. \tag{8.113}$$

This holds with a probability

$$p_\ell \sim e^{ns(x)} = \left(\frac{\ell}{\ell_0}\right)^{-s(x)/\ln 2}. \tag{8.114}$$

In other words, in the nesting construction of cubes, the fraction of the space such that the dissipation behaves as ℓ^x rarefies as $\ell^{-s(x)/\ln 2}$. Identifying, the exponent x with $\alpha - 1$ of (8.80) and the exponent $-s(x)/\ln 2$

with $3 - F(\alpha) = 1 - f(\alpha)$ of Section 8.6.1, we find that for the random cascade model,

$$F(\alpha) = 3 + \frac{s(\alpha - 1)}{\ln 2}, \qquad f(\alpha) = 1 + \frac{s(\alpha - 1)}{\ln 2}. \tag{8.115}$$

Hence, the $f(\alpha)$ function is essentially the Cramér function and should be called so. Furthermore, it is easily checked that (8.81), which states that $F(\alpha)$ and τ_q are related by a Legendre transformation, is equivalent to (8.108).

We observe that the β-model of Section 8.5.1 is recovered when the random variable W takes only two values: $1/\beta$ (with probability β) and zero (with probability $1 - \beta$). By (8.112), the corresponding values for m are $\ln_2 \beta$ and $+\infty$. This gives a somewhat pathological character to the large-deviations result. Indeed, the normalized sums S_n, defined in (8.99), also take only two values: $\ln_2 \beta$ (with probability β^n) and $+\infty$ (with probability $1 - \beta^n$). Hence, there is a *single* value for the 'deviation' x and a *single* value for the dimension, given by (8.21).

The dimension $F(\alpha)$ given by (8.115) may well become *negative* for some range of αs. As already explained in Section 8.5.1, this means that the nested set of cubes on which such an α holds terminate, because it rarefies too quickly with the generation index n. However, $F(\alpha)$ (or, equivalently, $s(x)$) is meaningful as a quantity controlling the probability of such rarefications (through (8.105)).

Large-deviations theory was here presented in its simplest form, for independent random variables. Like the law of large numbers and the central limit theorem, large-deviations theory has extensions to random variables with correlations, when they decrease sufficiently fast. Large deviations can also occur in deterministic chaos, but need not. Indeed, Biferale, Blank and Frisch (1994) have constructed deterministic cascade models in which the factors W_n of (8.92) are obtained by iterating a deterministic map. These models can display chaotic fluctuations but do not deviate from the K41 scaling because they have no large deviations.

8.6.5 *The lognormal model and its shortcomings*

In Section 8.6.3 the distribution of the multiplicative factors W was taken in a quite general form (with the restrictions stated at the beginning of that section). For mostly historical reasons, to which we shall return in Section 8.8, there has been a particular interest in the case where W has a lognormal law. Specifically, let us assume $W = 2^{-m}$, where m is a

Gaussian random variable, with mean

$$\overline{m} = \langle m \rangle \tag{8.116}$$

and variance

$$\sigma^2 = \langle (m - \overline{m})^2 \rangle. \tag{8.117}$$

The constraint $\langle W \rangle = 1$ gives

$$2\overline{m} = \sigma^2 \ln 2. \tag{8.118}$$

Since the sums $m_1 + \ldots + m_n$, like m, obey a Gaussian law, it is an elementary exercise to calculate the Cramér function:

$$s(x) = -\frac{(x - \overline{m})^2 \ln 2}{4\overline{m}}. \tag{8.119}$$

Hence, by (8.115),

$$F(\alpha) = 3 - \frac{(\alpha - 1 - \mu/2)^2}{2\mu}, \quad f(\alpha) = 1 - \frac{(\alpha - 1 - \mu/2)^2}{2\mu}. \tag{8.120}$$

Here, instead of the parameter $\overline{m}$, we have used

$$\mu \equiv 2\overline{m}, \tag{8.121}$$

a notation which is now traditional in turbulence (see, e.g., Monin and Yaglom 1975). From (8.81) and (8.88), it is easily shown that the lognormal model gives the following values for the exponents of moments of the dissipation and for the exponents of structure functions, respectively:

$$\tau_q = \frac{\mu}{2}(q - q^2), \quad \zeta_p = \frac{p}{3} + \frac{\mu}{18}(3p - p^2). \tag{8.122}$$

Hence, in the lognormal model, the second moment of the dissipation scales as $\ell^{-\mu}$ and the second order structure function has a correction $\mu/9$ to the K41 value 2/3.

It follows from (8.122) that ζ_p is a *decreasing* function of p when

$$p > p_\star = \frac{3}{2} + \frac{3}{\mu}. \tag{8.123}$$

Hence the lognormal model violates the condition of Section 8.4 that ζ_{2p} be a nondecreasing function of p, the condition needed to avoid supersonic velocities.

Furthermore, the lognormal model violates an inequality of Novikov (1970), which states that

$$\tau_q + 3q \geq 0 \quad \text{for } q \geq 0 \qquad \text{and } \tau_q + 3q \leq 0 \quad \text{for } q \leq 0. \tag{8.124}$$

To prove this relation, we start from the definition of ε_ℓ given in Section 8.6, as the local space average over a ball of radius ℓ of the dissipation. Since the quantity being integrated is positive, it follows that $\ell^3\varepsilon_\ell$ (unaveraged) is a nondecreasing function of ℓ. Raising to the qth power ($q \geq 0$) and averaging, we obtain that $\ell^{3q}\langle(\varepsilon_\ell)^q\rangle \propto \ell^{3q+\tau_q}$ is a nondecreasing function of ℓ. For $q \leq 0$, it should be a nonincreasing function of ℓ. This implies (8.124). Note that, if we had started from a definition of ε involving one- rather than three-dimensional averages, as is the case in the experimental measurements of dissipation fluctuation, we would have obtained instead of (8.124):

$$\tau_q + q \geq 0 \quad \text{for } q \geq 0 \qquad \text{and } \tau_q + q \leq 0 \quad \text{for } q \leq 0. \tag{8.125}$$

The reason why the lognormal model and many other random cascade models violate the Novikov inequality is the nonconservative character of the cascade as defined in Section 8.6.3 (Mandelbrot 1972). Furthermore, Orszag (1970) pointed out that the lognormal distribution is not uniquely defined by its moments.

Another property of the lognormal model is that the dimension $F(\alpha)$, predicted by (8.120), becomes *negative* for

$$\alpha > 1 + \frac{\mu}{2} + (6\mu)^{1/2}. \tag{8.126}$$

As noticed by Mandelbrot (1990, 1991), this is *not a shortcoming* of the lognormal model. Indeed, as already pointed out in Section 8.5.1, a negative dimension $F(\alpha)$ just means that the codimension $3 - F(\alpha)$ is greater than 3 and that the probability p_ℓ of encountering the corresponding scaling goes to zero with ℓ faster than ℓ^3.

Finally, we observe that the lognormal law is sometimes derived in an incorrect way. If the individual variables m_i in (8.113) are not Gaussian, the fact that there are *many* terms in the sum of the m_is does not justify using a Gaussian approximation. This would amount to replacing the Cramér function by just the first two terms in the Taylor expansion around the mean value of the m_is.[27] The Gaussian approximation is justified only when the sum of the m_is (minus its mean) has been divided by the square root of the number of terms. A mere product of a large number of independently and equally distributed positive random variables does not have an approximately lognormal distribution.

[27] In this particular respect the otherwise very useful and readable books by Lumley (1970) and Papoulis (1991) are somewhat misleading.

8.7 Shell models

Random cascade models give rise to nontrivial scaling laws which almost unavoidably display intermittency, but such models have only the remotest contact with the original Navier–Stokes equation. In contrast, shell models are deterministic, may or may not display chaos and intermittency and have a bit more of a Navier–Stokes flavor. The simplest shell model, of Desnyansky and Novikov (1974), is governed by the following infinite set of coupled nonlinear ordinary differential equations, labeled by an index $n = 0, 1, 2\ldots$ (called the shell index):

$$(d/dt + \nu k_n^2)u_n = \alpha\left(k_n u_{n-1}^2 - k_{n+1}u_n u_{n+1}\right) + \beta\left(k_n u_{n-1}u_n - k_{n+1}u_{n+1}^2\right) + f_n . \quad (8.127)$$

Here, the dynamical variables $u_n(t)$ are real numbers; u_{-1} which appears in the equation for $n = 0$ is taken equal to zero, the forcing term f_n is prescribed, time-independent and restricted usually to a single shell: $f_n = f\delta_{n,n_0}$ where δ_{n,n_0} is a Kronecker delta. The 'wavenumbers' k_n are given by

$$k_n = k_0 2^n, \quad (8.128)$$

where $k_0 > 0$ is a reference wavenumber. The ratio of a factor 2 between the wavenumbers of successive shells is an arbitrary choice, easily modified. The parameter $\nu > 0$ is a viscosity and the control parameters α and β are real numbers, one of which may vanish.

It is immediately seen that this model (and all other shell models to be discussed subsequently) has a number of properties in common with the Navier–Stokes equation. The nonlinear term is quadratic and has dimension [velocity]^2/[length]; it conserves the energy[28] $(1/2)\sum_n u_n^2$. It is invariant under time translations and under a discrete form of scaling transformations: when $f = \nu = 0$ and the 'boundary condition' for u_{-1} is ignored, the equations are invariant under

$$n \to n+1, \quad u_n \to (1/2)^h u_n, \quad t \to (1/2)^{1-h}t, \quad (8.129)$$

where h is an arbitrary scaling exponent. Finally, the shell model has exact *static* (time-independent) K41 solutions: at those scales where the force and the viscosity are negligible, it is readily seen that the nonlinear term vanishes when

$$u_n = Ck_n^{-1/3}, \quad (8.130)$$

for arbitrary C.

[28] In two dimensions, enstrophy conservation is also required; this can lead to unexpected effects (Aurell *et al.* 1994).

Shell models may also be viewed as severe truncations of the Navier–Stokes equation, retaining only one or a few modes as representatives of the (O (k_n^3)) Fourier modes in the nth octave of wavenumbers. In order to mimic the supposed 'local' (in scale) character of nonlinear interactions, only couplings to the nearest or next-nearest shells are usually kept.

Shell models originated with Lorenz (1972) and the Russian school (Desnyansky and Novikov 1974; Gledzer 1973 and the review in Gledzer, Dolzhansky and Obukhov 1981). Interest in shell models grew considerably when it was found that some of them can have a chaotic time-dependence and display intermittency with non-K41 scaling exponents (Pouquet, Gloaguen, Léorat and Grappin 1984; Ohkitani and Yamada 1989). Of particular interest is the Gledzer–Ohkitani–Yamada (GOY) model which has one complex mode per shell and next-nearest shell interactions and may be viewed as a complex version of a model introduced by Gledzer (1973). The governing equations are

$$(d/dt + \nu k_n^2)u_n = i\,(a_n u_{n+1}u_{n+2} + b_n u_{n-1}u_{n+1} + c_n u_{n-1}u_{n-2})^* + f_n\,, \quad (8.131)$$

where the star denotes complex conjugation,

$$a_n = k_n, \quad b_n = -k_{n-2}, \quad c_n = -k_{n-3}, \quad k_n = k_0 2^n, \quad (8.132)$$

and the force is applied to the fourth shell:

$$f_n = f\delta_{4,n}. \quad (8.133)$$

Ohkitani and Yamada (1989) gave qualitative and quantitative numerical evidence that the GOY model is chaotic; they also measured various statistical quantities indicating that the model is intermittent. Jensen, Paladin and Vulpiani (1991) calculated the 'structure functions', defined for shell models as

$$S_p(n) = \langle |u_n|^p \rangle \quad (8.134)$$

and found that: (i) at inertial-range scales,

$$S_p(n) \propto k_n^{-\zeta_p}; \quad (8.135)$$

(ii) the exponents ζ_p have a nontrivial dependence on the order p, suggesting *multifractality*. This multifractal behavior is indeed supported by further simulations (Pisarenko, Biferale, Courvoisier, Frisch and Vergassola 1993), involving up to 250×10^6 time steps. Figs. 8.18 and 8.19, taken from this reference show plots of the structure functions of order 2–10 (from top to bottom) and of their exponents, respectively. The

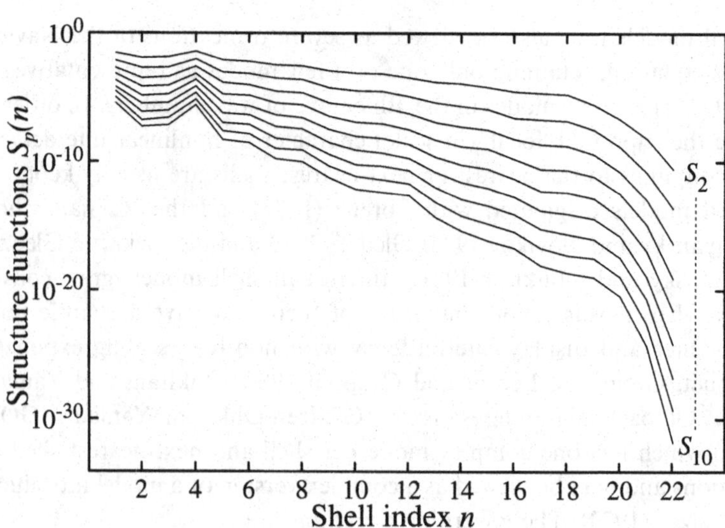

Fig. 8.18. Structure functions of order 2–10 (top to bottom) for the GOY model. The shell index n is $\log_2$ wavenumber. The data are averaged over 250×10^6 time steps.

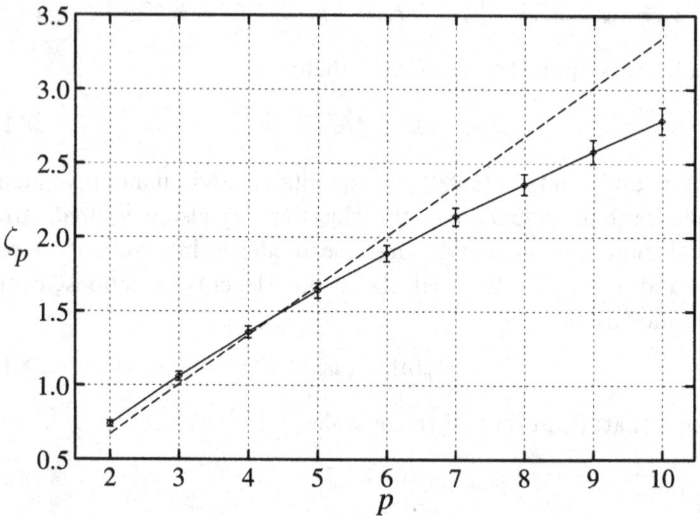

Fig. 8.19. Exponent ζ_p for the structure function vs order p. The dashed line corresponds to K41 ($\zeta_p = p/3$). The error bars are from a least squares fit of the structure functions shown in Fig. 8.18 to a power-law in the interval $6 \leq n \leq 18$.

calculations used 22 shells. The parameters were

$$f = (1+i) \times 5 \times 10^{-3}, \quad k_0 = 2^{-4}, \quad \nu = 10^{-7}. \tag{8.136}$$

The averaging was over a sufficiently long time span (about 5000 eddy turnover times at the forcing scale) to ensure that the data were essentially free of noise. (For example the oscillations seen in Fig. 8.18 are genuine.) The error bars shown in Fig. 8.19 were generated from a least squares fit of the logarithms of the structure functions to power-laws in the inertial range (from shell $n = 6$ to shell $n = 18$).[29]

Intermittency in shell models is the subject of much current work and is quite poorly understood at the moment. A few salient results will now be mentioned. All shell models possess 'pulse solutions' of the form (Siggia 1978):

$$u_n(t) = k_n^{-h} g_h \left[k_n^{1-h}(t - t_\star)\right], \tag{8.137}$$

where $t_\star$ and the scaling exponent $h < 1$ are arbitrary and where the functions $g_h(\cdot)$ are obtained by substitution into the dynamical equations (ignoring forcing and dissipation terms). When the function $g_h(\cdot)$ is localized, i.e. decreases rapidly to zero at small and large arguments, the temporal evolution of $u_n(t)$ has the form of a pulse of width $O\left(k_n^{h-1}\right)$ centered around the time $t_\star$. If well-separated pulses are generated by an instability at random times $t_\star$, it follows that the structure function of order p scales with an exponent $\zeta_p = 1 + (p-1)h$. Indeed, there are p factors k_n^{-h} (the prefactor in (8.137)) and a probability $\propto k_n^{1-h}$ of being within a pulse in the shell number n. To obtain a finite non-vanishing energy flux, one must take $h = 0$; then $\zeta_p = 1$, for all p, an extreme form of intermittency, similar to that obtained for Burgers' model (see Section 8.5.2). An attempt to develop a statistical mechanics of interacting pulses with different hs has been made by Parisi (1990).

A very different approach to intermittency in the GOY model is due to Benzi, Biferale and Parisi (1993). They constructed a multiplicative process of the sort discussed in Section 8.6.3 and required it to be consistent with certain exact moment relations derived from the GOY model. They were able to satisfy simultaneously to a good accuracy a number of constraints in excess of the number of free parameters in their multiplicative model and to compute analytically from this a graph of ζ_p which is consistent with the one presented in Fig. 8.19.

[29] Gat, Procaccia and Zeitak (1994) have suggested that discrepancies from pure K41 scaling in the GOY model are caused by the presence of a second quadratic invariant, in addition to the energy, which was identified by Kadanoff, Lohse, Wang and Benzi (1995).

One aspect of shell models which can be investigated systematically is the stability of K41 solutions and the bifurcations away from K41 solutions. Since K41 solutions emerge only in the limit $\nu \to 0$, the control parameter must be other than the viscosity, for example the ratio β/α in the Desnyansky–Novikov model (8.127), or a similar parameter which can be introduced into the GOY model (Biferale, Lambert, Lima and Paladin 1995). In addition to the technical problems in studying linear stability, there is a conceptual difficulty: an intermittent solution cannot be uniformly 'close' to K41 and thus amenable to small perturbation techniques. Some kind of nonperturbative method may thus be needed.

8.8 Historical remarks on fractal intermittency models

In 1961, at the *Colloque International de Mécanique de la Turbulence* in Marseille, Kolmogorov (1962) presented his theory of intermittency, often referred to as K62. Landau is given considerable credit:

> *But quite soon after they* (the K41 hypotheses) *originated, Landau noticed that they did not take into account a circumstance which arises directly from the assumption of the essentially accidental and random character of the mechanism of transfer of energy from the coarser vortices to the finer: with increase of the ratio ℓ_0/ℓ, the variation of the dissipation of energy*
>
> $$\varepsilon = \frac{\nu}{2}\sum_{\alpha}\sum_{\beta}\left(\frac{\partial u_\alpha}{\partial x_\beta} + \frac{\partial u_\beta}{\partial x_\alpha}\right)^2 \qquad (8.138)$$
>
> *should increase without limit.*

Curiously, nowhere in Landau's remarks (as quoted in Section 6.4), do we find any reference to 'fine' scales. However, it has become a tradition to accept Kolmogorov's view crediting Landau (see, e.g., the discussion in Section 25.1 of Monin and Yaglom 1975). As we have shown, Landau's remarks in no way imply that the K41 theory (in its scale-invariant version) is inconsistent.

Actually, a reference to Landau's remark seems to have been introduced at an advanced stage of the writing of the K62 paper, after Yaglom drew his attention to the possible relevance of Landau's footnote remark (A. Yaglom, private communication). The context of Kolmogorov's work on intermittency was the following. There had been experimental work by Gurvitch (1960) at the Institute of Atmospheric Physics in Moscow[30]

[30] Founded by Obukhov as an extension of a laboratory headed by Kolmogorov.

which indicated a certain variability of the turbulence intensity, a variability which Obukhov (1962) pointed out 'may be explained by variance of the dissipation rate ε'.

Following Obukhov (1962), Kolmogorov (1962)[31] considered fluctuations of the local space averages of the dissipation $\varepsilon_\ell(r)$, defined in (8.78), and stated (in our notation) that:

...it is natural to assume that when $\ell_0/\ell \gg 1$ the dispersion of the logarithm ... has the asymptotic behavior

$$\sigma_\ell^2 = A + \mu \ln(\ell_0/\ell), \tag{8.139}$$

where μ is some universal constant.

This is precisely the lognormal model already discussed in Section 8.6.5.

Kolmogorov had been interested in the lognormal law in 1941 when he proposed an interpretation for the approximate lognormality of the size distribution in pulverization of mineral ore (Kolmogorov 1941d). He described this process essentially as a multiplicative random cascade, the similarity of which to the Richardson cascade must have been obvious to him or became so at some point. Obukhov (1962) is the first to make the lognormal assumption for the fluctuations of the dissipation. He provides no particular justification, but quotes Kolmogorov's (1941d) 'lognormal' paper.

Neither Kolmogorov nor Obukhov have tried to justify the lognormal model for turbulence by invoking a multiplicative random cascade. This would indeed have been an incorrect application of the theory of large deviations (Section 8.6.4).[32] Actually, Kolmogorov (1962) formulates a *third hypothesis*, which is more or less equivalent to postulating a multiplicative process with independent factors, and then states:

Naturally the formulation of this hypothesis must be refined if mathematical rigor is desired and if it is to be used to derive the logarithmic normality of the distribution of velocity differences ...

Anyway, the lognormal model presents the two known inconsistencies discussed in Section 8.6.5. It violates the Novikov (1970) inequality and implies supersonic velocities at very high Reynolds numbers since ζ_p is not a monotonic function of p.

[31] The french version of the paper (Kolmogorov 1961) begins with a few lines in which he gives most of the credit to A.M. Obukhov.

[32] Large deviations were not considered in Kolmogorov's (1941d) 'lognormal' paper; thus, he would not have been able to predict correctly moments of the size distribution, had he tried.

It is paradoxical that despite all the aforementioned difficulties with Kolmogorov's 1962 paper, it nevertheless led to many fruitful further developments. Theoretical developments were mostly concerned with random cascade models of the sort considered in Section 8.6.3. The black and white model of Novikov and Stewart (1964) was extended by Yaglom (1966), who defined the general class of random cascade models. More detailed information on the contribution of the Russian school to this question may be found in Monin and Yaglom (1971, 1975).

In the late 1960s Mandelbrot (1968, 1974b) observed that random cascade models, when continued indefinitely (no finite viscous cutoff), lead to an energy dissipation generally concentrated on a set of non-integer (fractal) Hausdorff dimension. He also stressed that the relation between the 'dimension' of the dissipation and the correction to the K41 spectrum is different for the black and white case and for the general case (called by him 'weighted curdling'). The intrusion of what was considered at that time unnecessary exotic mathematics met with some resistance, but the author of this book became one active propagandist.[33]

In one respect the random cascade models appeared contrived: they were intended to predict scaling laws at *inertial-range* scales, yet they worked with the *dissipation* which is not an inertial-range quantity, as pointed out by Kraichnan (1974). He also gave a hint as to how this difficulty could be overcome (Kraichnan 1972, p. 213). A reformulation of the Novikov–Stewart model, the β-model of Section 8.5.1, using only inertial-range quantities (velocity increments and energy fluxes), was proposed by Frisch, Sulem and Nelkin (1978). This model became perhaps excessively popular. Indeed, the β-model was intended to be a minimally complex toy model and not a predictive model.

A few years later, Anselmet, Gagne, Hopfinger and Antonia (1984) obtained experimental data on high order structure functions of significantly better quality than previously feasible (see Section 8.3). Not surprisingly, the values of the exponents ζ_p agreed neither with the β-model nor with the lognormal model. The curvature they observed in the graph of ζ_p was interpreted by Parisi and Frisch (1985) in terms of the multifractal model of Section 8.5.3. Although the manifold of possible scaling laws derived in multifractal models is the same as in the random cascade models, the use of a Legendre transformation in the Parisi–Frisch approach made manifest the *multi*fractal character of such models, which was hardly

[33] Actually, as pointed out by Mandelbrot (1977), Richardson had pioneered the use of fractals in studying convoluted coastlines; also Lorenz (1963) had shown that they arise in simple nonlinear dynamical systems.

appreciated before.[34] Benzi, Paladin, Parisi and Vulpiani (1984) then constructed a particular instance of the multifractal model, the random β-model and also proposed using multifractals to describe the invariant measures of *chaotic dynamical systems.*

This problem differs from the one considered by Parisi and Frisch in several respects. The quantity with multifractal properties is not an (unsigned) function (the velocity field) but a positive measure, the invariant measure of a dynamical system. In this respect it is more like studying the multifractality of the dissipation. However, the dissipation is a random measure residing in the physical space and possessing the same symmetries as the turbulence (translations, rotations, etc.), whereas the invariant measure resides in the phase space, is deterministic and usually does not possess any symmetries. Grassberger (1983) and Hentschel and Procaccia (1983) have shown that the infinite set of Renyi dimensions D_q (Section 8.6.1) can be used to characterize the fine-scale properties of invariant measures. Halsey, Jensen, Kadanoff, Procaccia and Shraiman (1986) then developed a formalism for dynamical systems incorporating the same Legendre transform approach as in Parisi and Frisch, which was then also called 'multifractal'. Meneveau and Sreenivasan (1987, 1991; see also references therein) carried the formalism of Halsey, Jensen, Kadanoff, Procaccia and Shraiman back to turbulence and studied the multifractal properties of the dissipation (Section 8.6.1). Wu, Kadanoff, Libchaber and Sano (1990), while studying the temporal spectrum of temperature fluctuations in thermal convection, found that the data for different high Rayleigh numbers would not easily collapse onto a single curve (after shifts) when using log–log coordinates. Much better collapse was observed if they first divided the logarithms of both the temperature spectrum and the frequency by the logarithm of the Rayleigh number, a procedure which they (rightly) called 'multifractal'. Frisch and Vergassola (1991) then found that the Parisi–Frisch multifractal model of fully developed turbulence predicts precisely this kind of 'multifractal universality' when finite-viscosity effects are included (Section 8.5.5). Nelkin (1990) investigated the consequences of the multifractal model for moments of velocity-derivatives (Section 8.5.6).

A much deeper understanding of multifractality was achieved when it was realized how random multifractal measures are related to the theory of large deviations discussed in Section 8.6.4 (Mandelbrot 1989, 1991;

[34] Polyakov (1972) studied the scale-invariance of lepton–hadron interactions using a kind of multifractal formalism.

Oono 1989; Collet and Koukiou 1992).[35] This led to the probabilistic definition of multifractality discussed in Section 8.5.4, the velocity-analog of Mandelbrot's (1991) *Cramér renormalization.* One advantage is that one need not work with actual singularities. Concepts such as *negative dimensions* (Mandelbrot 1990, 1991; see also Meneveau and Sreenivasan 1991) thereby lose their magical character. The connection with large-deviations theory also makes clear what is wrong with the lognormal assumption (Section 8.6.5).

Finally, we mention that multifractals appear in many areas of theoretical physics (see, e.g., Paladin and Vulpiani 1987b) and that there is now considerable interest in multifractal measures and multifractal sets among mathematicians and theoretical physicists (Kahane and Peyrière 1976; Kahane 1993; Hentschel 1994; Waymire and Williams 1994; Blank 1995; see also references therein).

8.9 Trends in intermittency research

Half a century after Kolmogorov's work on the statistical theory of fully developed turbulence, we still wonder how his work can be reconciled with Leonardo's half a millennium old drawings of eddy motion in the study for the elimination of rapids in the river Arno (the example given in Fig. 8.20 is of particular interest in the context of Section 8.9.1).[36] Indeed, Kolmogorov's work on the fine-scale properties ignores any structure which may be present in the flow. In Section 7.4 we pointed out that many turbulent flows are known to possess 'coherent structures'. Their (re)discovery by Crow and Champagne (1971) and Brown and Roshko (1974) has led to questioning the relevance of the traditional statistical theory of turbulence. However, as far as inertial-range properties are concerned, coherent structures do not matter if they are confined to the large scales of the flow. But, is this really the case?

Traditional ways of analyzing turbulent flows, based on the velocity signal recorded by a probe, do not reveal any small-scale structures. Indeed, by visual inspection it is almost impossible to distinguish the properties at inertial-range scales of the kind of signal shown in Figs. 3.1(a) and (b) from that of a Gaussian noise with a $k^{-5/3}$ spectrum. However, the discrepancies from K41 scaling have led to the suggestion that the small

[35] In earlier work Mandelbrot (1974a) had already used large-deviations theory and introduced an $f(\alpha)$ function without interpreting it as a dimension.

[36] Lumley (1992) points out that Leonardo was probably aware of the need to describe turbulent flow as a mixture of coherent and random motion.

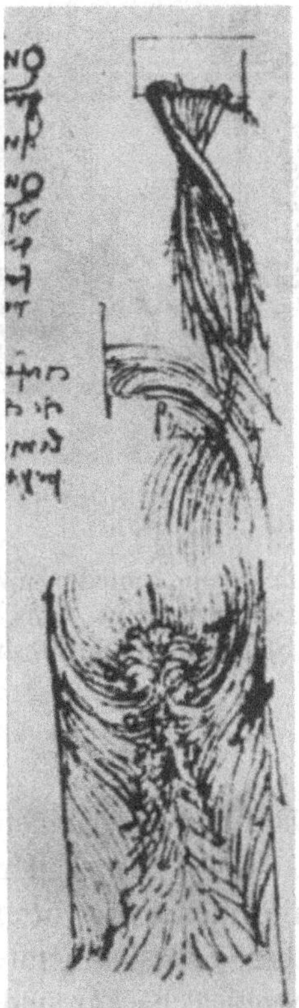

Fig. 8.20. Leonardo da Vinci: Study of liquid filaments of jets discharging from orifice (Ms. F, fo. 47 v., Bibliothèque de l'Institut de France, Paris).

scales have fractal or multifractal properties, as discussed earlier in this chapter. We have seen in Section 8.5.4 that multifractal scaling, in the probabilistic sense, does not require any fractal structures to be present in *individual* realizations. The fractal/multifractal description of turbulence is essentially of probabilistic nature and does not assume much about the *geometry* of fine scales.

It is remarkable that what may be called Leonardo's vision of turbulence is in the focus of current research on intermittency: there is growing evidence that there are structures with nontrivial geometry down to very fine scales, maybe of the order of the Kolmogorov scale.

Recent results indeed suggest that the fine scales of turbulent flow include a tangle of very intense and slender *vortex filaments*, a property which is already well established for *superfluid turbulence* (Donnelly and

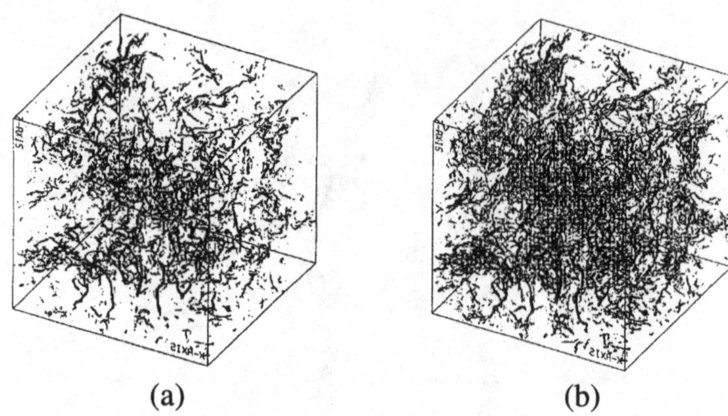

Fig. 8.21. View of the vorticity field, represented by a vector of length proportional to the vorticity amplitude at each grid point. Only vectors larger than a given threshold value are shown; (b) has a lower threshold than (a) (Vincent and Meneguzzi 1991).

Swanson 1986; Schwarz 1988) and for magnetic flux tubes in the solar photosphere (Stenflo 1973, 1994). We shall now review some of the numerical and experimental evidence for vortex filaments and attempts to incorporate them and other vortical structures in a statistical description of turbulence. For theoretical background on vortex filaments, see Saffman (1992).

Figs. 8.21(a) and (b) give a striking example of the tangle of vortex filaments obtained in turbulent flow at moderately high Reynolds number ($R_\lambda \approx 150$). The flow was simulated numerically using 240^3 grid points. A snapshot of the vorticity field is represented. The vorticity is shown only when its modulus exceeds a given threshold (chosen higher in (a) than in (b)). Similar figures, suggesting the proliferation of vortex filaments, may be found in Hosokawa and Yamamoto (1990) and She, Jackson and Orszag (1990; see also Fig. 7.4). The earliest simulation revealing such filaments seems to be by Siggia (1981; see also Kerr 1985). Additional information on the structure of individual vortex filaments comes from simulations of the Taylor–Green vortex (see Section 5.2) by Brachet *et al.* (1983) and Brachet (1990, 1991). Such simulations indicate that the vortex filaments are regions of high vorticity and low dissipation and thus, by (2.32), regions of low pressure.

Douady, Couder and Brachet (1991) took advantage of this property to make direct experimental observations of vortex filaments using a turbulent liquid seeded with bubbles (which migrate to the regions of low pressure).[37] Fig. 8.22, taken with a video camera monitoring the flow, shows an example of what is seen in such experiments. Further work was then done by Douady and Couder (1993) and by Bonn, Couder, van Dam and Douady (1993). The experiments, done *after* the simulations, revealed much about the birth and death of vortex filaments and their possible role in organizing the large-scale flow.

Concerning the vortex filaments, the picture emerging from the combined numerical and experimental work is as follows. The filaments are actually tubes with an approximately circular cross-section; their diameter is of the order of the Kolmogorov dissipation scale and their length is somewhere between the Taylor-scale λ and the integral scale ℓ_0. This length is, however, hard to define precisely. Indeed, when a vortex tube opens up, the strength of the vorticity decreases (so as to conserve the circulation) and may fall below the threshold for visualization; this phenomenon is easily observed when comparing Figs. 8.21(a) and (b). Jiménez (1993) observes that the vortex-tube Reynolds number $R_\gamma = \gamma/\nu$, based on the circulation γ of the tubes, is typically in the range 150–400. (He finds similar numbers for compact vortices observed in simulations of homogeneous shear flow, channel flow and plane mixing layers.) As for the dynamics of the filaments, Bonn, Couder, van Dam and Douady (1993) observe the following: 'The filaments appear abruptly by the rolling-up of thin layers where both shearing and stretching coexist ...They are unstable and undergo vortex breakdown (Leibovich 1978) by formation of helical distortions. The longest filaments are then observed to be transformed into large, long-lived eddies.'

It is then natural to ask the following questions:

- How are vortex filaments formed, what is their structure and what role do they play in the overall dynamics of the flow?
- What is the signature of vortex filaments in the statistical properties of the flow?

8.9.1 Vortex filaments: the sinews of turbulence?

Concerning the formation of vortex filaments, Bonn, Couder, van Dam and Douady observe that they appear wherever a large structure has created a thin layer with both shearing and stretching. The filaments which

[37] Fauve, Laroche and Castaing (1993) showed that low-pressure events can be used to detect vortex filaments.

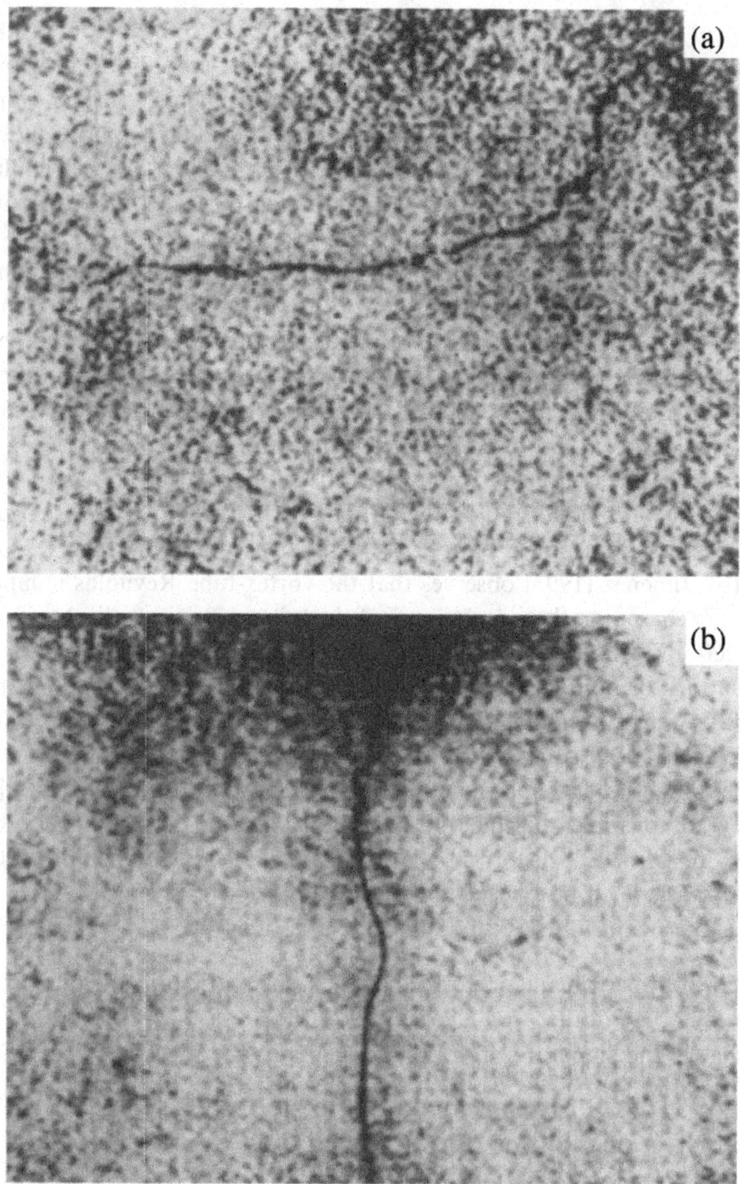

Fig. 8.22. Two images of high concentrations of vorticity obtained in water seeded with small bubbles for visualization. The tank is lit with diffusive light from behind: the bubbles appear dark on a light background: (a) a vorticity filament in a turbulent flow (exposure time 0.001 s); (b) the core of an axial vortex below a rotating disk (Bonn, Couder, van Dam and Douady 1993).

are seen in the simulations of Brachet, Meneguzzi, Vincent, Politano and Sulem (1992) and Passot, Politano, Sulem, Angilella and Meneguzzi (1995) originate from instabilities of pancake structures similar to those present in Fig. 7.7. A modulational perturbation analysis by Passot, Politano, Sulem, Angilella and Meneguzzi (1995), extending previous work by Neu (1984) and Lin and Corcos (1984) on braids in turbulent mixing layers, explained the formation of vortex filaments by the focusing of the vorticity within vortex layers with a finite (small) thickness. Jiménez (1993) proposed a mechanism of axial straining of an otherwise two-dimensional flow (see also Jiménez, Wray, Saffman and Rogallo 1993). Lundgren (1982, 1993; see also Pullin and Saffman 1993) has indeed observed that the strained flow is related to a fictitious unstrained two-dimensional flow by a simple scaling transformation. Where Lundgren and Jiménez differ is that the former stresses the spiral character of two-dimensional flow (needed to produce a $k^{-5/3}$ spectrum[38]), whereas the latter stresses the tendency of two-dimensional flow to form compact coherent vortices (see Section 9.7). It is also possible that some of the vortex filaments are born near solid boundaries, the only places where vorticity can be generated in an incompressible flow in the absence of buoyancy effects.

Clearly, more remains to be done to understand the formation of vortex filaments.

The internal structure of the vortex filaments can be understood, following Moffatt, Kida and Ohkitani (1994), by studying steady-state two-dimensional solutions of the Navier–Stokes equation with the vorticity of the form $(0, 0, \omega(x_1, x_2))$ in the presence of a three-dimensional large-scale straining field $\boldsymbol{U} = (\alpha x_1, \beta x_2, \gamma x_3)$, with $\gamma > 0$ and $\alpha + \beta + \gamma = 0$. The simplest case is obtained when $\alpha = \beta$ (axial strain). It is then easily checked that an exact solution with vanishing nonlinearity is the Burgers vortex (Burgers 1948; Townsend 1951; Saffman 1992; Moffatt 1994):

$$\omega(x_1, x_2) = \frac{\gamma\Gamma}{4\pi\nu} \exp\left[-\frac{\gamma(x_1^2 + x_2^2)}{4\nu}\right], \tag{8.140}$$

where Γ is the total circulation associated with the vortex. As soon as $\alpha \neq \beta$, nonlinearities reappear and solutions must be found perturbatively. An interesting feature of such nonaxisymmetric solutions, also observed by Brachet (1990) and Kida and Ohkitani (1992), is that vortic-

38 A spiral model with a $k^{-5/3}$ spectrum has also been proposed by Moffatt (1984).

ity contours remain approximately circular, whereas dissipation can be strongly anisotropic.

As to the dynamical role of vortex filaments a very appealing picture, albeit somewhat tentative, is the suggestion by Moffatt, Kida and Ohkitani (1994) that the vortex filaments are the 'sinews' of turbulence: 'Just as sinews serve to connect a muscle with a bone or other structure, so the concentrated vortices of turbulence serve to connect large eddies of much weaker vorticity; and just as sinews can take the stress and strain of muscular effort, so the concentrated vortices can accommodate the stress associated with the low pressure in their cores and the stress imposed by relative motion of the eddies into which they must merge at their ends.'

8.9.2 Statistical signature of vortex filaments: dog or tail?

Having identified 'simple' geometric objects, the vortex filaments, in turbulent flows, it is natural to ask if any of the known statistical properties of turbulence can be thus explained. Are the vortex filaments the *dog* or the *tail*? In the former case, they would be essential to explain the energetics and the scaling properties of high-Reynolds-number flow. In the latter case, they would have only marginal signatures, for example on the tails of p.d.f.s of various small-scale quantities and on the exponents ζ_p for large *p*s.

Before discussing various (speculative) dog theories, let us state some objections. In the opinion of Brachet (1990, 1991) and Bonn, Couder, van Dam and Douady (1993), the vortex filaments cannot be the dog insofar as they contain only a small fraction of the total dissipation of energy. This is supported by the numerical study of Jiménez, Wray, Saffman and Rogallo (1993) who have reconstructed faked velocity fields from (lower) truncated vorticity fields in which the vorticity is set equal to zero unless its magnitude exceeds a given fraction of the maximum vorticity, these truncated fields being then projected onto divergenceless vector fields. High thresholds select but the strongest vortex filaments. There is experimental evidence — albeit controversial — that vortex filaments may become unimportant at high Reynolds numbers: Abry, Fauve, Flandrin and Laroche (1994) observe that the frequency of occurrence of vortex filaments (in units of the inverse large eddy turnover time) decreases with R_λ; Belin, Maurer, Tabeling and Willaime (1995) find that, around $R_\lambda = 700$, the filaments may undergo a metamorphosis into less structured objects.

A major advocate of the dog theory has been Chorin (1988, 1990, 1994; Chorin and Akao 1991 and references therein). He assumes a collection of vortex tubes which have been stretched and folded into some complicated vortex tangle. He then introduces an analogy with a polymer, considered as a self-avoiding random walk. (Conservation of helicity is used to justify the nonintersecting character of vortex lines.) In an ordinary random walk, the typical distance traveled after N steps goes as $N^{1/2}$, while in a self-avoiding walk it goes as N^{μ} with $\mu = 3/5$ (de Gennes 1971). Equivalently, the vortex tangle has dimension $D = 1/\mu$. If, following Chorin, we now consider D to be a correlation dimension (in the sense of Grassberger and Procaccia 1984) and ignore the vector character of the vorticity, we obtain, from (8.83) that the vorticity correlation function scales as $\ell^{D-3} = \ell^{-4/3}$. This is equivalent to Kolmogorov's $\ell^{2/3}$ law for the second order structure function, here obtained from a highly intermittent model! Actually, Chorin points out that: (i) the energy constraint forces the vortices to fold and thus raises the dimension of the 'polymer' above 2, (ii) the vector character of the vorticity and the folding permit considerable cancellation, thereby invalidating the previous calculation. From there on, he has to resort to numerical calculation which will not be described here.

Passot, Politano, Sulem, Angilella and Meneguzzi (1995) have tried to calculate the contribution of the vortex filaments to the energy spectrum. Their argument, reexpressed in the language of the multifractal model, is as follows. Each vortex filament produces, by the Biot–Savart law, a $1/r$ velocity field which thus has scaling exponent $h = -1$. It is assumed that the vortex filaments are simple lines with dimension $D = 1$. We thus obtain a contribution $\propto \ell^{ph+3-D} = \ell^{2-p}$ to the structure function of order p (cf. (8.40)). The exponent $\zeta_p = 2 - p$ vanishes for $p = 2$, yielding a k^{-1} contribution to the energy spectrum (Townsend 1951). This is shallower than the K41 spectrum, but there may be a viscosity-dependent constant in front of the k^{-1} contribution, making this term relevant only at sufficiently small scales,[39] as Passot, Politano, Sulem, Angilella and Meneguzzi (1995) suggest. Our form of the argument, when applied to the structure function of order p, suggests that, at very small scales there may be a contribution $\propto \ell^{2-p}$, which overwhelms the $\ell^{p/3}$ K41 contribution, particularly at large ps. One could then argue that in trying to fit a single power-law with exponent ζ_p to this composite

[39] In this sense Passot, Politano, Sulem, Angilella and Meneguzzi (1995) is a 'tail theory'.

function, ζ_p will be underestimated, thereby leading to a 'multifractal artifact'.[40]

If multifractality is not an artifact, can we explain the measured values of the exponents ζ_p as reported in Section 8.3? A signature of the vortex filaments may be present in the numerical results of Vincent and Meneguzzi (1991) for *high-order* structure functions. We have already stressed in Section 8.3 that such structure functions are very difficult to measure accurately. Still, if we accept the exponents ζ_p of order 14–30, given in their Figure 10, we notice that they fall approximately on a straight line $\zeta_p = ap + b$ with $a \approx 0.9$ and $b \approx 2$. According to the multifractal model, a is the minimum scaling exponent $h_{\min}$ and b is the codimension $3 - D(h_{\min})$. Hence, $D(h_{\min}) \approx 1$, the dimension of filaments.

Another signature of vortex filaments may be found in the flatness of the velocity derivative: a collection of vortex filaments with a finite core, assumed to be approximately straight and independent, can produce a large flatness but gives a vanishing contribution to odd order moments such as the skewness.

More ambitiously, is it possible to explain the *whole set* of exponents ζ_p in terms of vortex filaments?

She and Lévêque (1994) proposed a phenomenology involving a hierarchy of fluctuation structures associated with the vortex filaments. From this they derive a relation with no adjustable parameters, namely:

$$\zeta_p = p/9 + 2 - 2(2/3)^{p/3}, \tag{8.141}$$

which is in remarkable agreement with experimental results processed by ESS (Benzi, Ciliberto, Tripiccione *et al.* 1993), as discussed in Section 8.3. The corresponding τ_q exponents, given by (8.88), are consistent with Novikov's inequality (8.124) for both positive and negative qs. The slope of ζ_p for large p is $1/9$. Hence, the most singular (i.e. smallest) scaling exponent is $h_{\min} = 1/9 \approx 0.11$. Note that Gagne's (1987) experimental data suggest an asymptotically linear behavior of ζ_p with an exponent $h_{\min} \approx 0.18$. Dubrulle (1994) observed that (8.141) corresponds to a log–Poisson distribution: in the notation of Section 8.6.4, the variables m_i have a Poisson distribution. Dubrulle also noted that (8.141) may be obtained as a suitable limit of the random β-model of Benzi, Paladin, Parisi and Vulpiani (1984). She and Waymire (1995) independently noted the log–Poisson property. A variable with a Poisson distribution is an instance of an *infinitely divisible random variable*, i.e. a variable which

[40] This argument is easily adapted to the case in which the vortex filaments form a fractal tangle of dimension D (change $\propto \ell^{2-p}$ into $\propto \ell^{3-D-p}$).

can be written as the sum of an arbitrary number of independently and identically distributed variables (see, e.g., Lévy 1965; Feller 1968b). Using the Lévy-Khinchine representation of infinitely divisible laws, She and Waymire (1994) proposed a generalization of (8.141) which may be realized in a context different from incompressible turbulence.

Reconciling the statistical picture of multifractal turbulence with the geometric picture of vortex filaments may be easier if we resort to the *circulation*[41]

$$C_\ell = \oint_{\partial D_\ell} \boldsymbol{v}(\boldsymbol{r}) \cdot d\boldsymbol{r}, \tag{8.142}$$

rather than to velocity increments. Here, ∂D_ℓ is the boundary of a disk D_ℓ of radius ℓ (or a square of side ℓ). Instead of moments of velocity increments, we suggest using here moments of C_ℓ/ℓ. This quantity has the same dimension and maybe the same exponents ζ_p as velocity increments. It is conjectured that a single vortex filament (say, with unit circulation), if sufficiently convolved, can give rise to multifractal scaling of the circulation.

The circulation around ∂D_ℓ is, of course, equal to the flux of the vorticity through the disk D_ℓ. In contrast to the approach based on the dissipation, which uses three-dimensional integrals of a positive measure, one is here working with two-dimensional integrals of a *signed measure*. It is therefore possible to study aspects other than multifractality, such as the *cancellation index* which measures how much cancellation of opposite-sign measure takes place on very fine scales. The importance of this in studying magnetic field production by the dynamo effect has been recognized by Ott, Du, Sreenivasan, Juneja and Suri (1992) and Du and Ott (1993).

The calculation of the circulation (or of the vorticity flux) appears to be much simpler from numerical simulations than from experiments. There is, however, a magnetohydrodynamic variant of this problem which, in our opinion, can be studied via high-resolution observations of solar magnetic fields: one then uses, instead of the vorticity flux, the magnetic flux, obtained from the Zeeman effect (see, e.g., Stenflo 1994).

Whether or not the vortex filaments are the 'dog', there are a number of interesting research topics which should be explored, for example the *statistical mechanics of three-dimensional vortex filaments*. This is, of course, much more difficult than the corresponding two-dimensional problem

[41] Migdal (1994) has shown how to develop a functional formalism for turbulence in terms of the p.d.f. of the circulations around all possible contours.

with point vortices, pioneered by Onsager (1949; see also Section 9.7.2). Indeed, in three dimensions the individual entities, being strings, already have infinitely many degrees of freedom. One of the open issues is the possibility of Debye-type screening in three dimensions (Ruelle 1990).

8.9.3 The distribution of velocity increments

In this section on 'trends' in intermittency we have chosen to discuss mostly vortex filaments. Among the other topics of current interest, there is the study of probability densities (p.d.f.s) of velocity increments and velocity gradients at high Reynolds numbers. Measurements of p.d.f.s of increments and gradients are obtained from experimental and numerical data (Van Atta and Park 1972; Gagne, Hopfinger and Frisch 1990; Vincent and Meneguzzi 1991; Noullez, Wallace, Lempert, Miles and Frisch 1996). An example is given in Fig. 8.23.

The main features observed can be summarized as follows. For $\ell \sim \ell_0$, the integral scale, the p.d.f. of increments is essentially indistinguishable from a Gaussian. At inertial-range separations, the p.d.f. develops approximately exponential wings.[42] At even smaller scales, the p.d.f. takes the form of a 'stretched exponential', i.e. the exponential of (minus a constant times) a fractional power of the absolute value of the velocity increment; the exponent is less than unity, so that the p.d.f. decreases slower than exponentially.

The p.d.f. of the increment δv_ℓ for the smallest separation $\ell_{\rm min}$ available is generally used to determine the p.d.f. of the velocity gradient, assumed to be approximately given by $(\delta v_\ell/\ell)_{|\ell=\ell_{\rm min}}$. This may be difficult to circumvent, but it is a questionable procedure. Indeed, inspection of Fig. 8.23 reveals that, even when $\ell_{\rm min}$ is comparable to the Kolmogorov dissipation scale η, very large velocity increments (which can become comparable to v_0) occur with significant probabilities.[43] For $\ell \sim \eta$, the tail of the p.d.f. of velocity increments is thus not necessarily representative of the tail of the p.d.f. of velocity gradients. Furthermore, the present quality of data does not easily permit us to distinguish between (at least) two possibilities: (i) different functional forms for the p.d.f.s of increments, associated respectively with the energy range, the inertial range and the

[42] For longitudinal increments, the negative wing is higher than the positive wing, as dictated by Kolmogorov's four-fifths law (Section 6.2) which implies a negative skewness of the velocity increment.

[43] This phenomenon is much more marked for transverse than for longitudinal increments; transverse increments are indeed not constrained by the condition of incompressibility.

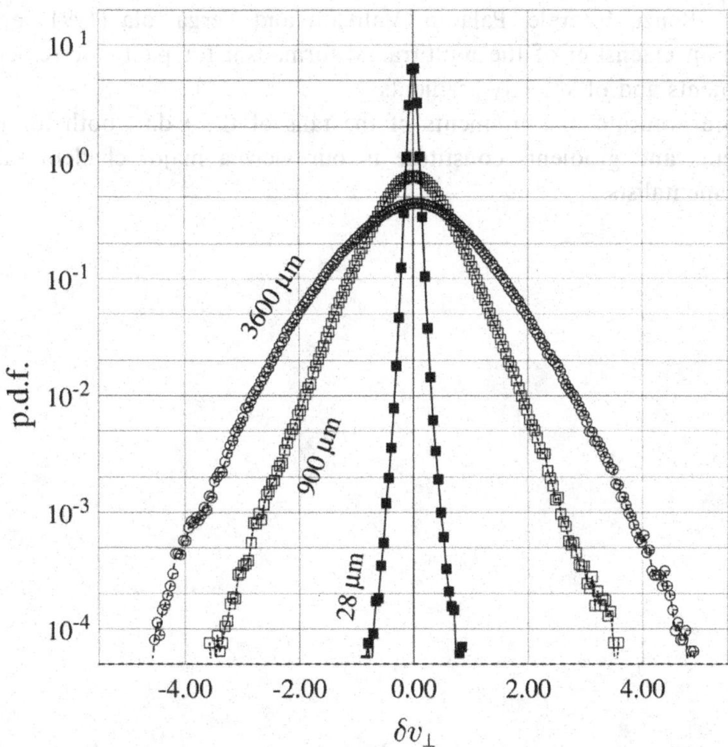

Fig. 8.23. Probability densities of transverse velocity increments obtained by the RELIEF flow tagging technique in a turbulent jet at $R_\lambda \approx 240$ for various separations ℓ, as labeled. The Kolmogorov dissipation scale is $\eta = 16\,\mu$m. The velocity increment scale is linear and in units of v_0, the r.m.s. transverse velocity fluctuation. Lower levels of p.d.f. (not shown) are too noisy to be significant (Noullez, Wallace, Lempert, Miles and Frisch 1996).

dissipation range and (ii) a functional form varying continuously with the scale.

Various theoretical interpretations have been proposed, mostly for the latter possibility. Castaing, Chabaud and Hébral's (1990) interpretation is based on a variational formulation for fully developed turbulence, introduced by Castaing (1989). She (1991; see also She and Orszag 1991; She, Jackson and Orszag 1991) proposed a nonlinear transformation relating fluctuations of the velocity gradient and fluctuations of the large-scale velocity. This is based on a picture of fully developed turbulence as a collection of weakly correlated random eddies and strongly correlated structured eddies. It also makes use of a 'mapping function' (Kraichnan

1990). Benzi, Biferale, Paladin, Vulpiani and Vergassola (1991) proposed an extension of the multifractal formalism for p.d.f.s of velocity increments and of velocity gradients.[44]

More accurate measurements of the tails of the p.d.f.s both for increments and gradients constitute in our view a major challenge for experimentalists.

[44] The multifractal model, in its probabilistic formulation of Section 8.5.4, puts some constraints on the p.d.f. of increments, but does not determine it in a unique way.

9

Further reading: a guided tour

9.1 Introduction

This chapter contains supplementary material beyond the scope of the previous chapters. The choice of topics is, of course, determined largely by the author's own interests and knowledge (or lack thereof). No attempt is made, for example, to discuss supersonic and magnetohydrodynamic turbulence or turbulent convection. As for numerical simulation of turbulence, one of the most important current tools in turbulence research, it defies being summarized and requires a book of its own. Some topics are presented rather briefly, either because there is no need to elaborate them (Section 9.2 on further reading in turbulence and fluid mechanics) or because very good reviews are easily found in the literature (Section 9.3 on mathematics and Section 9.4 on dynamical systems). Other topics require more detailed presentation for lack of suitable review or just because the author's viewpoint is somewhat unusual. Section 9.5 is an introduction, occasionally rather critical, to closure, functional and diagrammatic methods. Section 9.6 is devoted to eddy viscosity, multiscale methods and renormalization; it includes some little known historical material on nineteenth century turbulence research. Finally, Section 9.7 deals mostly with recent developments in two-dimensional turbulence.

9.2 Books on turbulence and fluid mechanics

One of the earliest reviews of (mostly homogeneous) turbulence, was written by von Neumann (1949) in the form of a report to the Office of Naval Research, after he had attended a conference in Paris on *Problems of Motion of Gaseous Masses of Cosmical Dimensions*, organized jointly by the International Union of Theoretical and Applied Mechanics and the International Astronomical Union. This report was only privately circulated for many years. It was and remains a most remarkable paper, containing many original ideas. Von Neumann was the first to recognize that, when the Reynolds number is increased, the separation of microscopic and macroscopic scales increases also (cf. Section 7.5), thereby ruling out any 'ultraviolet catastrophe'. He noted that analyticity of the flow suggests an exponential law for the behavior of the energy spectrum at high wavenumbers. Thoroughly convinced that a 'high-speed computing program ... should be undertaken as soon as feasible', he made near the end of the paper the following observations which deserve being quoted *in extenso*:

> *These considerations justify the view that a considerable mathematical effort towards a detailed understanding of the mechanism of turbulence is called for. The entire experience with the subject indicates that the purely analytical approach is beset with difficulties, which at this moment are still prohibitive. The reason for this is probably as was indicated above: That our intuitive relationship to the subject is still too loose — not having succeeded at anything like deep mathematical penetration in any part of the subject, we are still quite disoriented as to the relevant factors, and as to the proper analytical machinery to be used.*
>
> *Under these conditions there might be some hope to 'break the deadlock' by extensive, but well-planned, computational efforts. It must be admitted that the problems in question are too vast to be solved by a direct computational attack, that is, by an outright calculation of a representative family of special cases. There are, however, strong indications that one could name certain strategic points in this complex, where relevant information must be obtained by direct calculations. If this is properly done, and the operation is then repeated on the basis of broader information then becoming available, etc., there is a reasonable chance of effecting real penetrations in this complex of problems and gradually developing a useful, intuitive relationship to it. This should, in the end, make an attack with analytical methods, that is truly more mathematical, possible.*[1]

Batchelor's (1953) *Homogeneous Turbulence* was a landmark in the field, written after considerable experimental and theoretical work had been done in Cambridge in the years after World War II. This monograph contributed to bringing the idea of a 'universal equilibrium' theory of turbulence (one aspect of Kolmogorov's contribution) to a wide scientific audience. It even included some discussion of intermittency (for velocity-derivatives) and some ideas about reverse energy flow in two-dimensional turbulence (see Section 9.7).

Ten years later, two important review papers were published in *Handbuch der Physik.* Lin and Reid's (1963) review of theoretical work includes

[1] It took over 20 years until von Neumann's dream started coming true (Orszag and Patterson 1972).

considerable material on 'early' closures of the quasi-normal type which had severe pitfalls (see Section 9.5). It also contains a clear exposition of the problems encountered with the so-called Loitsyansky invariant (see Section 7.7). Corrsin's (1963) review of experimental work remains a classic on how to set up and measure homogeneous tubulence (see also the very readable book by Bradshaw 1971).

The currently most complete book on homogeneous turbulence, by Monin and Yaglom (1975) was first published in Russian in 1967, the substantially revised English version being edited by John Lumley. It contains a wealth of information on theory and experiments, including considerable material on intermittency but not on fractal and multifractal models. It has about 1000 references (fairly distributed between East and West), most of which are discussed at length in the text. This 'big red book' is the constant companion of anyone working seriously in the field. Many important topics not covered in the present book are discussed, for example the statistical properties of passive scalars, such as temperature fields.[2] Written around the same time, and first published in Bulgarian and Russian, Panchev's (1971) book contains a nice presentation of Kolmogorov's four-fifths law and much material on early attempts by Obukhov and others to derive simple closed spectral equations by heuristically relating the energy transfer to the energy spectrum. Considerable material on compressible turbulence may be found in Favre, Kovasznay, Dumas, Gaviglio and Coantic (1976).

After ways were found to circumvent the more obvious pitfalls of closure (see Section 9.5) books and reviews tended to devote considerable space to what may be called 'advanced closures'. Leslie's (1973) book is mostly concerned with Robert Kraichnan's direct interaction approximation, a topic we shall come back to in Section 9.5.3. Orszag (1977) and Rose and Sulem (1978) are among the best reviews of the analytic theory of homogeneous turbulence published in the 1970s. Their emphasis is on closure and statistical mechanics. Lesieur's (1990) book contains similar material with more details, particularly on two-dimensional turbulence; in addition it has some material about 'real world turbulence'. The reader will appreciate this well-organized book and find that it has relatively little overlap with the present one.[3] For example, it does not mention

[2] More recent useful references on this are: Sreenivasan (1991), Pumir, Shraiman and Siggia (1991), Gollub, Clarke, Gharib, Lane and Mesquita (1991), Holzer and Pumir (1993), Pumir (1994), and Tong and Warhaft (1994).

[3] The same author has also produced a very readable adaptation for broad scientific audiences, including many beautiful color pictures (Lesieur 1994).

Kolmogorov's four-fifths law, an omission it shares with most books and reviews, Landau and Lifshitz (1987), Monin and Yaglom (1975) and Panchev (1971) being among the exceptions. Many topics of current interest in turbulence research are discussed in the *Tentative Dictionary* of turbulence, edited by Tabeling and Cardoso (1994).

Let us also mention Zeldovich, Ruzmaikin's and Sokoloff's (1990) very enjoyable essay *The Almighty Chance*, which contains ample material on fractals and homogeneous turbulence. Concepts such as intermittency are presented in an elementary way and illustrated by many nonhydrodynamic examples involving, for instance, mushrooms, urban centers, nuclear accidents, etc. Finally, it is very enlightening to watch the film on turbulence by Stewart (1972).

Shear flow turbulence, also called 'inhomogeneous' turbulence, is not within the scope of this book. Recommended reading includes Hinze (1959), Townsend (1976) and Monin and Yaglom (1971). The last is by far the most comprehensive treatment of the subject. A new edition is expected around 1996. Tennekes and Lumley's (1972) *A First Course in Turbulence*, which covers both homogeneous and shear flow turbulence, deserves a particular mention: it has become very popular among the many users of turbulence, scientists, engineers, etc. Although more than 20 years old, this very readable book is far from being obsolete. One could even say that, for much of the material that one regretted not finding in it 10 years ago, maybe there is less regret now.

Students and scientists interested in turbulence do not necessarily have much background in fluid mechanics. Classical textbooks which can provide the required background include Batchelor (1970), Landau and Lifshitz (1987), Tritton (1988) and Guyon, Hulin and Petit (1991). Physicists will love the short but penetrating introduction to fluid mechanics found in Chapters 40 and 41 of the second volume of Feynman (1964). A remarkably intuitive elementary introduction to fluid mechanics is Vogel (1981). More specialized reading on mathematical aspects of fluid mechanics, hydrodynamic stability, thermal convection, magnetohydrodynamics and vortex dynamics may be found in Chandrasekhar (1961), Lions (1969), Temam (1977), Chorin and Marsden (1979), Drazin and Reid (1981), Bayly, Orszag and Herbert (1988), Cross and Hohenberg (1993), Siggia (1994), Moffatt (1978), Saffman (1992) and Chorin (1994). Finally, special mention should be made of the whole set of Geophysical Fluid Dynamics Lecture Notes published for over 30 years by the Woods Hole Oceanographic Institute (Woods Hole, Massachusetts); these lecture notes have always been at the forefront of research in turbulence and hydrodynamic stability and their applications to astro/geophysical sciences.

9.3 Mathematical aspects of fully developed turbulence

Since we are presently still lacking a *systematic* theory of fully developed turbulence, is anything to be gained by focusing on *rigorous* approaches? A.N. Kolmogorov, a leading mathematician of the twentieth century, certainly did not restrict his investigations of turbulence to what he could handle rigorously. His scientific activities, as he liked to point out, encompassed rather unrelated fields, including history, mathematics and turbulence (V.I. Arnold, private communication). Some questions relating to high-Reynolds-number flow, which have been identified only after the completion of Kolmogorov's work on turbulence, are however likely to benefit from a close collaboration between mathematicians and physicists.

The two central issues are *singularities* and *depletion*, both of which are beyond simple phenomenology (see Section 7.8) and must be tackled by exquisitely precise numerical simulations or by rigorous methods. If finite-time singularities are present in viscous flow, such flow is not analytic and the dissipation-range spectrum should be a power-law. This is highly unlikely, but cannot be completely ruled out. If inviscid flow blows up (i.e. develops singularities) after a finite time, the nature of the singularities may be reflected in the scaling properties of fully developed turbulence. We pointed out in Section 8.5.4 that the converse need not be true: (multi)fractality does not imply finite-time singularities. Furthermore, even if blow-up takes place, the first singularity to occur could be different in nature from those taking place after blow-up, when the limit of zero viscosity is taken.[4] As for depletion or reduction of nonlinearity, it provides a possible way to avoid singularities and is related to the intriguing tendency of high-Reynolds-number flow to organize itself into coherent structures such as vortex sheets, pancakes and filaments.

Let us now briefly list what has been proven rigorously about singularities in viscous and inviscid three-dimensional flow. For viscous (Navier–Stokes) flow no real breakthrough has occurred since Leray

[4] This is indeed the case for Burgers' equation, where the first singularity is locally as $(x - x_\star)^{1/3}$ while later singularities are shocks, i.e. discontinuities (Fournier and Frisch 1983b).

(1934) showed the *existence* of a *weak* solution. This solution does not have enough smoothness to rule out singularities and need not be unique. In fact, the lack of uniqueness provided at that time a possible explanation of the turbulent behavior. Scheffer (1976, 1977), using a functional estimate established by Leray (1934), proved that the Hausdorff dimension of the instants of time where the viscous flow is singular cannot exceed one-half. The best result in this direction was obtained by Cafarelli, Kohn and Nirenberg (1982), who proved that, in the four-dimensional space-time, the Hausdorff dimension of singularities cannot exceed 1. In other words, such hypothetical singularities are on a very small set. It is most likely, of course, that this set is empty. Some details on these results, putting them into a modern perspective for both mathematicians and physicists, may be found in the reviews of Constantin (1991) and Gallavotti (1993). For viscous flow, Ladyzhenskaya (1969) has proven that regularity for all times (global regularity) holds when the viscous term $\nu\nabla^2\boldsymbol{v}$ is replaced by $-\mu(-\nabla^2)^\alpha$, where the *dissipativity* $\alpha \geq 2$. As observed by Rose and Sulem (1978), K41 phenomenology suggests that regularity holds as soon as $\alpha > 1/3$, while the best proven result is $\alpha \geq 5/4$ (Lions 1969).

We turn now to inviscid (Euler) flow. For lack of a smoothing mechanism, the solutions never become more regular than the initial data. Even for sufficiently smooth initial data, for example with a vorticity which is Hölder continuous,[5] regularity is proven only up to a finite time. The first result of this kind was obtained by Lichtenstein (1925) who assumed the initial vorticity to be Hölder continuous and zero outside of a bounded set. A number of improvements, obtained in the 1970s, may be found, for example, in Ebin and Marsden (1970), Kato (1972), Bardos and Frisch (1976) and the review by Rose and Sulem (1978). As already pointed out in Section 7.8, none of these results significantly improves on what would have been obtained by assuming that the (Lagrangian) rate of change of the vorticity modulus is of the order of the squared vorticity. Such a relation implies blow-up of the supremum of the vorticity. Actually, an important result by Beale, Kato and Majda (1989) ensures that no finite-time loss of regularity can occur without blow-up of the supremum of the vorticity.[6]

In the early 1980s, when the first high-resolution simulations of three-

[5] Increments of the vorticity over a small distance ℓ are bounded by a constant times ℓ^α for some $0 < \alpha \leq 1$.

[6] In fact, if the first loss of regularity occurs at time $T_\star$, the time-integral of the spatial supremum of the vorticity modulus from 0 to $T_\star$ is infinite.

dimensional Euler flow did not reveal the expected singularities (Brachet *et al.* 1983), it became clear that all the existing rigorous bounds were overestimating the growth of nonlinearities because they failed to take into account the development of structures with strongly depleted nonlinearities (see, e.g., Frisch 1983, p. 689). As we now know, depleted structures are ubiquitous: pancakes (Section 7.8), vortex filaments with approximately circular cross-sections (Section 8.9), circular two-dimensional vortices (Section 9.7), etc. It is a major challenge to take into account the depleted *geometry* of inviscid or turbulent flow and thereby to improve on existing rigorous results. This question is now being tackled seriously. A reasonably detailed and very enjoyable account of what is known may be found in the review by Constantin (1994, Sections 4 and 5). The key result is the Constantin relation[7] for the Lagrangian derivative of the vorticity modulus $|\omega|$, *viz*:

$$D_t|\omega| \equiv (\partial_t + \boldsymbol{v} \cdot \nabla)|\omega| = \alpha|\omega|, \tag{9.1}$$

where α is given by

$$\alpha(\boldsymbol{r}) \equiv \frac{3}{4\pi}\,\mathrm{P.V.}\int D(\hat{\boldsymbol{r}}', \hat{\boldsymbol{\omega}}(\boldsymbol{r}+\boldsymbol{r}'), \hat{\boldsymbol{\omega}}(\boldsymbol{r}))\,|\boldsymbol{\omega}(\boldsymbol{r}+\boldsymbol{r}')|\,\frac{d\boldsymbol{r}'}{|\boldsymbol{r}'|^3}. \tag{9.2}$$

Here, the hat on a vector designates the unit vector parallel to this vector, the symbol P. V. stands for 'principal value' and

$$D(\boldsymbol{a}, \boldsymbol{b}, \boldsymbol{c}) \equiv (\boldsymbol{a} \cdot \boldsymbol{b})\,\mathrm{Det}\,(\boldsymbol{a}, \boldsymbol{b}, \boldsymbol{c}), \tag{9.3}$$

where $\mathrm{Det}\,(\boldsymbol{a}, \boldsymbol{b}, \boldsymbol{c})$ is the determinant of the three vectors $\boldsymbol{a}$, $\boldsymbol{b}$ and $\boldsymbol{c}$.

The quantity α has the dimension of a vorticity. Yet, as pointed out by Constantin, it is geometrically depleted because of the presence of the determinant: if the *direction* of the vorticity changes very slowly in the neighborhood of point $\boldsymbol{r}$, the determinant is going to be very small. If the time-integral of the spatial supremum of α remains finite, so will the supremum of the vorticity and blow-up may be avoided altogether. Of course, the previous sentence has a big 'if'. What has been established so far are various strengthened versions of the Beale, Kato and Majda (1989) result (Constantin 1994; Constantin, Fefferman and Majda 1995).[8]

[7] Called by him the 'alpha and omega' relation.

[8] Other relations, possible starting points for mathematical analysis, may be found in Ohkitani and Kishiba (1995).

Some hope of obtaining much stronger results comes from related work on 'active scalars', i.e. a scalar quantity which is being advected by a two-dimensional incompressible flow, the stream function of which is a prescribed function of the local intensity of the scalar (Constantin 1994 and references therein). This problem has features resembling those of the three-dimensional Euler flow and global regularity and/or blow-up results seem at hand (A. Majda and P. Constantin, private communication). Another two-dimensional problem in which a careful control of the geometry has led to a real breakthrough is the two-dimensional vortex patch, i.e. the inviscid motion of a domain of uniform vorticity bounded by a curve $\mathscr{C}$, outside of which the vorticity vanishes. The initial data are already discontinuous, and the issue is whether or not an initially smooth curve $\mathscr{C}$ will remain so for ever or develop a singularity (cusp or other) after a finite time (Majda 1986). The topic was the subject of some controversy among physicists. This was settled when Chemin (1993) proved the regularity for all times. A more elementary proof is given by Bertozzi and Constantin (1993; see also Constantin 1994 and references therein).

It is now clear that progress on the 'big' Euler and Navier–Stokes problems in three dimensions requires more than just better functional analysis. Some geometry is needed. But 'how much geometry is enough geometry'? The work in progress described above uses only some information on the three-dimensional geometry of vortex lines. Actually, there is *infinitely* more geometry in the Euler flow: incompressible Euler flow, in any dimension, when formulated in Lagrangian coordinates, may be viewed as a geodesic flow in a suitable infinite-dimensional space. Indeed, let $\boldsymbol{r}(t, \boldsymbol{a})$ denote the position at time t of the fluid particle initially at $\boldsymbol{a}$, it is elementary to check that the extremals of the space-time integral of $|\partial_t \boldsymbol{r}(t, \boldsymbol{a})|^2$, subject to $\boldsymbol{r}(0, \boldsymbol{a}) = \boldsymbol{a}$, to prescribed 'final positions' $\boldsymbol{r}(T, \boldsymbol{a})$ and to the incompressibility constraint $\mathrm{Det}\, \partial r_i/\partial a_j = 1$, are just the solutions of the Euler equation, in which the pressure appears as a Lagrange multiplier (see, e.g., Serrin 1959, Chapter IV).

In more mathematical language, the Euler flow in a domain M is described by the geodesics on the group $\mathrm{S\,Diff}(M)$ of volume-preserving diffeomorphisms of M, endowed with the right-invariant Riemannian metric defined by the kinetic energy. From there, using Lie algebra methods, Arnold (1966, 1978) discovered that the Euler flow is actually an infinite-dimensional analog of the motion of a solid body around a fixed point. This has many interesting consequences for fluid mechanics, which are being actively investigated (see, e.g., Arnold and Khesin 1992;

Marsden and Weinstein 1983; Zeitlin 1991, 1992; Morrison 1993 and other contributions in the same Woods Hole proceedings). Associated to such Lagrangian formulations of the Euler equation, there is a symplectic structure and various Hamiltonian formulations. This includes the famous representation in terms of Clebsch variables (see, e.g., Lamb 1932), where the vorticity is written as $\nabla\lambda \wedge \nabla\mu$ and λ and μ are advected scalars (material invariants).[9] Another Hamiltonian representation, which is useful when dealing with helical flow is due to Kuz'min (1983) and Oseledets (1989; see also Gama and Frisch 1993).

For seriously tackling some of the major open mathematical problems of the Euler flow, it may thus be necessary to learn symplectic geometry. (Remember that physicists had to learn Riemannian geometry to do general relativity !) It is not obvious that such methods can be extended to viscous flow, i.e. to true turbulence.

9.4 Dynamical systems, fractals and turbulence

Kolmogorov, in addition to being the founder of modern probability theory and modern turbulence theory, was also the founder of modern dynamical systems theory. We do not intend here to review his numerous contributions to this field and just mention the work on invariant tori for slightly perturbed Hamiltonian systems, which eventually led to what is called KAM theory (Kolmogorov 1954; Arnold 1963; Moser 1962). Since the 1970s, physicists all over the world have become strongly interested in nonlinear dynamics and in fractals. Key references may be found in textbooks, reviews and collection of reprints such as Mandelbrot (1977), Bergé, Pomeau and Vidal (1984), Guckenheimer and Holmes (1986), Schuster (1988), Cvitanović (1989), Devaney (1989), Ruelle (1989, 1991), Falconer (1990), Manneville (1990), Evertsz and Mandelbrot (1992), Iooss and Adelmeyer (1992), and Cross and Hohenberg (1993).

A frequently asked questions is: What have we learned from looking at turbulence with a dynamical systems viewpoint? This was one of the key topics included in the meeting *Wither Turbulence?*, organized by Lumley (1990; see, in particular, the contribution by H. Aref; see also Dracos and Tsinober 1993). It should be clear, from our Chapter 3, that turbulence (including fully developed turbulence) is formally a problem in dynamical systems. Conceptually, this is helpful. For example,

9 The corresponding vortex lines, being at the intersections of surfaces of constant λ and constant μ, have a rather constrained topology; this can be avoided by increasing the number of Clebsch variables.

we know that there is no need to invoke singularities to explain the unpredictable behavior of turbulent flow. Neither is there any need for external noise to explain random behavior of turbulent flow.[10] Also, dynamical systems theory allows us to think of homogeneous turbulence as a flow in which translational and other symmetries are restored through chaos.

On the other hand, the tools which have been developed mostly for low order dynamical systems (bifurcations, Lyapunov exponents, dimension of attractors, etc.) have been rather useless for fully developed turbulence, which has a very large number of degrees of freedom (see Section 7.4). There are also severe practical limitations on the measurements of dimensions of attractors when these are too high (Atten, Caputo, Malraison and Gagne 1984; Ruelle 1989). However, there has been some success in modeling the large scales of wall-bounded turbulent flow by low-dimensional dynamical systems (Aubry, Holmes, Lumley and Sone 1988; Holmes 1990).

One of the most fruitful applications of dynamical systems to fluid mechanics, if not exactly turbulence, is *chaotic advection*, i.e. the chaotic motion of passive tracer particles in a prescribed flow (Aref 1984; Dombre *et al.* 1986; Ottino 1989). Here, the phase space of the dynamical system is just the usual two- or three-dimensional physical space. When the flow is incompressible the dynamics are conservative (Hamiltonian). Chaotic advection, in the simplest three-dimensional case, exploits the observation made by Arnold (1965; see also Hénon 1966) that very simple steady flow, which is a solution of the Euler equation, can have nonintegrable streamlines, so that the Lagrangian motion is chaotic. One of the simplest examples, called the Arnold–Beltrami–Childress flow or, simply, ABC flow is given by

$$\left.\begin{aligned} u_1 &= A\sin x_3 + C\cos x_2 \\ u_2 &= B\sin x_1 + A\cos x_3 \\ w_3 &= C\sin x_2 + B\cos x_1. \end{aligned}\right\} \tag{9.4}$$

If, instead of passive particles, one has a passive magnetic field in a prescribed conducting flow, chaotic advection can help the *dynamo effect*, i.e. the unbounded growth of magnetic fields (Moffatt 1978). Indeed, in the absence of molecular diffusion, the magnetic field would behave just like a pair of infinitesimally close tracer particles, so that its growth would

[10] Some external noise is useful insofar as it prevents the system from being trapped in one of many possible attractors with small basins.

be controlled by the maximum Lyapunov exponent.[11] Diffusion, even in very small amounts, makes the problem considerably more complex (Arnold, Zeldovich, Ruzmaikin and Sokoloff 1981; Zeldovich, Ruzmaikin and Sokoloff 1983; Childress *et al.* 1990; Childress and Gilbert 1995 and references therein).

If turbulence has benefitted, occasionally, from dynamical systems, the converse is also true. In Chapter 8 we saw that the concept of *multifractal*, which first appeared in fully developed turbulence, has turned out to be very useful in dynamical systems. Paradoxically, the multifractal character of fully developed turbulence is still a controversial matter, whereas attractors of chaotic dissipative dynamical systems are known, and sometimes proven rigorously, to be multifractal.

This brings us finally to fractals and turbulence, a subject already discussed at length in Chapter 8. Here, we just want to add a word of warning: fractals with no particular dynamical significance may appear when processing data of fully developed turbulence. This has to do with *fractal level crossing*, a phenomenon which may be understood by considering the case of the ordinary one-dimensional Brownian motion $W(t)$ (Kahane 1985). A simple consequence of the fact that the increments of $W(t)$ scale as the square root of the time increments is that the graph of $W(t)$ has fractal dimension $3/2$. Consequently, the instants of time where the graph intersects any smooth curve, for example the crossings of a given level, form a set of fractal dimension $D = 1/2$. Similarly, a one-dimensional random function with Hölder exponent h or a self-similar random function with scaling exponent h will produce sets of dimension $D = 1 - h$ under level crossing, whether or not the random function is Gaussian (Falconer 1990). Even pure K41 turbulence will look fractal in this way.[12] Similar artifacts are even more likely to occur in higher dimensions since graphical representation by level crossing is one of the simplest ways to plot a multidimensional field.[13] For further information on the geometry of fractal graphs in connection with turbulence, see Constantin and Procaccia (1993).

[11] Arnold (1972) already noticed that chaotic advection may favor instability of nonconducting flow.

[12] Praskovsky, Foss, Kleiss and Karyakin (1993) analyzed the fractal properties of level crossing for high Reynolds number turbulence data and failed to find a wide range of scales exhibiting constant fractal dimension; they nevertheless quoted a value $D \approx 0.4$; this is not consistent with $D = 1 - h$ for $h \approx 1/3$.

[13] For example, the fractals appearing in a solar magnetogram need not be indicative of any genuine fractal small-scale magnetic activity.

9.5 Closure, functional and diagrammatic methods

After the work of Wiener (1930) and Taylor (1935) on the energy spectrum of (stationary or homogeneous) random functions, a major challenge in turbulence appeared to be the prediction and measurement of its energy spectrum. For many scientists in the field this became *the* major challenge. Over a period of three decades, starting with the work of Millionshchikov (1941a, b) on the fourth-cumulant-discard approximation, which has grave inconsistencies, there were many attempts to give a contracted description of the Navier–Stokes dynamics involving only the energy spectrum or a *finite* number of statistical functions. This culminated in the work of Kraichnan (1958, 1961) on the direct interaction approximation (DIA) and its companions, the Lagrangian history direct interaction approximation (LHDI; Kraichnan 1965, 1966) and the eddy damped quasi-normal Markovian approximation (EDQNM; Orszag 1966, 1977) both of which are compatible with the K41 theory.

To the best of our knowledge, Kolmogorov never attempted to derive such 'closed' equations.[14] His 1941 work dealt with the global probabilistic structure of the random velocity field and cannot be reduced to the prediction of the $k^{-5/3}$ energy spectrum.

A number of more ambitious methods have been developed since the 1950s, intended precisely to cope with the global probabilistic structure. They will be referred to here as 'functional' methods. Indeed, they make use of the characteristic functional (Section 4.2) of the velocity field or some generalization thereof, and not just of a finite set of moments. Functional methods were pioneered by Hopf (1952). Their usage grew very rapidly when it became clear, around 1960, that the diagrammatic and functional methods developed for quantum field theory by Dyson, Feynman, Schwinger and others (see also Section 9.6.4) could be applied, at least formally, to the Navier–Stokes equation with random initial conditions and/or random forces (Kraichnan 1958, 1961; Rosen 1960; Wyld 1961; Lewis and Kraichnan 1962; Tatarskii 1962; see also the references in Chapter 10 of Monin and Yaglom 1975). The paper by Martin, Siggia and Rose (1973) 'Statistical dynamics of classical systems' became a landmark in this subject, not only for the Navier–Stokes problem, but also for a whole class of nonlinear statistical problems, arising, for example, in critical dynamical phenomena.

[14] With the possible exception of his work on modeling of inhomogeneous turbulence (Kolmogorov 1942).

Functional and diagrammatic methods are sometimes referred to as 'field-theoretic methods' in turbulence. This is somewhat misleading since what they borrowed from field theory (e.g., the so called Dyson equation) was mostly introduced in the early 1950s. Developments using more recent field-theoretic methods such as conformal invariance for two-dimensional turbulence (Polyakov 1993) are generating considerable interest (Falkovich and Hanany 1993; Falkovich and Lebedev 1994; Benzi, Legras, Parisi and Scardovelli 1995).

Section 9.5.1 is a brief introduction to one of the most useful functional methods, the Hopf equation. Functional methods using Feynman integrals and subsequent diagrammatic expansions are discussed in Section 9.5.2. The author is rather critical of such methods, which seem so far to have been mostly useless in understanding turbulence. An exception to this is Kraichnan's DIA (Section 9.5.3). Closures, whether based on heuristic approximations or on functional and diagrammatic methods, are discussed in Section 9.5.4, with special emphasis on their shortcomings.

9.5.1 The Hopf equation

An almost trivial example can be used to give an idea of what is the essence of Hopf's (1952) functional method. Consider the following ordinary differential equation:

$$\frac{dv}{dt} = v^2, \tag{9.5}$$

and let us introduce

$$K_{\text{unav.}}(t,z) \equiv e^{izv(t)}, \tag{9.6}$$

where 'unav.' means 'unaveraged'. It is trivial to check that

$$\partial_t K_{\text{unav.}} = iz\left(\frac{1}{i}\frac{\partial}{\partial z}\right)^2 K_{\text{unav.}}. \tag{9.7}$$

Now, let the initial condition $v(0)$ be chosen randomly, and define

$$K(t,z) \equiv \langle K_{\text{unav.}}(t,z)\rangle = \langle e^{izv(t)}\rangle, \tag{9.8}$$

namely the characteristic function of the random variable $v(t)$. Since (9.7) is linear, it is also satisfied by $K(t,z)$. With the label 'unav' removed (9.7) is the Hopf equation associated with (9.5). Since knowledge of the characteristic function is equivalent to that of the p.d.f., of which it is the Fourier transform, the Hopf equation indirectly determines the (single-time) p.d.f.

The Hopf equation associated with the Navier–Stokes equation is obtained in basically the same way. We start from the Navier–Stokes equation in the form (2.13), obtained after elimination of the pressure term, and observe that it has the general structure:

$$\partial_t v = \gamma(v, v) + Lv, \tag{9.9}$$

where L and $\gamma(.,.)$ are suitable linear and bilinear operators[15] representing the viscous and nonlinear terms, respectively. All time arguments are equal to t and are not written. It is, of course, possible to write these objects in explicit form as differential and/or integral operators (remember that the solution of the Poisson equation, necessary to eliminate the pressure, involves an integral operator). Since differential operators can always be written formally as convolutions with suitable derivatives of Dirac distributions, it may be assumed that both L and $\gamma(.,.)$ are integral operators.

Mathematicians like to use coordinate-free notation, while physicists tend to prefer somewhat more coordinate-explicit notation. Given that the velocity field (at any time t) depends on a three-dimensional position vector $\boldsymbol{r}$ and has three components, a coordinate-explicit form of (9.9) reads

$$\begin{aligned}\partial_t v_{i_1}(\boldsymbol{r}_1) = &\int\!\!\int d\boldsymbol{r}_2\, d\boldsymbol{r}_3\, \gamma_{i_1 i_2 i_3}(\boldsymbol{r}_1, \boldsymbol{r}_2, \boldsymbol{r}_3) v_{i_2}(\boldsymbol{r}_2) v_{i_3}(\boldsymbol{r}_3)\\ &+ \int d\boldsymbol{r}_2\, L_{i_1 i_2}(\boldsymbol{r}_1, \boldsymbol{r}_2) v_{i_2}(\boldsymbol{r}_2).\end{aligned} \tag{9.10}$$

Such notation makes subsequent manipulations leading to the Hopf equation quite cumbersome, so that one cannot see the forest for the trees. Hence, theoretical physicists have invented a shorthand notation where arguments such as i_1 and $\boldsymbol{r}_1$ are lumped into just a symbol 1. Eq. (9.9) then reads:

$$\partial_t v(1) = \gamma(1, 2, 3) v(2) v(3) + L(1, 2) v(2), \tag{9.11}$$

where it is understood that repeated arguments are summed/integrated over. In this notation, it is straightforward to show that the time-dependent characteristic functional

$$K(t, z) \equiv \left\langle e^{[iz(1)v(t,1)]} \right\rangle \tag{9.12}$$

[15] Without loss of generality $\gamma(.,.)$ may be assumed symmetrical in its arguments.

satisfies the Hopf (1952) equation

$$\partial_t K(t,z) = iz(1)\left[\frac{\gamma(1,2,3)}{i^2}\frac{\partial^2}{\partial z(2)\partial z(3)} + \frac{L(1,2)}{i}\frac{\partial}{\partial z(2)}\right]K(t,z). \quad (9.13)$$

When the Navier–Stokes equation is discretized in the space variables by using, for example, finite differences for derivatives and Riemann sums for integrals, operators such as $\partial/\partial v(2)$ are partial differential operators; without discretization, they are functional (or Fréchet) derivatives. One can also work with the spatial Fourier components of the velocity field. It is then necessary to separate real and imaginary parts, so that the $v(.)$s are real. Knowing $K(t,z)$, from (4.12), one can calculate all single-time moments of the velocity field.

When a deterministic forcing term $f(t,1)$ is added to the r.h.s. of (9.11) a term $iz(1)f(t,1)K(t,z)$ must clearly be added to the r.h.s. of the Hopf equation. When the force is *random*, the average of $iz(1)f(t,1)$ times $e^{iz(1)v(t,1)}$ cannot in general be expressed in terms of $K(t,z) = \langle e^{iz(1)v(t,1)}\rangle$. There is, however, an important exception, when $f(t,1)$ is a stationary Gaussian random function of zero mean and it is rescaled into $f_\epsilon(t,1) \equiv (1/\epsilon)f(t/\epsilon^2,1)$, so that, as $\epsilon \to 0$, the random function f_ϵ becomes δ-correlated (white noise) in the time variable. It may then be shown that, in the limit $\epsilon \to 0$, the effect of the random force is just to add to the r.h.s. of the Hopf equation (9.13) a term $i^2 z(1)z(2)F(1,2)K(t,z)$, where $F(1,2) = \int_0^\infty dt\,\langle f(t,1)f(0,2)\rangle$ (Novikov 1964). Novikov's proof uses Gaussian integration by parts (p. 43); alternatively, multiscale techniques of the kind discussed in Section 9.6.2 may be used (see, e.g., Gama, Vergassola and Frisch 1994, Appendix D).

The Hopf equation has been rather useful. Hopf (1952) showed that when $f = v = 0$, it possesses Gaussian solutions with an energy spectrum $E(k) \propto k^2$, corresponding to the equipartition of the kinetic energy among all spatial Fourier modes. Such solutions appear already to some extent in the work of Burgers (1939, Section 21). Kraichnan (1967) refers to them as 'absolute equilibrium solutions'[16] (see also Orszag 1977; Rose and Sulem 1978; Lesieur 1990). Absolute equilibrium solutions seem highly unphysical in view of the approximately $k^{-5/3}$ spectrum of three-dimensional turbulence. Actually, they are appropriate at the very smallest wavenumbers of turbulent flow maintained by forcing at intermediate wavenumbers (Forster, Nelson and Stephen 1977). We also note that the Hopf equation has been used for a proper mathematical

[16] They played a key role in his theory of the inverse cascade in two dimensions; see also Section 9.7.1.

formulation of the problem of homogeneous turbulence (Vishik and Fursikov 1988).

One of the nicest applications of the Hopf equation is the derivation of the hierarchy of (single-time) *cumulants.* From the definition (9.12) of the characteristic functional it follows that it is the generating functional of (single-time) moments of the velocity field:

$$K(t,z) = 1 + iz(1)\langle v(t,1)\rangle + \frac{i^2}{2!}z(1)z(2)\langle v(t,1)\,v(t,2)\rangle + \cdots. \tag{9.14}$$

As is well known, moments beyond the second order need not become small when some of their spatial arguments are widely separated. For example, if the distances $\boldsymbol{r}_1 - \boldsymbol{r}_2$ and $\boldsymbol{r}_3 - \boldsymbol{r}_4$ are small compared to the integral scale ℓ_0, while $\boldsymbol{r}_1 - \boldsymbol{r}_3$ is large compared to ℓ_0, then

$$\langle v(t,1)\,v(t,2)\,v(t,3)\,v(t,4)\rangle \approx \langle v(t,1)\,v(t,2)\rangle\,\langle v(t,3)\,v(t,4)\rangle, \tag{9.15}$$

which is independent of $\boldsymbol{r}_1 - \boldsymbol{r}_3$. Cumulants $\langle v(t,1)\cdots v(t,p)\rangle_c$ are defined by subtracting from moments all possible factorized terms involving lower order moments. It may be checked that the generating function of cumulants is just the logarithm of the characteristic functional:

$$H(t,z) \equiv \ln K(t,z) = iz(1)\langle v(t,1)\rangle_c + \frac{i^2}{2!}z(1)z(2)\langle v(t,1)\,v(t,2)\rangle_c + \cdots. \tag{9.16}$$

Substituting $K = e^H$ into the Hopf equation (9.13) with the forcing term added, we immediately obtain what could be called the *log–Hopf* equation:

$$\begin{aligned}\partial_t H(t,z) = &-iz(1)\gamma(1,2,3)\left[\frac{\partial^2 H}{\partial z(2)\partial z(3)} + \frac{\partial H}{\partial z(2)}\frac{\partial H}{\partial z(3)}\right]\\ &+ z(1)L(1,2)\frac{\partial H}{\partial z(2)} - z(1)z(2)F(1,2).\end{aligned} \tag{9.17}$$

Expanding (9.17) in powers of z and assuming that the mean velocity (which is also the first order cumulant) vanishes, we obtain a hierarchy of cumulant equations. The first two nonvanishing read (all time arguments are equal and omitted):

$$\begin{aligned}&\partial_t\langle v(1)v(2)\rangle_c =\\ &+\gamma(1,3,4)\,\langle v(2)v(3)v(4)\rangle_c + \gamma(2,3,4)\,\langle v(1)v(3)v(4)\rangle_c\\ &+ L(1,3)\,\langle v(2)v(3)\rangle_c + L(2,3)\,\langle v(1)v(3)\rangle_c + 2F(1,2),\end{aligned} \tag{9.18}$$

$$
\begin{aligned}
&\partial_t \langle v(1)v(2)v(3)\rangle_c = \\
&+ 2[\gamma(1,4,5)\,\langle v(2)v(4)\rangle_c\,\langle v(3)v(5)\rangle_c \\
&+ \gamma(2,4,5)\,\langle v(1)v(4)\rangle_c\,\langle v(3)v(5)\rangle_c \\
&+ \gamma(3,4,5)\,\langle v(1)v(4)\rangle_c\,\langle v(2)v(5)\rangle_c] \\
&+ \gamma(1,4,5)\,\langle v(2)v(3)v(4)v(5)\rangle_c \\
&+ \gamma(2,4,5)\,\langle v(1)v(3)v(4)v(5)\rangle_c \\
&+ \gamma(3,4,5)\,\langle v(1)v(2)v(4)v(5)\rangle_c + L(1,4)\,\langle v(2)v(3)v(4)\rangle_c \\
&+ L(2,4)\,\langle v(1)v(3)v(4)\rangle_c + L(3,4)\,\langle v(1)v(2)v(4)\rangle_c. \qquad (9.19)
\end{aligned}
$$

Subsequent equations in this hierarchy are too cumbersome to be shown here. In symbolic form, the equation for the nth cumulant ($n \geq 4$) C_n may be written:

$$
\partial_t C_n = \sum_{\substack{r+s=n+1,\\ r,s\geq 2}} \gamma\, C_r C_s + \gamma\, C_{n+1} + L\, C_n. \qquad (9.20)
$$

It is worth noting that, since the random force is Gaussian, it contributes only to (9.18) for the second order cumulant. In the isotropic case, this equation reduces to the energy transfer relation (6.37), when written in the Fourier space.

The cumulant hierarchy has often been used as a starting point for closure approximations (Section 9.5.4). One can also look for statistically stationary scaling solutions in the limit of vanishing viscosity (so that all terms involving L-operators disappear). In the absence of a force, or at scales at which the force is negligible, it is easily checked that the cumulant hierarchy is invariant under the rescaling $\boldsymbol{r} \to \lambda \boldsymbol{r}$ and $C_n \to \lambda^{nh} C_n$, where h is an arbitrary scaling exponent. This result is, of course, just a restatement of the scale invariance of the Navier–Stokes equation, discussed in Sections 2.2 and 6.1. Using (9.18), which is basically the Kármán–Howarth–Monin relation (6.8), one can then obtain scale-invariant solutions with the K41 value $h = 1/3$, just as in Section 6.3.1.

It is actually possible to do somewhat better and to use the cumulant hierarchy to show that a K41 ansatz does not lead to any blatant contradiction. Orszag and Kruskal (1966, 1968) investigated the cumulant hierarchy (written in the Fourier space), made such a K41 ansatz and looked for possible infrared (small wavenumber) or ultraviolet (large wavenumber) divergences in the ensuing Fourier integrals which might signal some inconsistencies not captured by naive scaling arguments. They found good evidence that all Fourier-integrals are convergent, implying the localness of interactions in the assumed K41 inertial range. The main problem is to show that there is no infrared divergence, in spite of the divergence of the total energy when a $k^{-5/3}$ energy spectrum is used. The above convergence stems mostly from cancellations among the three terms in the r.h.s. of (9.19) involving products of second order cumulants. These are the same cancellations as required for *random Galilean invariance* (Section 6.2.5). Indeed, when a uniform isotropic Gaussian random velocity $\boldsymbol{U}$ is added to the velocity field $\boldsymbol{v}$, the second order cumulant changes from $\langle v_i(\boldsymbol{r})v_j(\boldsymbol{r}')\rangle$ to $\langle v_i(\boldsymbol{r})v_j(\boldsymbol{r}')\rangle + (1/3)\langle U^2\rangle\delta_{ij}$, while all other cumulants are unchanged. Such cancellations take place only when working with *single-time* statistical averages, because all space arguments are then shifted by the same amount, $-\boldsymbol{U}t$. Methods of the sort discussed in Sections 9.5.2 and 9.5.3 use multiple-time formalisms and thus lead to more serious infrared difficulties.

9.5.2 Functional and diagrammatic methods

While the solution of the Hopf equation contains all the information about single-time moments, it may be desirable to have a formalism dealing with the full space-time statistical properties of the velocity field. This can be done by representing the solution of the Navier–Stokes equation in terms of functional integrals of the type first used by Wiener and Feynman (for an introduction to functional integrals, see Kac 1959, Chapter IV).

As with the Hopf equation, we shall begin with a very elementary example which involves only ordinary integrals. Suppose we want to solve the ordinary algebraic equation

$$v + h(v) = s. \tag{9.21}$$

Here, s is a prescribed real number, which is deterministic at this point (but will eventually become random) and $h(v)$ is a polynomial in the real variable v chosen such that (9.21) has a single real solution. For example, we could take $h(v) = v^3$. Let us denote by $v(s)$ the solution of (9.21) and by $\delta(v - v(s))$ the Dirac distribution at $v(s)$. From the relation

$\delta(\lambda x) = \delta(x)/|\lambda|$, it follows that

$$\delta(v - v(s)) = \delta(s - v - h(v))J(v), \quad J(v) \equiv |1 + h'(v)|. \tag{9.22}$$

Using the exponential representation of δ-functions, this may also be written:

$$\delta(v - v(s)) = \frac{1}{2\pi} \int dp\, J(v)\, e^{ip(s-v-h(v))}. \tag{9.23}$$

Any function of the solution (observable) $F(v(s))$ may now be represented as a double integral:

$$F(v(s)) = \frac{1}{2\pi} \int\!\!\int dv\, dp\, F(v)\, J(v)\, e^{ip(s-v-h(v))}. \tag{9.24}$$

At this point, let s become a centered random Gaussian variable; we then have $\langle e^{ips} \rangle = e^{-(1/2)\langle s^2 \rangle p^2}$. Averaging (9.24), we obtain:

$$\langle F(v(s)) \rangle = \frac{1}{2\pi} \int\!\!\int dv\, dp\, F(v) \mathscr{P}(p, v), \tag{9.25}$$

where

$$\left.\begin{aligned} \mathscr{P}(p, v) &\equiv J(v) e^{-A(p,v)}, \\ A(p, v) &\equiv \frac{1}{2} \langle s^2 \rangle p^2 + p[v + h(v)]. \end{aligned}\right\} \tag{9.26}$$

Similarly, we can calculate the mean Green's function[17] g, i.e. the mean of $\partial v/\partial s$. Since the multiplication of the integrand in (9.24) by ip is equivalent to a derivative with respect to s, we obtain:

$$\langle g \rangle \equiv \left\langle \frac{\partial v}{\partial s} \right\rangle = \frac{1}{2\pi} \int\!\!\int dv\, dp\, ip\, v\, \mathscr{P}(p, v). \tag{9.27}$$

It quite straightforward to extend this formalism to the full Navier–Stokes equation (6.6) with a Gaussian random force, which may be written

$$v(1) + \gamma(1, 2, 3) v(2) v(3) = s(1). \tag{9.28}$$

Here, symbols such as 1, 2 and 3 are defined as in Section 9.5.1, but with the additional inclusion of the time variable. Eq. (9.28) is obtained by applying the inverse of the heat operator $\partial_t - \nu \nabla^2$ to (6.6), after elimination of the pressure. By going through essentially the same steps

[17] Also called the *response function.*

as before, we obtain the following expression for the mean of an arbitrary functional $F(v(.))$ of the solution of the Navier–Stokes equation:

$$\langle F(v)\rangle = \int Dv\, Dp\, F(v)\mathscr{P}(p,v), \tag{9.29}$$

where

$$\left.\begin{aligned} \mathscr{P}(p,v) &\equiv J(v)\, e^{-A(p,v)}, \\ A(p,v) &\equiv \frac{1}{2}\langle s(1)s(2)\rangle\, v(1)v(2) \\ &\quad + ip(1)\,[v(1)+\gamma(1,2,3)v(2)v(3)]\,. \end{aligned}\right\} \tag{9.30}$$

Here, $J(v)$ is the Jacobian of the map from v to s defined by (9.28):[18]

$$J(v) \equiv \det\{\delta(1,2) + 2\gamma(1,2,3)v(3)\}\ ; \tag{9.31}$$

for discrete variables Dv and Dp are products of discrete differentials. In the continuous limit, (9.29) becomes a functional integral. Similarly, the mean Green's function is expressed as

$$\langle g(1,2)\rangle \equiv \left\langle \frac{\partial v(1)}{\partial s(2)} \right\rangle = \int Dv\, Dp\, v(1)\, ip(2)\, \mathscr{P}(p,v). \tag{9.32}$$

Eqs. (9.29)–(9.32) are superficially similar to Feynman integral representations used in quantum field theory with a cubic 'Lagrangian' (Feynman 1950). It is clear that, except for the $\gamma(1,2,3)v(2)v(3)$ term, (9.29) and (9.32) are Gaussian integrals (i.e. over exponentials of quadratic forms) which may be calculated explicitly. Feynman graphs are generated by expanding these integrals in powers of the 'bare vertex' $\gamma(1,2,3)$, using diagrams analogous to those in Fig. 4.1 for bookkeeping. Alternatively, functional differential equations à la Schwinger (1951b) may be obtained. Because it is not easy to present such matters in a concise way, we shall not dwell on the technical aspects of the subject (see, e.g., Machlup and Onsager 1953; Martin, Siggia and Rose 1973; Phythian 1977; Frisch, Fournier and Rose 1978).[19]

Diagrammatic methods have led repeatedly to false hopes in the theory of turbulence. A caricature of the situation would be to state that either

[18] For the Navier–Stokes equation, one can use a discretization where the Jacobian becomes a constant which may be absorbed into the definition of $Dv\,Dp$ (Phythian 1977). Otherwise, the Jacobian may be written as the integral over auxiliary *anticommuting ghost fields* of the exponential of a quadratic form in these fields (Parisi and Sourlas 1979).

[19] In Frisch (1968) the reader will find an elementary presentation of the diagrammatic machinery for the case of linear equations with random coefficients, which is somewhat simpler than the nonlinear Navier–Stokes equation with random forces.

they produced wrong results or that, when the results were right, they could be obtained by much simpler methods. We shall now try to give a more balanced view of the situation.

One of the difficulties with any multiple-time method is that compatibility with K41 is much less straightforward than in a single-time formalism such as the Hopf formalism (see end of Section 9.5.1). In Chapter 6 (p. 90) we observed that K41 is tantamount to having finite limits for the structure functions as the integral scale $\ell_0 \to \infty$. Structure functions are single-time quantities. Multiple-time quantities become singular in this limit because the Eulerian correlation time for any fixed spatial separation ℓ goes to zero. Indeed, this is basically the time for an eddy of size ℓ to be moved a distance $O(\ell)$ due to the sweeping action of the most energetic eddies present. There is worse: the action of very large eddies on much smaller ones, which in the K41 theory is akin to a random Galilean transformation (Section 7.3), may be misrepresented because Galilean invariance is difficult to preserve in a multiple-time diagrammatic formalism. Indeed, this invariance involves a cancellation between the *linear* term $\partial_t \boldsymbol{v}$ and the *nonlinear* term $\boldsymbol{v} \cdot \nabla \boldsymbol{v}$. In diagrammatic methods, the linear term is kept as such, whereas an expansion is performed in the nonlinear term. Hence, almost any simplification of the expansion will destroy the cancellation. We use the word 'simplification' because: (i) the complete problem is intractable and (ii) there is no known *approximation* procedure with a clearly identified expansion parameter.[20] Instead, one resorts typically to 'loop expansions' involving resummations of diagrams of increasing topological complexity (see, e.g., Amit 1984).

Scientists interested in using field-theoretical tools for turbulence have been aware of the difficulty with random Galilean invariance since the late 1950s (Kraichnan 1958, 1959). The difficulty is usually overcome by resorting to some form of Lagrangian coordinates, i.e. one tries to follow the trajectory of fluid particles to avoid sweeping effects. One successful attempt, which is not directly linked to diagrammatic methods is Kraichnan's LHDI (see next section). The Russian school has developed an alternative method, called 'quasi-Lagrangian', in which the true Lagrangian displacement of a reference point is subtracted from all spatial arguments (Belinicher and L'vov 1987; L'vov 1991).

Once random Galilean invariance is recovered, one can look for possible remaining divergences which would be indicative of a breakdown of

[20] With some exceptions, such as the problem studied by Forster, Nelson and Stephen (1977), discussed in Section 9.6.4.1.

K41. None has been found so far. This negative result should certainly not be taken as evidence that K41 is the correct theory. If there is a mechanism leading to intermittency somewhat similar to what is found in multiplicative random cascade models (Section 8.6.3), it is unlikely that it can be captured by perturbation or even renormalized perturbation theory.

There is at least one instance for which one can prove that diagrammatic methods fail to give the correct result *even when resummation is performed to all orders*, namely Burgers' model. It has indeed all the required invariance and conservation properties to make it an acceptable candidate for K41 scaling. However, the presence of shocks leads to a different scaling (Section 8.5.2). Furthermore, Fournier and Frisch (1983b) showed that formal resummation to all orders leads to an energy spectrum at high wavenumbers $\propto k^{-3}$ which is neither K41 nor the correct $\propto k^{-2}$ spectrum.

To end on a more optimistic note, we mention recent work on anomalous scaling of (local) dissipation correlations. We have seen in Sections 8.6.1 and 8.6.2 that the correlation of the dissipation is found experimentally to scale as $\ell^{-\mu}$ with $\mu \approx 0.2$. Golitsyn (1962) showed that naive application of K41 scaling arguments gives an exponent[21] $\mu = 8/3$. Lebedev and L'vov (1994) and L'vov and Procaccia (1995) suggest that K41 scaling in the inertial range could be consistent with the anomalous value of ≈ 0.2 for μ. Since the (local) dissipation is quadratic in velocities, they worked with the four-point velocity correlation functions expressed via diagrammatic techniques. To obtain the dissipation correlation function, four gradient operators are applied and then two pairs of points are made to coincide.[22] This theory predicts also corrections to inertial-range scaling involving small powers of the viscosity and which, therefore, may stay relevant up to very high Reynolds numbers. Such corrections could be mistaken for 'intermittency' corrections to K41.[23] Note that in the traditional fractal/multifractal models the dissipation correlation function involves $(\ell/\ell_0)^{-\mu}$, where ℓ_0 is the integral scale. In the alternative theory it involves $(\ell/\eta)^{-\mu}$, where η is the dissipation scale, so that, for fixed ℓ, the dissipation correlation function goes to zero with the viscosity.

21 Frisch, Lesieur and Sulem (1976) pointed out that such correlations involve both dissipation-range and inertial-range quantities; with this taken into account, K41 predicts $\mu = 0$.

22 When this kind of situation occurs in field theory, one speaks of 'composite operators' (see, e.g., Itzykson and Zuber 1985, Section 8-2-6); this can lead to anomalous scaling.

23 It may be hard to fully reconcile this vision of intermittency with the results on anomalous scaling obtained by the extended self-similarity technique applied to experimental data (see p. 131).

The derivation of this anomalous scaling has a number of delicate points. Fortunately, it has been found that similar arguments can be applied to the anomalous scaling of the dissipation correlation function for a passive scalar advected by a prescribed Gaussian random velocity field with a very short correlation time (L'vov, Procaccia and Fairhall 1994).

For the same passive scalar problem, Kraichnan (1994) predicted anomalous scaling of structure functions at inertial-range separations,[24] confirmed by two-dimensional numerical simulations using 8192^2 grid-points (Kraichnan, Yakhot and Chen 1995).

Both of these questions can be approached via nondiagrammatic methods. This requires the study of linear partial differential equations in several space variables with N-particle Schrödinger-like operators, satisfied by the single-time multiple-point correlation functions of the scalar (Kraichnan 1968b). Anomalous scaling is caused by the presence of zero modes (i.e. elements of the null-space) of these operators. The equation for the two-point function can be solved analytically. Higher-order equations require perturbation techniques (Chertkov, Falkovich, Kolokolov and Lebedev 1995; Fairhall, Gat, L'vov and Procaccia 1996; Gawędzki and Kupiainen 1995) or a mixture of analytical and numerical techniques (Shraiman and Siggia 1995). The results confirm anomalous scaling for both the dissipation correlation function and the structure functions of order 4 and beyond. For passive vectors, such as magnetic fields, anomalous scaling is already present in the structure function of order 2 (Vergassola 1996).

9.5.3 The direct interaction approximation

The direct interaction approximation (DIA) may be viewed either as a closure or as the *asymptotically exact* solution of a sequence of stochastic models closely resembling the Navier–Stokes equation, when written in the abstract form (9.28). The models are obtained by introducing random couplings among N replicas of the Navier–Stokes equation, labeled by Greek indices, running from 1 to N. Specifically, (9.28) is changed into

$$v_\alpha(1) + \frac{1}{N}\Phi_{\alpha\beta\delta}\,\gamma(1,2,3)\,v_\beta(2)\,v_\delta(3) = s_\alpha(1). \qquad (9.33)$$

Here, the $s_\alpha(.)$s are independent copies of the centered Gaussian random force[25] and the $\Phi_{\alpha\beta\delta}$s are centered Gaussian random variables of unit

[24] In Kraichnan's theory, the corrections to normal scaling involve powers of ℓ/ℓ_0.

[25] More precisely, of the inverse of the heat operator $\partial - \nu\nabla^2$, applied to the random force.

variance, which are all taken to be independent, except for the constraint that the tensor $\Phi_{\alpha\beta\delta}$ be completely symmetric in α, β and δ. It is easily checked that this constraint ensures that the *random coupling model* (9.33) has the same quadratic invariants (e.g., energy and helicity) as the original equation (9.28).

It may be shown that, in the limit $N \to \infty$: (i) any *finite* subset of the $v_\alpha(.)$s of different αs becomes a system of independent Gaussian centered random functions with common correlation function $\langle v(1)v(2)\rangle$, (ii) the Green's function $g_{\alpha\beta}(1,2) \equiv \partial v_\alpha(1)/\partial s_\beta(2)$ tends to a deterministic limit $\langle g(1,2)\rangle\,\delta_{\alpha\beta}$. The correlation function and the (mean) Green's function satisfy the so-called DIA equation, which can be written in various ways, the most symmetrical being

$$\left.\begin{aligned}
&\langle g(1,2)\rangle = \\
&\delta(1,2) + 4\gamma(1,3,4)\,\gamma(5,6,7)\langle v(3)v(6)\rangle\langle g(4,5)\rangle\langle g(7,2)\rangle, \\
&\langle v(1)v(1')\rangle = \\
&\langle g(1,2)\rangle\langle g(1',2')\rangle\,\langle s(2)s(2')\rangle \\
&+2\langle g(1,2)\rangle\langle g(1',2')\rangle\gamma(2,3,4)\gamma(2',3',4') \\
&\times\langle v(3)v(3')\rangle\langle v(4)v(4')\rangle,
\end{aligned}\right\} \qquad (9.34)$$

where $\delta(1,2)$ denotes the identity.

The derivation of the DIA equations from the random coupling model may be found in Kraichnan (1961). It uses a diagrammatic method. The key observation is that, when the $\Phi_{\alpha\beta\delta}$-factors are introduced, the weight put on most diagrams in the limit $N \to \infty$ goes to zero; the surviving diagrams[26] can be summed and lead to closed equations for the correlation and mean Green's functions. A nondiagrammatic derivation may be found in Lesieur, Frisch and Brissaud (1971).

The random coupling model shares a number of properties with the Navier–Stokes equation: invariance under space- and time-translations, under rotations, under scaling transformations, the same linear and quadratic conserved quantities, etc. When Gaussian initial conditions are assumed and the correlation function is expanded in powers of the Reynolds number R, there is agreement up to terms $O(R^3)$. One major discrepancy is the lack of invariance of the random coupling model

[26] Which may technically characterized by the absence of 'vertex corrections'.

(and of the DIA equations) to random Galilean transformations, from which it follows that the $k^{-5/3}$ law for the energy spectrum does not hold (Kraichnan 1958). Instead, in the inertial range, the DIA gives $E(k) = \text{const}\,(\varepsilon v_0)^{1/2}\,k^{-3/2}$, so that the energy spectrum depends not only on the wavenumber k and on the mean dissipation rate ϵ, but also on the r.m.s. velocity v_0. Although experimental data give a clear advantage to a $k^{-5/3}$ law over a $k^{-3/2}$ law, the relation between invariance under random Galilean transformations and localness (see p. 105) is not clear. Kraichnan (1959, p. 536) pointed out that 'A prominent feature of turbulence at high Reynolds numbers is the presence of sharply defined, extended, and tangled vortex sheets and filaments.' If filaments and sheets have an extension comparable to the integral scale ℓ_0, then the effect of the energy-containing eddies on such structures clearly cannot be represented just as a quasi-uniform sweep.[27] However, there may be other, less coherent, inertial-range structures which contribute more than filaments to the energy spectrum and which are approximately swept by the energy-containing eddies.

Kraichnan (1964, 1965, 1966) found a way to overcome the difficulty with random Galilean transformations and developed the Lagrangian history direct interaction (LHDI).[28] This makes use of a generalized velocity $\boldsymbol{v}(\boldsymbol{r}, t \mid t')$ which has both Eulerian and Lagrangian characteristics. It is defined as the velocity measured at time t' in that fluid element which passes through $\boldsymbol{r}$ at time t. When the DIA equations are written in explicit form for the Navier–Stokes equation, there appear integrals over the past history of the flow, which can be altered when working with the generalized field in such a way as to recover random Galilean invariance.

9.5.4 Closures and their shortcomings

Insofar as Landau's objection to the universality of the Kolmogorov constant (Section 6.4) is ignored or considered a marginal issue, it is a reasonable goal to develop a closure theory able to predict the Kolmogorov constant and other similar dimensionless constants, not given by scaling arguments. At least, one would like to reduce the number of constants which must be independently adjusted from experimental and/or numerical data. If a closure theory fails to give the $k^{-5/3}$ energy

[27] This perhaps explains why some data from the simulation of decaying turbulence at $R_\lambda < 200$ agree with the DIA better than they agree with closures that are consistent with random Galilean transformations (Herring and Kerr 1993).

[28] This is quoted freely from Kraichnan (1975).

spectrum, it is most likely that it is inconsistent with one or several of the basic properties of the Navier–Stokes equation, usually Galilean invariance. No closure approximation has so far managed to capture any genuine intermittency corrections to K41.

Closures (for homogeneous and isotropic turbulence) are obtained by heuristically modifying an infinite set of equations, such as the single-time cumulant hierarchy (9.18)–(9.20). The simplest way is to work only with the energy transfer relation (6.37) and to express the energy flux Π_k in terms of the energy spectrum $E(k)$. If the relation involves only $E(k)$ and k and is local (in the Fourier space), consistency with K41 is inescapable. A somewhat more sophisticated way is to work with the first two cumulant relations (9.18)–(9.19) and to express the unknown term, involving the fourth order cumulant, in terms of second and third order cumulants. Even more sophisticated closures use partial resummations of the Feynman diagrams generated, for example, from the functional representation discussed in Section 9.5.2. No attempt will be made here to review the subject of closure and its numerous applications (see, e.g., Panchev 1971; Orszag 1977; Rose and Sulem 1978; Lesieur 1990).

Some of the early closures, particularly the quasi-normal approximation of Millionshchikov (1941a, b) in which the fourth order cumulant is discarded, had difficulties coping with probabilistic constraints, such as the positivity of the energy spectrum (Ogura 1963; see also Tatsumi 1957). Positivity is automatically guaranteed for *realizable* closures; realizability refers to the existence of a stochastic model of which the closure is the (asymptotically) exact solution, just as the DIA is the solution of the random coupling model (Section 9.5.3).

A very frequently used realizable closure is the eddy damped quasi-normal Markovian approximation (EDQNM; Orszag 1966, 1977). The corresponding stochastic model is a modification of Kraichnan's random coupling model (9.33): the random coupling factors $\Phi_{\alpha\beta\delta}$ depend now on the triad of interacting wavevectors $\boldsymbol{k}$, $\boldsymbol{p}$ and $\boldsymbol{q}$ which appear in the nonlinear term (when written in the Fourier space); they are taken to be Gaussian and δ-correlated in the time variable. The coefficient θ_{kpq}, in front of the δ-function, which has the dimension of time, is generally based on (a suitably symmetrized version of) the local eddy turnover time. This ensures compatibility with K41, but at the price of introducing at least one adjustable numerical constant. The somewhat more cumbersome LHDI (Section 9.5.3), which is free of any adjustable constants, appears in practice to be realizable, but has never been proven so.

Closure approximations, because they work only with the second order moments of the velocity,[29] ignore any structure which may be present in the flow, such as vortex sheets or filaments. As a consequence, they cannot correctly represent a statistical distribution of structures with strongly depleted nonlinearities (see pp. 119 and 201). The resulting overestimation of the strength of nonlinearities can result in spurious finite-time blow-up for the initial value problem with zero viscosity. We know that for the three-dimensional Euler problem the blow-up issue is far from settled (Sections 7.8 and 9.3); as for the EDQNM closure, it unequivocally predicts blow-up (see, e.g., Lesieur 1990). A more dramatic discrepancy is obtained for the two-dimensional equations of magnetohydrodynamics (MHD). Catastrophic stretching of the vorticity by the Lorentz force[30] leads to blow-up of the mean square vorticity in closure calculations (Pouquet 1978). Actually, Frisch, Pouquet, Sulem and Meneguzzi (1983) showed that two-dimensional ideal MDH flow develops current sheets with strongly depleted nonlinearities; gradients (of the velocity and of the magnetic field) grow to very large values, but the growth is only exponential, at least for the moderately long times accessible in the simulations. They also showed that this depletion cannot be captured by a closure, unless it includes (at least) fourth order moments.

There are some cases where closed equations for the energy spectrum can be derived *systematically*. This requires the presence of an expansion parameter other than the Reynolds number. One instance is the problem studied by Forster, Nelson and Stephen (1977) by renormalization group techniques (see Section 9.6.4). Another instance, with many applications in geophysics, plasma physics and solid state physics, is the theory of *resonant wave interactions*, developed by Benney and Saffman (1966) and Benney and Newell (1969; see also Zakharov, L'vov and Falkovich 1992; L'vov 1994). The general setting is a nonlinear equation of the following form:

$$\partial_t v = \epsilon N(v) + \mathscr{D} v. \tag{9.35}$$

Here, $N(v)$ is some nonlinear partial differential operator, $\mathscr{D}$ is a purely dispersive linear operator with eigenvalues $i\omega_k$ (functions of the wavevector $\boldsymbol{k}$) and ϵ is a small parameter. To leading order, only resonant wave interactions survive, i.e. nonlinear interactions between wavevectors such that the phase-shifts due to the dispersive terms cancel out.[31] The original

[29] And partially with third order moments.
[30] In the absence of a magnetic field the vorticity is conserved; see Section 9.7.
[31] Otherwise, there is suppression of nonlinearity by phase-mixing.

theory was worked out in terms of the single-time cumulant hierarchy. Depending on the form of the nonlinearity and of the dispersion relation ω_k, cumulants beyond some order may be phase-mixed. In the simplest cases, the quasi-normal approximation may become asymptotically exact.[32]

9.6 Eddy viscosity, multiscale methods and renormalization

9.6.1 Eddy viscosity: a very old idea

The concept of eddy viscosity and the related concept of eddy diffusivity (for the transport of scalar quantities) have become perhaps the most frequently used tools in the modeling of inhomogeneous turbulent flow, including in the work of Kolmogorov (1942). The idea originated in the nineteenth century.

As we have know since the work of Maxwell on kinetic theory, one manifestation of molecular motion and collisions is momentum transport (diffusion of momentum): in a fluid, shear (gradients of the macroscopic velocity) produces stress (momentum fluxes). Locally, the stress is proportional to the strain (symmetric part of the velocity gradient). The coefficient of proportionality, called the viscosity, is typically of the order of the velocity of thermal molecular motion multiplied by the mean-free-path.

Turbulent transport of momentum might be regarded in an analogous way to molecular transport, with the small-scale eddies playing the role of molecules and the correlation length (integral scale) playing the role of the mean-free-path. This is roughly the vision of Prandtl (1925) who introduced the concept of mixing length.[33] The analogy between microscopic and turbulent transport is hinted at in Lord Kelvin (1887; p. 352) and appears clearly in Lamb (1932; already in the fourth edition of 1916) when presenting Reynolds' (1894) theory of the 'Reynolds stresses'. Lamb also attributes to Reynolds the introduction of the eddy viscosity.

Actually, the concept of eddy viscosity emerged before Maxwell's work. We shall now try to give a brief account of what we found by searching mostly through publications of the French Academy of Sciences. After

[32] It was found by Frisch, Legras and Villone (1996) that the cumulant hierarchy approach may give erroneous results when the wavevectors are discrete rather than continuous. The correct approach makes use of averaging and normal form techniques which have been used for a very long time in celestial mechanics (see, e.g., Arnold, Kozlov and Neishtadt 1988).

[33] The concept of mixing length for the transport of vorticity was already present in the work of Taylor (1915).

the introduction of the viscous term in the equations of hydrodynamics by Navier (1823), the question arose of whether the viscosity coefficient (usually denoted ϵ at that time) is the same everywhere. Saint-Venant[34] (1843) worried briefly about this and a few years later wrote the following (Saint-Venant 1851), which is quoted by Boussinesq (1877, p. 7)[35]:

Si l'hypothèse de Newton, reproduite par Navier et Poisson, et qui consiste à prendre le frottement intérieur proportionnel à la vitesse des filets glissants les uns devant les autres, peut être appliquée approximativement pour les divers points d'une même section fluide, tous les faits connus portent à inférer qu'il faut faire croître le coefficient de cette proportionnalité avec les dimensions des sections transversales; ce qui s'explique jusqu'à un certain point en remarquant que les filets ne marchent pas parallèlement entre eux avec des vitesses régulièrement graduées de l'un à l'autre, et que les ruptures, les tourbillonnements et les autres mouvements compliqués ou obliques, qui doivent beaucoup influer sur la grandeur des frottements, se forment et se développent davantage dans les grandes sections.[36]

Boussinesq, a former student of Saint-Venant, carried these ideas further and stressed that turbulence will greatly increase the viscosity. After recalling the work of Poiseuille on the measurement of viscosity for laminar flow, he writes the following, which contains a number of pioneering ideas on turbulence (Boussinesq 1870):

Mais je fais voir au §IX de ce même Mémoire [Boussinesq 1868; pp. 402–403] qu'il n'en est pas ainsi lorsqu'il s'agit de canaux découverts ou de tuyaux de conduite d'un certain calibre. Le liquide, n'étant plus alors aussi resserré latéralement, possède toujours des mouvements oscillatoires rapprochant ou éloignant brusquement des parois le fluide qui en est voisin. L'action tangentielle qu'exerce la paroi sur ce fluide change donc sans cesse, et, par ses variations combinées avec la vitesse générale de translation du même fluide, imprime à ce dernier des mouvements rotatoires. Ceux-ci se transmettant aux couches liquides plus intérieures, toute la masse fluide est bientôt sillonnée de tourbillons dont la matière glisse, avec une vitesse relative finie, sur celle qui l'environne. La moyenne des vitesses observées en un même point durant un petit instant n'est donc plus sensiblement égale à chacune d'elles, et la force moyenne tangentielle exercée à travers un petit élément plan fixe doit dépendre, non seulement de la manière dont varie cette vitesse moyenne aux points environnants,

34 His full last name is Barré de Saint-Venant, but his name is mostly quoted as Saint-Venant.

35 A somewhat briefer version of this may be found in Saint-Venant (1850).

36 In English: If Newton's assumption, reproduced by Navier and Poisson, which consists in taking interior friction proportional to the speed of the fluid elements sliding against each other, can be applied approximately to the set of points of a given fluid section, all the known facts lead us to infer that the coefficient of this proportionality [the viscosity] should increase with the size of transverse sections; this may be explained up to a point by noticing that the fluid elements are not progressing parallel to each other with regularly graded velocities, and that ruptures, eddies and other complex and oblique motions, which must strongly affect the magnitude of frictions, are formed and develop more in large sections.

c'est-à-dire des dérivées du premier ordre par rapport aux coordonnées x, y, z de ses trois composantes u, v, w suivant les axes, mais encore de la grandeur et du nombre des discontinuités dont les vitesses vraies y sont affectées. En effet, les frottements produits dans ce cas étant dus à des glissements finis entre couches adjacentes, doivent être bien plus grands que si les vitesses variaient avec continuité de chaque point aux points voisins.[37]

In the same paper Boussinesq (1870) proposes an expression for the friction coefficient (eddy viscosity) for pipe flow, namely Au_0h, where A is a dimensionless constant depending on wall roughness, u_0 is the speed of the liquid near the wall and h is the mean radius. In the language of Taylor and Prandtl this would be called a mixing-length expression. Further expressions are presented in the *Memoir* (Boussinesq 1877).[38]

Note that the remarkable work of Saint-Venant and Boussinesq on turbulence had its experimental roots in the careful observation of *water* flow in canals, the practical importance of which was still considerable in the nineteenth century. This may also explain why it fell somewhat into oblivion. Eventually, the needs of the emerging science of aerodynamics at the beginning of the twentieth century led to the rediscovery of many of the ideas of Saint-Venant and Boussinesq.

From a more fundamental viewpoint, it seems that the idea of eddy viscosity emerged because the natural state of many 'real' flows is turbulent, so that the concept of pure molecular viscosity seemed inappropriate. It must be stressed that, early in the nineteenth century, the distinction between true molecules and fictitious 'fluid molecules' was a bit blurred. Eddies were probably considered just one particular form of molecules.

[37] In English: But I show in §IX of said *Memoir* [Boussinesq 1868; pp. 402–403] that it is not so when open channels or pipes of a sufficient diameter are involved. The liquid, being much freer of lateral constraints, possesses always oscillatory motions which will suddenly move neighboring fluid towards the wall or away from it. The tangential stress exerted by the wall on the fluid thus varies constantly, and, by its variations combined with the general translation speed of that very fluid, communicates rotary motions to the fluid. These being transmitted to liquid layers further inside, the whole mass of the fluid is soon threaded by eddies the substance of which slides, with a finite relative velocity, over the surrounding substance. The mean of the velocities observed at a given point during a small interval of time thus differs significantly from individual velocities, and the mean tangential force exerted through a given small planar element should depend not only on the manner of variation of the velocity at nearby points, i.e. on the derivatives of the first order with respect to the coordinates x, y, z of its three components u, v, w along the axes, but also on the magnitude of the velocity and the number of discontinuities it suffers. Indeed, the friction experienced, being caused by finite sliding between adjacent layers, will be much larger than would be the case should the velocities vary in a continuous way from each point to neighboring points.

[38] This *Memoir* is quoted, e.g., by Hinze (1959) and Monin and Yaglom (1971) as originating the idea of eddy viscosity. It is also famous for its interpretation of Scott Russell's solitary wave.

Further investigation of the interplay of ideas in kinetic theory and in fluid dynamics during the nineteenth century would be of considerable interest.[39]

The analogy with kinetic theory can help us to understand the limitations of the concept of eddy viscosity and also why this concept has been regarded by some theoreticians of turbulence as (at best) a pedagogical device. Hilbert, Chapman and Enskog, at the beginning of the twentieth century, were able to present systematic derivations of the equations of hydrodynamics from kinetic theory (for references, see Brush 1976). They started from the Boltzmann equation and used singular expansion methods in which the expansion parameter, the ratio of the mean-free-path to the scale of hydrodynamical motion, known as the Knudsen number, has to be small.

In contrast, when eddy viscosity ideas are applied to turbulence modeling, there is usually no clear separation of scales: the integral scale (the analog of the mean-free-path) is comparable to the scale of inhomogeneities. It thus becomes hard to justify the use of an eddy viscosity. Indeed, in many instances where an eddy viscosity argument gives a substantially correct answer, alternative more systematic arguments can be used. Two examples will suffice.

Landau and Lifshitz (1987) used an eddy viscosity presentation of the Kolmogorov 1941 theory of homogeneous turbulence. It gives the 'right' result, i.e. $v_\ell \sim (\varepsilon\ell)^{1/3}$. Such a result is, however, uniquely constrained by dimensional analysis (if the only ingredients are ε and ℓ). Hence, any dimensionally consistent approach gives the same result. Actually, Kolmogorov's derivation by a similarity argument is more fundamental. Note also that, in this example, the eddy viscosity depends on the scale under consideration.

Von Kármán (1930; see also Goldstein 1969, p. 21) and Prandtl (1932) derived the logarithmic law of the variation of the mean velocity with respect to the distance from the wall for a turbulent boundary layer. Again, this can be done by an eddy-viscosity argument or by a somewhat deeper similarity argument (both may be found in Section 7-4 of Hinze 1959). Note that, this time, the eddy viscosity invoked depends on the point under consideration.

Fortunately, there are many problems in turbulence where a clear separation of scale is present and justifies the use of an eddy viscosity and, more generally, of *multiscale methods*. These are among the few

[39] An exposition of the early history of kinetic theory may be found in Brush (1976).

truly systematic tools which exist in turbulence and deserve to be studied carefully. In Section 9.6.2, we give a general idea of what is involved and a guide to the existing literature. In Section 9.6.3, we give an overview of the variety of problems which can be treated by such techniques. In Section 9.6.4, we discuss the renormalization group approach to fully developed turbulence, which may be viewed as an extension of multiscale methods with admixture of heuristic elements.

9.6.2 Multiscale methods

For pedagogical reasons, we shall begin with an elementary example, the heat equation with a diffusivity varying periodically in space. Our treatment is based on Bensoussan, Lions and Papanicolaou (1978) where the method is known as *homogenization*, i.e. the replacement of a heterogeneous material by one that is homogeneous at large scales. Consider a one-dimensional material, say a metal rod, with a diffusivity (conductivity) $\kappa(x)$ which is 2π-periodic. The temperature field $\theta(x,t)$ satisfies the heat equation:

$$\partial_t\theta = \partial_x\kappa(x)\partial_x\theta. \tag{9.36}$$

(Here, and in what follows, the leftmost derivative ∂_x acts on the entire expression on its right.) We wish to show that on scales large compared to 2π the diffusion of heat may be described by a heat equation with a uniform diffusivity $\kappa_{\rm eff}$, called the *effective (or eddy) diffusivity* which we intend to calculate.

We first give a very simple heuristic argument. Suppose that at two points x_1 and x_2, separated by a distance large compared to the period 2π, the temperature is maintained at two fixed values $\theta(x_1)$ and $\theta(x_2)$. A constant (in space and time) heat flux Φ will be established after relaxation of transients. Locally, the temperature gradient and the heat flux are related by

$$\Phi = -\kappa(x)\partial_x\theta. \tag{9.37}$$

Hence,

$$\begin{aligned}\theta(x_2)-\theta(x_1) &= -\,\Phi\int_{x_1}^{x_2}\frac{dx}{\kappa(x)}\\ &\simeq -\,(x_2-x_1)\left\langle\frac{1}{\kappa(x)}\right\rangle\Phi,\end{aligned} \tag{9.38}$$

where

$$\left\langle \frac{1}{\kappa(x)} \right\rangle \equiv \frac{1}{2\pi} \int_0^{2\pi} \frac{dx}{\kappa(x)}. \tag{9.39}$$

This may be rewritten as

$$\Phi = -\frac{1}{\langle 1/\kappa(x) \rangle} \frac{\theta(x_2) - \theta(x_1)}{x_2 - x_1} = -\kappa_{\text{eff}} \frac{\theta(x_2) - \theta(x_1)}{x_2 - x_1}, \tag{9.40}$$

which demonstrates the existence of an effective diffusivity equal to the *harmonic mean* of the diffusivity, i.e. the inverse of the average (over the period) of the inverse diffusivity.

Now, we rederive this result using a multiscale method. We assume that the temperature field θ depends not just on the *fast* variables x and t, but also on the *slow* variables $X = \epsilon x$ and $T = \epsilon^2 t$, where ϵ is a small parameter.[40] We assume that θ has the same 2π-periodicity in the fast variable x as the diffusivity $\kappa(x)$ and no dependence on the fast time t.[41] Since the temperature field depends on x and t both through the fast variables and through the slow variables, space- and time-derivatives must be decomposed as follows:

$$\partial_x \to \partial_x + \epsilon \partial_X, \qquad \partial_t \to \partial_t + \epsilon^2 \partial_T. \tag{9.41}$$

We now expand the temperature field in powers of ϵ:

$$\theta = \theta^{(0)} + \epsilon \theta^{(1)} + \epsilon^2 \theta^{(2)} + \cdots, \tag{9.42}$$

where all the $\theta^{(n)}$s depend, in principle, on x, t, X and T. We then use (9.41) and (9.42) in (9.36) and identify the various powers of ϵ. Only the first three equations are needed. They read

$$\partial_t \theta^{(0)} = \partial_x \kappa(x) \partial_x \theta^{(0)}, \tag{9.43}$$

$$\partial_t \theta^{(1)} = \partial_x \kappa(x) \partial_x \theta^{(1)} + \partial_x \kappa(x) \partial_X \theta^{(0)} + \partial_X \kappa(x) \partial_x \theta^{(0)}, \tag{9.44}$$

and

$$\begin{aligned} \partial_t \theta^{(2)} + \partial_T \theta^{(0)} = {} & \partial_x \kappa(x) \partial_x \theta^{(2)} + \partial_x \kappa(x) \partial_X \theta^{(1)} \\ & + \partial_X \kappa(x) \partial_x \theta^{(1)} + \partial_X \kappa(x) \partial_X \theta^{(0)}. \end{aligned} \tag{9.45}$$

Eq. (9.43) expresses that $\theta^{(0)}$ is in the null-space of the heat operator

[40] The rationale for rescaling the time variable by a factor ϵ^2 is that diffusion on spatial scales $O(1/\epsilon)$ is expected to occur on times scales $O(1/\epsilon^2)$; any other choice would lead to mathematical inconsistencies.

[41] A nontrivial dependence on the fast time is obtained only if the diffusivity κ itself is time-dependent.

$\partial_t - \partial_x \kappa(x) \partial_x$. With the assumed 2π-periodicity, $\theta^{(0)}$ will relax to a constant, independent of x and t. In other words, except for transients, $\theta^{(0)}$ does not depend on the fast variables. It follows that, in (9.44), the term containing $\partial_x \theta^{(0)}$ vanishes. Eq. (9.44) may be considered as an equation for $\theta^{(1)}$, containing an inhomogeneous term $\partial_x \kappa(x) \partial_X \theta^{(0)}$, which is independent of t. Hence, $\partial_t \theta^{(1)}$ may be assumed to vanish (after relaxation of transients). The remaining terms all begin with a ∂_x and may be integrated to give

$$\kappa(x) \left(\partial_x \theta^{(1)} + \partial_X \theta^{(0)} \right) = C. \tag{9.46}$$

Dividing by $\kappa(x)$ and observing that $\langle \partial_x \theta^{(1)} \rangle$ vanishes by the assumed periodicity, we find that

$$C \left\langle \frac{1}{\kappa(x)} \right\rangle = \partial_X \theta^{(0)}, \tag{9.47}$$

where the averaging is now only over fast variables. A most interesting phenomenon happens with (9.45): it cannot be solved for $\theta^{(2)}$ unless an additional constraint is imposed on $\theta^{(0)}$. Indeed, taking the average of (9.45), we obtain (note that $\langle \theta^{(0)} \rangle = \theta^{(0)}$)

$$\partial_T \theta^{(0)} = \partial_X \langle \kappa(x) \rangle \partial_X \theta^{(0)} + \partial_X \left\langle \kappa(x) \partial_x \theta^{(1)} \right\rangle. \tag{9.48}$$

This condition, known as the *solvability condition*, occurs in all singular perturbation problems.[42] We now use the value of $\theta^{(1)}$ from (9.46) and (9.47) in (9.48) and obtain, after a cancellation of the term $\partial_X \langle \kappa(x) \rangle \partial_X \theta^{(0)}$

$$\partial_T \theta^{(0)} = \kappa_{\text{eff}} \partial_X \partial_X \theta^{(0)}, \tag{9.49}$$

with κ_{eff} given by the same expression (9.40) as in the previous heuristic derivation.

The multiscale method just described is easily extended to more than one dimension, when κ is periodic in several variables. There is one major difference: the analog of (9.44) can no longer be solved explicitly. However, $\theta^{(1)}$ still depends linearly on $\partial_X \theta^{(0)}$, the large-scale temperature gradient. Therefore, the solvability condition for (9.45) again takes the form of a (generally anisotropic) diffusion equation with constant coefficients. The expression for the effective diffusivity (now a second order tensor) thus involves the solution of an auxiliary problem. This solution must be determined numerically or perturbatively by taking advantage of an expansion parameter other than ϵ. It is intuitively clear

[42] It expresses a Fredholm alternative, namely that the inhomogeneous term in the equation for $\theta^{(2)}$ is orthogonal to the null-space of the adjoint of the operator acting on $\theta^{(2)}$.

that, even in the isotropic case where the effective diffusivity is just a scalar, its value is not the harmonic mean of the periodically varying diffusivity. Indeed, in one dimension, regions with a very low diffusivity strongly deplete the effective diffusivity by blocking the heat flux, whereas in more than one dimension, the heat may diffuse around the regions of low diffusivity.

In the example of the heat equation we have encountered all the key ingredients of multiscale methods for linear problems: One starts with a linear partial differential equation $A\phi = 0$ where the field ϕ may be scalar- or vector-valued and the operator A possesses a nontrivial null-space $\phi^{(0)}(x,t)$. One then looks for solutions in which the null-space function is slowly modulated in space and time, that is, depends on the slow variables $X = \epsilon x$ and $T = \epsilon^s t$. The form of the large-scale (or homogenized) equation and the value of the exponent s depend mostly on the *symmetries* of the problem. This is best illustrated by the examples discussed in Section 9.6.3. Technically, the solution ϕ is extended into a function of both the fast and the slow variables. Derivatives are decomposed according to $\partial_x \to \partial_x + \epsilon \partial_X$ and $\partial_t \to \partial_t + \epsilon^s \partial_T$. The solution is expanded in powers of ϵ and substituted into the equation (with decomposed derivatives). Equations for the various orders in ϵ are then obtained. The large-scale equation always emerges as a solvability condition to some order in ϵ. The values of the coefficients in the large-scale equation are obtained in terms of the solution of the lower order equations. Analytic expressions for these are available only in special cases.

The technique is easily extended to nonlinear equations, of the form $A\phi + B(\phi) = 0$, where $B(\phi)$ is a nonlinear functional of ϕ. The main new difficulty is that it is necessary to find the order in ϵ of the leading term in ϕ: this is often done by a dominant balance argument in the large-scale equation (the form of which can usually be guessed by symmetry arguments). The final equation, which emerges again as a solvability condition, is nonlinear; still, the auxiliary problems to be solved at intermediate orders are usually linear in the unknown fields.

The reader interested in learning about multiscale methods (a good investment in this author's view) can study the general theory of homogenization in the book by Bensoussan, Lions and Papanicolaou (1978). It is written for the mathematician rather than for the physicist and the emphasis is on problems related to the heat equation. An introduction to multiscale methods for the transport of passive scalars in a turbulent flow may be found in Section 3.1 of Frisch (1989) and in Section 2 of Fauve

(1991). Among the research papers which are quoted in the next section, some may be used as introductory material. These include Fannjiang and Papanicolaou (1994), Dubrulle and Frisch (1991) and Gama, Vergassola and Frisch (1994).

9.6.3 Applications of multiscale methods in turbulence

Multiscale methods are now quite extensively used in turbulence. Typically, the questions which can be addressed by them are of the following form: a basic 'small-scale' flow with some form of translation invariance (periodicity, quasi-periodicity, random homogeneity) is prescribed. This flow is 'perturbed' by the introduction of either an advected scalar or a modification of the flow itself. The basic flow has spatial scales $O(1)$ and the perturbation is at much larger scales, which are $O(1/\epsilon)$. In other words, one considers the basic flow as a kind of 'microscopic'[43] material and one tries to find its macroscopic properties. Depending on how weak the perturbation is (or remains), linear or nonlinear theory has to be used.

9.6.3.1 Turbulent transport of scalars

The simplest and oldest problem which can be solved by multiscale techniques is the *turbulent transport of passive scalars*, such as a dye or a passive temperature field. This basic equation is an advection diffusion equation:

$$\partial_t \theta + \boldsymbol{v} \cdot \nabla \theta = D \nabla^2 \theta. \tag{9.50}$$

Here, $\theta(\boldsymbol{r}, t)$ is the concentration of the passive scalar at the position $\boldsymbol{r}$ and the time t, D is the molecular diffusivity and $\boldsymbol{v}(\boldsymbol{r}, t)$ is the prescribed velocity field, which can be either space-time periodic or random homogeneous and stationary. It is essential that $\langle \boldsymbol{v} \rangle = 0$. Otherwise, advection by the mean velocity dominates the large-scale dynamics on time scales $O(1/\epsilon)$. Formal application of the multiscale techniques (see, e.g., Frisch 1989) leads to an anisotropic diffusion equation for the leading order term in the slow variables $X = \epsilon x$ and $T = \epsilon^2 t$:

$$\partial_T \theta^{(0)} = D_{ij} \nabla_i \nabla_j \theta^{(0)}, \tag{9.51}$$

where ∇ denotes space-derivatives in the slow variable and D_{ij} is an effective diffusivity tensor, the expression of which involves the solution

[43] 'Mesoscopic' would be more appropriate, since no atomic scales are involved.

of an auxiliary equation. In those instances where the flow has enough symmetry to guarantee isotropy[44] of the tensor D_{ij}, we shall write

$$D_{ij} = \frac{1}{3} D_{\mathrm{E}} \delta_{ij}, \tag{9.52}$$

where D_{E} is the effective (or eddy) diffusivity. D_{E} is always greater or equal to the molecular diffusivity: turbulence can only enhance diffusion.

If we think of $\theta^{(0)}$ as representing the p.d.f. of a particle subject to turbulent diffusion, a consequence of (9.51) and (9.52) is that the mean square displacement of the particle grows linearly in time:

$$\langle R^2(T) \rangle \equiv \int |X|^2 \theta^{(0)}(X, T) d^3X = \langle R^2(0) \rangle + 2D_{\mathrm{E}} T. \tag{9.53}$$

This is characteristic of diffusive behavior. Actually, in a turbulent flow the long-time behavior may display *anomalous diffusion.* For example it may be superdiffusive(the mean square displacement grows faster than T) or subdiffusive (growth slower than T). Such questions and many others are discussed in the review by Isichenko (1992; see also Avellaneda and Majda 1990). In such instances, there is no effective diffusivity and the large-scale behavior is not governed by an ordinary diffusion equation.

When normal diffusion holds, it is of interest to find upper and lower bounds for D_{E} and to study its behavior as the molecular diffusivity tends to zero. For periodic flow, some of the best results may be found in Fannjiang and Papanicolaou (1994; see also references therein). For turbulent flow, the questions was first addressed by Taylor (1921) who actually ignored the molecular diffusion and observed that, for times long compared to the Lagrangian velocity autocorrelation time, the mean square displacement of a particle, starting at the Lagrangian location $\boldsymbol{a}$, is given by:

$$\langle R^2(T) \rangle \simeq 2T \int_0^\infty \langle \boldsymbol{v}^{\mathrm{L}}(\boldsymbol{a}, s) \cdot \boldsymbol{v}^{\mathrm{L}}(\boldsymbol{a}, 0) \rangle \, ds, \tag{9.54}$$

where $\boldsymbol{v}^{\mathrm{L}}(\boldsymbol{a}, s)$ is the velocity in Lagrangian coordinates. In view of (9.53), (9.54) implies that the limit of the effective diffusivity for vanishing molecular diffusivity is simply the time-integral of the Lagrangian autocorrelation function.[45] Taylor's formula may become invalid in situations where particles can get trapped. This can happen because of the presence of either closed stream-lines, Kolmogorov–Arnold–Moser

[44] This holds trivially if the flow is random and isotropic; it also holds if it has square symmetry in two dimensions or cubic symmetry in three dimensions.

[45] A simple argument of Saffman (1960) shows that this limit is mostly attained from below, but there are couterexamples (M. Vergassola, private communication).

tori (as in the ABC flows defined in Section 9.4) or closed pockets of recirculation (Pomeau, Pumir and Young 1988).

9.6.3.2 Transport of vector quantities and large-scale instabilities

When a prescribed flow $\boldsymbol{v}$ is subject to a weak perturbation $\boldsymbol{w}$, the perturbation satisfies, to leading order, the *linearized Navier–Stokes* equation:

$$\left.\begin{aligned} \partial_t \boldsymbol{w} + \boldsymbol{v}\cdot\nabla\boldsymbol{w} + \boldsymbol{w}\cdot\nabla\boldsymbol{v} &= -\nabla p' + \nu\nabla^2\boldsymbol{w}, \\ \nabla\cdot\boldsymbol{w} &= 0. \end{aligned}\right\} \tag{9.55}$$

Eq. (9.55) has some similarity with (9.50), the advection–diffusion equation for a passive scalar. The main difference, stemming from the vector character of the perturbation, is the third term on the l.h.s. of the first equation, whereby the perturbation couples to the gradient of the prescribed velocity field. This is why there is considerably more life in the large-scale transport of vector quantities.

One possibility is that the large-scale dynamics is first order in both time- and space-derivatives. Indeed, it has been shown that when the basic flow has no center of symmetry, that is, $\boldsymbol{v}$ lacks parity-invariance (cf. Section 2.2), the large-scale equation has the following form (Frisch, She and Sulem 1987; Sulem, She, Scholl and Frisch 1989):

$$\left.\begin{aligned} &\partial_T\langle w_i^{(0)}\rangle = \alpha_{ij\ell}\nabla_j\langle w_\ell^{(0)}\rangle - \nabla P, \\ &\nabla_j\langle w_j^{(0)}\rangle = 0. \end{aligned}\right\} \tag{9.56}$$

(In (9.56), the notation is the same as in Section 9.6.3.1.) In three dimensions, plane-wave solutions of (9.56) may grow exponentially with a rate proportional to the wavenumber. This is called the *anisotropic kinetic alpha* (AKA) effect and is the hydrodynamic analog of the alpha effect, known in magnetohydrodynamics since the 1960s (Steenbeck, Krause and Rädler 1969; see also Moffatt 1978). The alpha effect is frequently used to explain the growth of large-scale magnetic fields (dynamo effect) in conducting media lacking parity-invariance.[46] In contrast to the alpha effect, the AKA effect vanishes for isotropic flow, whence its name. The AKA effect arises basically from a modification of the Reynold stresses (the mean momentum flux) by a large quasi-uniform flow.[47] The

[46] The absence of parity-invariance is more fundamental than the presence of helicity, although the latter implies the former (Gilbert, Frisch and Pouquet 1988).

[47] This requires that Galilean invariance be broken, e.g., by the presence of a small-scale driving force.

nonlinear saturation of the AKA instability has been studied by Sulem, She, Scholl and Frisch (1989).

When the basic flow is parity-invariant, the large-scale equation has the following form (Dubrulle and Frisch 1991):

$$\left.\begin{aligned} \partial_T \langle w_i^{(0)} \rangle &= v_{ij\ell m} \nabla_j \nabla_\ell \langle w_m^{(0)} \rangle - \nabla_i P, \\ \nabla_j \langle w_j^{(0)} \rangle &= 0. \end{aligned}\right\} \tag{9.57}$$

This is a pressure-modified diffusion-like equation. For isotropic basic flow it reduces to:

$$\partial_T \langle \boldsymbol{w}^{(0)} \rangle = v_E \nabla^2 \langle \boldsymbol{w}^{(0)} \rangle, \qquad \nabla \cdot \langle \boldsymbol{w}^{(0)} \rangle = 0, \tag{9.58}$$

where v_E is the eddy viscosity. The calculation of $v_{ij\ell m}$ or of v_E in general requires the solution of two auxiliary problems. This can be done pertubatively when the Reynolds number of the basic flow is small (Dubrulle and Frisch 1991).[48]

The multiscale method thus provides a rational justification for the use of the old concept of eddy viscosity. A number of *caveat*s are needed however.

First, the derivation assumes a large separation of scales and this is not the case in most traditional applications of the concept of eddy viscosity.

Second, the eddy viscosity need not be positive: it has been shown, at least in two dimensions, that the eddy viscosity is frequently negative and thus leads to large-scale instabilities (Vergassola, Gama and Frisch 1993; Gama, Vergassola and Frisch 1994; see also Sivashinsky and Frenkel 1992).[49]

Third, when the basic flow is not isotropic, the quantities $v_{ij\ell m}$ in the large-scale equation (9.57) are the components of a fourth order tensor; the eddy viscosities are actually the eigenvalues of the linear operator which appears when plane-wave solutions are assumed in (9.57). In two dimensions they are always real, but in three dimensions they may become complex; when the imaginary part dominates, the large-scale dynamics is more Schrödinger-like than heat-like (Wirth, Gama and Frisch 1995).

[48] The exact expression of the eddy viscosity may be obtained by closure (Kraichnan 1976) in special instances, e.g., when the basic flow has a very short correlation time.

[49] The idea of negative eddy viscosity appears in Starr (1968) and has been proposed by Kraichnan (1976) as an interpretation of the inverse cascade of energy in two dimensions (see also Section 9.7). A negative eddy viscosity does not violate any thermodynamic principle, since there is usually a saturation mechanism, involving nonlinearities not represented in (9.55).

Fourth, at the same time as the basic flow induces a change in the viscosity from its molecular value to an eddy viscosity, it also usually *modifies the nonlinear term*: the coefficient in front of the advection term $\boldsymbol{v} \cdot \nabla \boldsymbol{v}$ changes from unity to another value which may vanish accidentally (Gama, Vergassola and Frisch 1994; Vergassola and Gama 1994);[50] furthermore, in the absence of mirror-invariance, a new 'chiral' nonlinear term appears (in two dimensions) which leads to drastic enhancement of vortical structures of a given sign (Vergassola 1993; Gama, Vergassola and Frisch 1994).

For parallel periodic flow, for example flow in the x_2-direction which depends only on x_1, the eddy viscosities can be calculated in explicit form. An instance is the Kolmogorov flow:

$$\boldsymbol{v} = (0, \sin x_1). \tag{9.59}$$

Nepomnyashchy (1976; see also Sivashinsky 1985) showed that large-scale perturbations perpendicular to the basic flow experience an eddy viscosity given by

$$\nu_{\mathrm{E}} = \nu - \frac{1}{2\nu}, \tag{9.60}$$

so that a large-scale instability[51] appears when $\nu < \nu_c = 1/\sqrt{2}$. Nepomnyashchy (1976) and Sivashinsky (1985) also studied the nonlinear régime in which the eddy viscosity is marginally negative. This is governed by an equation with cubic nonlinearity of the Cahn–Hilliard type, which has been investigated by She (1987; see also Frisch, Legras and Villone 1996). When the eddy viscosity is strongly negative, some transient aspects of the nonlinear régime are still within the scope of multiscale methods (E and Shu 1993).

Finally, we stress that what we have just discussed is certainly not exhaustive of the problems which can be tackled by multiscale methods (starting from the Navier–Stokes equation). For example, when Galilean invariance holds, the large-scale dynamics may be viscoelastic (Fauve 1983; Coullet and Fauve 1985; Frisch, She and Thual 1986).[52] A challenging problem is to find the large-scale description appropriate for a tangle of vortex filaments of the kind discussed in Section 8.9.

[50] In field-theoretic language, this is known as 'vertex renormalization'.

[51] This instability was actually discovered by Meshalkin and Sinai (1961).

[52] The possibility of viscoelastic behavior of turbulent flow was considered by Crow (1968) and, to some extent, by Lord Kelvin (1887).

9.6.4 Renormalization group (RG) methods

The term 'renormalization' was used originally to denote the elimination of infinities in quantum electrodynamics.[53] After the work of Dyson, Feynmann, Schwinger, Tomonoga and many more (see, e.g., Schwinger 1958), renormalization became one of the major tools of quantum field theory.

The applicability of similar ideas to the problem of *critical phenomena* was foreseen by Di Castro and Jona-Lasinio (1969). The 'renormalization group' (RG) method, which combines elements of quantum field renormalization with the spin-blocking ideas of Kadanoff (1966), was developed by Wilson and his colleagues (Wilson 1972; Wilson and Fisher 1972; Wilson and Kogut 1974; see also Toulouse and Pfeuty 1975; Amit 1984; Parisi 1988; Itzykson and Drouffe 1989; Le Bellac 1992). The resulting approach to scaling in critical phenomena is reminiscent of the Richardson cascade in turbulence: There is a hierarchy of scales of increasing length such that the dynamics at a given scale modifies the effective parameters describing the dynamics at the next scale. The renormalization group (actually, a semi-group) gives the transformation rules. The scaling laws are obtained from the asymptotic behavior after indefinite iteration.

As could be expected, attempts were quickly made to apply similar ideas to fully developed turbulence (Nelkin 1974; see also Rose and Sulem 1978; Eyink and Goldenfeld 1994). Twenty years later turbulence remains unsolved. However, RG methods stand a good chance of playing a role in the solution of the problem of turbulence. We shall therefore take the reader on a guided tour of what has been done so far. First, we should make it clear that none of the problems which have been solved in a *systematic way* by the use of RG methods are genuine turbulence problems; they would be better described as (nontrivial) statistical Navier–Stokes problems. Extensions of such work to the grand problem of turbulence involve a mixture of systematic and heuristic steps.

9.6.4.1 The Forster–Nelson–Stephen problem

The problem which Forster, Nelson and Stephen (1977; referred to hereafter as FNS) tackled by systematic RG methods is that of the

[53] As stressed by Coleman (1979), the first instance of renormalization actually appeared in hydrodynamics: a very light hollow sphere of finite radius, released at the bottom of a swimming pool, does not experience an enormous buoyancy-induced upward acceleration, since it acquires a *virtual mass* equal to one half of the mass of the water displaced (Stokes 1843; see also Landau and Lifshitz 1987, Section 11).

Navier–Stokes equation in the presence of a driving force (2.43)-(2.44), a problem closely parallel to dynamical (time-dependent) critical phenomena (see, e.g., Ma and Mazenko 1975). The force $\boldsymbol{f}$ is assumed to be Gaussian, white noise in time and such that the energy input to the fluid per unit mass and per unit wavenumber (forcing spectrum) is[54]

$$F(k) = Dk^{3-\epsilon}, \tag{9.61}$$

with $0 \leq \epsilon \ll 1$ and $D > 0$. Actually, FNS did not vary the exponent in the forcing power spectrum, but instead changed the dimension of the space, as in Wilson and Fisher (1972). Here, as in Frisch and Fournier (1978), the exponent is changed, while the space dimension is kept equal to three.

The resulting energy spectrum has the form

$$E(k) \propto k^{1-2\epsilon/3}. \tag{9.62}$$

This result can be predicted by a simple scaling argument. We observe that the random force $\boldsymbol{f}$, when written in the physical space, scales as follows:

$$\boldsymbol{f}(\lambda\boldsymbol{r}, \lambda^{1-h}t) \overset{\text{law}}{=} \lambda^{(\epsilon+h-5)/2}\boldsymbol{f}(\boldsymbol{r}, t), \tag{9.63}$$

where the symbol $\overset{\text{law}}{=}$, equality in law, is defined in Section 4.3. Eq. (9.63) follows from (9.61) and the observation that the distribution $\delta(t-t')$ which is present in the space-time correlation of the force has dimension -1 (it scales as $|t-t'|^{-1}$). We know that, in the limit of vanishing viscosity, the Navier–Stokes equation without a force is invariant under scaling transformations with arbitrary scaling exponent h (Section 2.2). The Navier–Stokes equation with force $\boldsymbol{f}$, is still invariant if $\boldsymbol{f}$ transforms like the other terms, namely is multiplied by λ^{2h-1}. This requires $h = -1+\epsilon/3$. Hence, the energy spectrum has the exponent $-1-2h = 1-2\epsilon/3$. This establishes (9.62). The corresponding eddy turnover time $t_\ell = \ell/v_\ell$ is proportional to $\ell^{1-h} = \ell^{2-\epsilon/3}$. We see that, for $\ell \to \infty$, this eddy turnover time becomes much shorter than the viscous diffusion time (proportional to ℓ^2). Hence, at large scales (also called the infrared (IR) domain), it is legitimate to ignore viscosity, as required for our scaling argument. At small scales (ultraviolet (UV) domain), we expect that the force and the viscous dissipation will just balance each other, thereby producing a

[54] The notation ϵ is borrowed from critical phenomena and is not related to the mean energy dissipation, denoted ε.

spectrum

$$E(k) = \frac{F(k)}{2\nu k^2} = \frac{D}{2\nu} k^{1-\epsilon}. \tag{9.64}$$

Observe that in this scaling argument we have used the positivity of ϵ, but not its smallness.

If the matter is so simple, why use the renormalization group at all? It is actually needed to obtain the constants in front of the power-laws and also for predicting the behavior at $\epsilon = 0$. Let us now try to give a flavor of how this is done. In the literature, RG methods make extensive use of field-theoretic machinery and cannot be easily approached unless one is familiar with this. Actually the RG, as implemented by FNS, may be viewed as an iterated multiscale method (Section 9.6.2) involving a scale-dependent eddy viscosity. It may indeed be shown that the eddy viscosity produced by a flow with an energy spectrum (9.64) has a UV divergence[55] at $\epsilon = 0$. It follows that, for small ϵ, the dominant interactions are between widely separated scales, such that the ratio M of their scales has a logarithm of the order of $O(1/\epsilon)$. The following construction then emerges: one considers a sequence of scales $\ell_n = \ell_0 M^n$ $(n = 0, 1, \ldots)$ and associated wavenumbers $k_n = \ell_n^{-1}$. The effect of scale ℓ_n on scale ℓ_{n+1} is a change in the (eddy) viscosity which is so small that, to leading order, the viscosity may be considered a smooth function of the wavenumber k. A perturbative calculation then gives

$$\frac{d\nu(k)}{dk} = -\frac{AD}{[\nu(k)]^2 k^{1+\epsilon}}, \tag{9.65}$$

with an explicitly evaluated constant A (Fournier and Frisch 1983a). Eq. (9.65), apart from the constant A, follows from dimensional analysis and the condition that $d\nu(k)/dk$ is evaluated to leading (linear) order in the forcing spectrum $F(k)$. Integration of (9.65) and substitution into (9.64) gives the desired result (9.62) and produces a log-corrected k^1 spectrum for $\epsilon = 0$.

Several comments are now made on the FNS approach. First, we stress that at each level the eddy viscosity can be calculated perturbatively because the Reynolds number remains small. Actually, the *bare* Reynolds number, based on the molecular viscosity, would grow without bound as $k \to 0$, but the *renormalized* Reynolds number, based on the eddy viscosity evaluated at the previous step, goes to a value $O(\epsilon^{1/2})$.

[55] A simple way to show this is to use (50) of Dubrulle and Frisch (1991) which gives the eddy viscosity at low Reynolds numbers.

Second, it is the separation of scales (also a consequence of the smallness of ϵ) which justifies a hydrodynamical (Navier–Stokes) description at all levels. In the traditional presentation of the RG, one distinguishes 'greater wavenumbers' in a *finite* shell, say from Λ to $\Lambda/2$ and 'lesser wavenumbers', from zero to $\Lambda/2$. One then manipulates the equations to eliminate the greaters in favor of the lessers, an operation called 'decimation'. The resulting equation for the lessers is somewhat of a nightmare; it contains, in addition to Navier–Stokes terms, infinitely many other terms which are eventually shown to be irrelevant, in the sense that they do not contribute to the leading order asymptotic behavior of the solution for small ϵ and large scales. Decimation of a very wide shell (justified by the near divergence of the eddy viscosity) circumvents this difficulty and allows us to work with only a finite number of coefficients.

Third, it may be asked why only the coefficient in front of the viscous term in the Navier–Stokes equation is renormalized (changes) at each step. We have already observed in Section 9.6.3.2 that the coefficient in front of the advection term $\boldsymbol{v} \cdot \nabla \boldsymbol{v}$ can be renormalized under the action of a small-scale flow. In the FNS problem, the force, having a white-noise-dependence on the time, preserves the Galilean invariance of the Navier–Stokes equation,[56] so that 'vertex corrections' are ruled out.[57] Also, the force itself may be renormalized. Indeed, the nonlinear interaction of two large wavenumbers (with nearly opposite wavevectors) produces a low wavenumber 'beating' input. This 'eddy noise' (Rose 1977) has a forcing spectrum proportional to k^4 in three dimension and is thus irrelevant at small wavenumbers, because the direct forcing (9.61) dominates. There are, however, other situations, also considered by FNS, where eddy noise is relevant.

9.6.4.2 Extension of RG to Kolmogorov 1941 scaling régimes

We have already observed that the exponent $1 - 2\epsilon/3$ for the spectrum of the solution to the problem with the forcing $\propto k^{3-\epsilon}$ is obtained by a simple scaling argument. As noticed by de Dominicis and Martin (1979), this implies that the result for the exponent is nonperturbative (in ϵ), contrary to the constant in front of the power-law. Hence, there should be some finite range of ϵs for which (9.62) holds. If $\epsilon = 4$ is in this range, the $k^{-5/3}$ energy spectrum is obtained. In other words, a forcing with a

[56] $\boldsymbol{f}(\boldsymbol{r} - \boldsymbol{U}t, t)$ has the same space-time correlation as $\boldsymbol{f}(\boldsymbol{r}, t)$ when $\boldsymbol{f}$ is δ-correlated in the time.

[57] They do, however, occur in magnetohydrodynamics (Fournier, Sulem and Pouquet 1982).

k^{-1} spectrum may produce the K41 energy spectrum (de Dominicis and Martin 1979).

One possible cause for worry is that, as soon as $\epsilon \geq 3$, the energy spectrum (9.62) has a total energy diverging at the IR end. This causes difficulties in diagrammatic calculations of the sort discussed in Section 9.5, but it is not obvious that anything drastic happens physically since the dynamics is much more sensitive to the large-scale mean square strain (enstrophy) than to the mean square velocity (Section 7.3).

A simple phenomenological argument suggests that $\epsilon = 4$ may not produce a $k^{-5/3}$ spectrum exactly. Let us assume that the forcing spectrum $F(k) = Dk^{-1}$ for $k > k_0$, where k_0 is an IR cutoff, and otherwise vanishes. The amount of energy injected at wavenumbers $k > k_0$ is then $\varepsilon(k) = D\ln(k/k_0)$. If we assume that this energy cascades to higher wavenumbers and use the K41 expression (6.65) for the energy spectrum, with $\varepsilon(k)$ substituted for the mean energy dissipation rate, we obtain

$$E(k) = C_{\text{Kol}} D^{2/3} k^{-5/3} \left(\ln \frac{k}{k_0} \right)^{2/3}, \tag{9.66}$$

i.e. a log-corrected K41 spectrum. If we then let $k_0 \to 0$, the energy spectrum becomes infinite. Eq. (9.66) could be invalidated by intermittency corrections: random Gaussian forcing at all scales need not suppress the build-up of intermittency since, after enough cascade steps, the cascade rate may overwhelm the direct input from the force (R. Kraichnan, private communication).

We also observe that, in the FNS régime, at small ϵs the statistical equilibrium results from a balance between the input due to the assumed forcing and a drain due to eddy viscosity, whereas in an inertial range, both the input and the output originate from nonlinear interactions.

Yakhot and Orszag (1986a,b; see also Dannevik, Yakhot and Orszag 1987) have extended RG ideas to make quantitative predictions about amplitudes in K41 scaling régimes. They assumed a random force à la FNS with a Dk^{-1} spectrum. This force should not be viewed as *external*, but rather as representing the eddy noise, i.e. the random input to a given wavenumber from the beating of larger wavenumbers (cf. above). The RG (to lowest order in ϵ) is used to determine the constant in front of the energy spectrum in terms of D, as explained in Section 9.6.4.1. It remains to relate D and the mean energy dissipation rate ε. For this, an expression of the energy flux based on the EDQNM closure is used (Section 9.5.4). This involves the rate of relaxation of triple correlations, which in standard closure must be expressed phenomenologically. Here, it is expressed self-consistently in terms of the RG eddy viscosity, thereby avoiding the use of adjustable constants. The use of adjustable constants would, indeed, be fatal to a theory which *imposes scaling* and only intends to determine constants. In spite of the somewhat arbitrary resort to closure, the RG[58] theory leads to numerical values in good agreement with experimental data. For example, it gives $C_{\mathrm{Kol}} \approx 1.6$ for the Kolmogorov constant and $\kappa \approx 0.37$ for the von Kármán constant. The method has also been extended to give explicit evaluation of turbulence transport coefficients for engineering modeling (Orszag *et al.* 1993).

Kraichnan (1987) has pointed out that it is not clear whether RG is conceptually superior to closure. Actually, there is more than one way in which RG can be used to 'calibrate' closure. For example, closures such as EDQNM or Kraichnan's (1971a) 'test field model', which are both consistent with K41, can be used to study the same problem as FNS for *small* ϵ and thereby to determine their adjustable constants (Fournier and Frisch 1983a). Other observations about the use of renormalization group methods in turbulence may be found in Lesieur (1990) and Eyink (1994b).

9.7 Two-dimensional turbulence

By 'two-dimensional turbulence', one generally understands the study of high-Reynolds-number solutions of the incompressible Navier–Stokes equation (1.2) which depend only on two cartesian coordinates, here denoted x and y. It is easily checked that the component of the velocity along the third coordinate axis satisfies a simple advection–diffusion equation without back-reaction on the horizontal (x–y) flow. Hence, without loss of generality one may assume that the velocity has only two components and work with a stream function ψ. It is then simplest to use the vorticity equation (2.15). The vorticity has a single (vertical) component, denoted ζ, which satisfies

$$\partial_t \zeta - J(\psi, \zeta) \equiv \partial_t \zeta + \boldsymbol{v} \cdot \nabla \zeta = \nu \nabla^2 \zeta + \eta, \tag{9.67}$$

$$\nabla^2 \psi = -\zeta, \tag{9.68}$$

where

$$J(\psi, \zeta) \equiv (\partial_x \psi)(\partial_y \zeta) - (\partial_x \zeta)(\partial_y \psi), \qquad \boldsymbol{v} \equiv (\partial_y \psi, -\partial_x \psi). \tag{9.69}$$

Note that (9.67) contains a forcing term η, the curl of the driving force.

[58] In the Yakhot–Orszag extension, it is known as RNG.

As is well known, (9.67) includes no vortex-stretching term, so that one essential aspect of three-dimensional turbulence is lost.[59] Two-dimensional turbulence might thus be viewed as just a toy model, somewhat easier to analyze and certainly easier to simulate and to visualize than three-dimensional turbulence.

Actually, there exist numerous situations, in natural flow and in laboratory experiments, which are constrained to quasi-two-dimensional or layer-wise motion. The most important examples arise in geophysical and planetary flow. For example, two-dimensional turbulence is relevant to the dynamics of oceanic currents, the motion of intense eddies such as tropical cyclones, the existence of the polar vortex and the mixing of chemical species in the polar stratosphere, a key factor in the production of the ozone hole, and in other large-scale motions of planetary atmospheres (see, e.g., Dritschel and Legras 1993; Waugh *et al.* 1994). Two-dimensional turbulence dominates the short- and medium-term internal variability of the climate system, and largely contributes to determining the average conditions and the probability of extreme events. The use of two-dimensional approximations in geophysical flow was proposed by Charney (1947). It is presented in detail in Lesieur (1990; see also Pedlosky 1979). In addition, the two-dimensional Euler (inviscid) equation applies to strongly magnetized plasmas in the 'guiding-center' approximation (see, e.g., Kraichnan and Montgomery 1980).

Two-dimensional turbulence is also a very active area of experimental investigation (see, e.g., Bondarenko, Gak and Dolzhansky 1979; Gledzer, Dolzhansky and Obukhov 1981; Couder 1984; Sommeria 1986; Gharib and Derango 1989; Cardoso, Marteau and Tabeling 1994).

Hence, there is a large body of literature, which we shall not attempt to review in detail here. After a brief overview of — now classical — material on cascades and vortices (Section 9.7.1), our emphasis will be on two recent (or recently revived) topics which could be also of relevance for three-dimensional turbulence (Sections 9.7.2–9.7.3).[60]

9.7.1 Cascades and vortices

Two-dimensional turbulence has been reviewed by Kraichnan and Montgomery (1980) and by Lesieur (1990) and others. What distinguishes it most from three-dimensional turbulence is the conservation of vorticity along fluid particle paths when viscosity and forcing are ignored. As a

[59] But stretching of vorticity gradients is definitely present.

[60] The reader is also referred to a very interesting essay by Pomeau (1994).

consequence, the enstrophy $\frac{1}{2}\langle\zeta^2\rangle$ cannot increase under the sole action of nonlinearity.

Kraichnan (1967a) conjectured that, if energy is injected into the flow at a constant rate ε at some intermediate scale ℓ_0, an *inverse cascade* of energy will take place until the largest scales available are attained. This inverse cascade should follow the same $k^{-5/3}$ law as the direct three-dimensional cascade. Kraichnan's conjecture was based on a 'absolute equilibrium' statistical mechanics argument using the Euler equation with finitely many Fourier modes (Galerkin truncation), no force and no viscosity. Later, he gave a negative eddy viscosity interpretation (Kraichnan 1976). The inverse cascade, possibly the most important result in fully developed turbulence since Kolmogorov's work, has been largely confirmed by simulations of increasing size (Lilly 1971; Siggia and Aref 1981; Frisch and Sulem 1984; Herring and McWilliams 1985; Smith and Yakhot 1993; Borue 1994) and, to a lesser extent, by experiment (Sommeria 1986). Once the inverse cascade reaches the 'size of the box' (when simulated with periodic boundary conditions), vortices are formed, with a size comparable to that of the forcing, and the $k^{-5/3}$ range is disrupted at large scales (Smith and Yakhot 1993, 1994; Borue 1994). The shape of the spectrum at the smallest wavenumbers depends on what kind of mechanism is or is not assumed for removing the energy. For example, with a constant 'Ekman' friction term $-\alpha\zeta$ added to the r.h.s. of (9.67), a k^{-3} range is observed at the *largest scales* (Smith and Yakhot 1994); it corresponds to a balance of nonlinear transfer with friction, which may be obtained by equating the eddy turnover time ℓ/v_ℓ with the time of damping by friction α^{-1}. This is one of the possible explanations of the paradoxical fact that in the atmosphere k^{-3} and $k^{-5/3}$ ranges are indeed observed, but the former is at larger scales (Lilly and Peterson 1983).

The analog of the K41 energy cascade towards small scales is an enstrophy cascade (Kraichnan 1967a; Batchelor 1969), for which the energy spectrum follows a k^{-3} law with a logarithmic correction. This correction stems from the highly non-local character of nonlinear interactions in the enstrophy cascade: most of the shear acting on a given small scale comes from much larger scales. Although closure theory and closure calculations completely support the dual picture of a direct enstrophy cascade and an inverse energy cascade in forced turbulence (see, e.g., Kraichnan 1971b; Pouquet, Lesieur, André and Basdevant 1975), direct numerical simulations do not very strongly support the k^{-3} law for forced turbulence. Legras, Santangelo and Benzi (1988) obtained considerably steeper spectra at small scales, but Borue (1994), using higher Reynolds numbers, obtained reasonably clean k^{-3} spectra. Unfortunately, most of these calculations do not integrate the Navier–Stokes equation (9.67), but a modified equation with a high power of the Laplacian as dissipation term.[61]

The Batchelor–Kraichnan theory also predicts a k^{-3} spectrum for unforced (decaying) two-dimensional turbulence. Again, the numerical evidence is not so strong. Brachet, Meneguzzi, Politano and Sulem (1988) obtained a k^{-3} spectrum in a moderately long but genuine Navier–Stokes simulation. Santangelo, Benzi and Legras (1989) found that, eventually, the energy spectrum becomes quite steep and does so in a fashion which depends on the initial conditions. They ascribe this to the formation of a hierarchy of *coherent vortices*. These have been observed in simulations by Fornberg (1977) and Basdevant, Legras, Sadourny and Béland (1981) and were studied systematically by McWilliams (1984): the long-time evolution of two-dimensional fields is dominated by small coherent vortices whose vorticities are much stronger than that of the well-mixed background. Individually, the vortices are well described (in a suitable frame of motion) as solutions of the two-dimensional time-independent Euler equation:

$$J(\psi, \nabla^2\psi) = 0. \tag{9.70}$$

This is an extremely broad class of functions.[62] It includes circular vortices, in which ψ is a function only of the distance to the vortex center. Circular vortices, as long as they are well separated, behave mostly as a system of point-vortices, a problem which has been studied for more than a century (see, e.g., Kirchhoff 1877; Lamb 1932; Aref 1983 and Section 9.7.2). Within such vortices, the nonlinearity is completely depleted, so that cascade arguments cease to be valid; in a sense coherent vortices constitute drops of 'laminar' fluid in an otherwise turbulent flow (Benzi, Paladin, Paternello, Santangelo and Vulpiani 1986).

9.7.2 Two-dimensional turbulence and statistical mechanics

Statistical mechanics, as developed by Maxwell, Boltzmann, Gibbs and their followers, has been amazingly successful in predicting the behavior of systems with very many degrees of freedom. Its main successes have, however, been in the area of equilibrium distributions for *conservative*

61 A procedure referred to as 'modified dissipativity' or 'hyperviscosity'.

62 We shall return to this matter in Section 9.7.2.

(Hamiltonian) dynamics. Three-dimensional turbulence is *dissipative*: as seen in Section 5.2, a very minute amount of viscosity suffices to produce a finite energy dissipation. In two dimensions, when the viscosity is small, so is the energy dissipation; ignoring the viscosity altogether may thus be of some relevance.[63] In the absence of viscosity the two-dimensional Navier–Stokes equation (9.67) reduces to the Euler equation. As we know, the Euler equation in any dimension may be written in Hamiltonian form (Section 9.3). In two dimensions a particularly simple formulation is obtained when the vorticity field $\zeta(\boldsymbol{r})$ is approximated by a large number N of point vortices of individual circulations c_i:

$$\zeta(\boldsymbol{r}) \approx \sum_i c_i \delta\left(\boldsymbol{r} - \boldsymbol{r}_i(t)\right), \tag{9.71}$$

where $\boldsymbol{r}$ stands now for (x, y). The canonically conjugate Hamiltonian variables are then simply the $c_i x_i$s and the $c_i y_i$s (no summation on i) and, in the absence of boundaries, the Hamiltonian is the energy of interaction of the vortices (see, e.g., Batchelor 1970):

$$H = \sum_{i<j} c_i c_j V(\boldsymbol{r}_i, \boldsymbol{r}_j). \tag{9.72}$$

Here,

$$V(\boldsymbol{r}, \boldsymbol{r}') \equiv -\frac{1}{2\pi} \ln |\boldsymbol{r} - \boldsymbol{r}'| \tag{9.73}$$

is the Green's function for the negative Laplacian. Note that, if we define the stream function

$$\psi(\boldsymbol{r}) = \int V(\boldsymbol{r}, \boldsymbol{r}')\zeta(\boldsymbol{r}')\, d\boldsymbol{r}', \tag{9.74}$$

with ζ given by the r.h.s. of (9.71), the Hamiltonian is (after removal of an infinite self-energy term) equal to $\frac{1}{2}\int \zeta\psi d\boldsymbol{r} = \frac{1}{2}\int \boldsymbol{v}^2 d\boldsymbol{r}$, i.e. the kinetic energy of the system.[64]

Onsager (1949) investigated the equilibrium statistics for such point-vortices and discovered the possibility of *negative temperature states* in which the entropy is a decreasing function of the energy and close clustering of equal-sign vortices is favored. Onsager's theory provides an attractive interpretation for the formation of coherent vortices. Actually, Fröhlich and Ruelle (1982) found that negative temperature states are ruled out if one considers the limit $N \to \infty$ and simultaneously assumes

[63] We shall come back to the difficulty that enstrophy dissipation cannot be neglected.

[64] In a bounded domain, the appropriate Green's function must be used and a term of interaction of the individual point vortices with the boundary is present.

that the mean energy per vortex remains finite. Eyink and Spohn (1993; see also Caglioti, Lions, Marchioro and Pulverenti 1992) were able to prove the existence of negative temperature states when $N \to \infty$ by considering a *mean field* theory: the Hamiltonian H in (9.72) is replaced, by H/N so that the mean energy per *vortex pair* remains finite. For the case of a bounded domain and when all the vortices have the same circulation, they obtained the following equation for the single-vortex distribution (which may be interpreted as a *mean vorticity* and is here denoted ζ):

$$-\nabla^2\psi(\boldsymbol{r}) = \zeta(\boldsymbol{r}) = F(\psi(\boldsymbol{r})), \tag{9.75}$$

$$F(\psi(\boldsymbol{r})) = \frac{e^{-\alpha-\beta\psi(\boldsymbol{r})}}{Z(\beta)}, \qquad \int F(\psi(\boldsymbol{r}'))\,d\boldsymbol{r}' = 1. \tag{9.76}$$

For the neutral case of an equal number of $+1$ and -1 vortices, the function $F(.)$ is often a hyperbolic sine[65] instead of an exponential and (9.75) is known as the 'sinh–Poisson' equation.

Actually, the sinh–Poisson equation has been known since the work of Joyce and Montgomery (1973; see also Montgomery and Joyce 1974, and Kraichnan and Montgomery 1980) who derived it by a maximum entropy principle.[66] The mere fact that there is a functional relation between the vorticity ζ and the stream function ψ (whatever the function $F(.)$) implies that they are steady-state solutions of the Euler equation. Montgomery, Matthaeus, Stribling, Martinez and Oughton (1992) have performed very long high-Reynolds-number integrations of the decaying two-dimensional Navier–Stokes equation and found that such flow relaxes to a configuration with vanishing nonlinearity which is well described by the sinh–Poisson equation.[67]

This brings us to a fundamental question: to what extent is it legitimate to infer the long-time behavior of the *slightly dissipative* two-dimensional Navier–Stokes equation by studying equilibrium properties of conservative systems such as point vortices? How can we take into account the dissipation of enstrophy, a phenomenon which is always observed in high-Reynolds-number simulations?

65 The hyperbolic sine law holds only if the 'chemical potentials' α_+ and α_- associated with each vortex species are equal; this may not be the case, even with an equal number of $+1$ and -1 vortices: the equilibrium state may break the parity symmetry of the system.

66 In the context of stellar dynamics, (9.76) has also been derived by Lynden-Bell (1967) for the Jeans–Vlasov–Poisson equation of gravitationally interacting stars. He also considered a variant which is the gravitational counterpart of the case of continuous vorticity distribution discussed below.

67 According to Pasmanter (1994) $F(\psi) = c\psi/(1-\psi^2)$ with $-1 < \psi < +1$ gives an equally satisfactory fit.

The beautiful work of Robert and Sommeria (Robert 1990, 1991; Robert and Sommeria 1991) and of Miller (1990) indicates that the use of conservative statistical mechanics may actually be justified for real fluids.[68] We shall now try to give a flavor of this work. For this, we shall borrow occasionally from Eyink and Spohn (1993) and from Pasmanter (1994). Let us first observe that inviscid evolution of a smooth (or piecewise smooth) vorticity field leads to extremely convoluted vorticity contours with steep vorticity gradients (Zabusky, Hughes and Roberts 1979; Dritschel 1989). Vorticity and area are conserved along fluid particle paths. As a consequence all the integrals

$$\Omega_p \equiv \int \zeta^p(\boldsymbol{r})\, d\boldsymbol{r} \tag{9.77}$$

are conserved, in addition to the energy

$$E = \frac{1}{2}\int \zeta(\boldsymbol{r})\psi(\boldsymbol{r})\, d\boldsymbol{r} = \frac{1}{2}\int\int V(\boldsymbol{r},\boldsymbol{r}')\zeta(\boldsymbol{r})\zeta(\boldsymbol{r}')\, d\boldsymbol{r}\, d\boldsymbol{r}'. \tag{9.78}$$

If the vorticity field is smoothed over a distance a, small compared to the size of large vortical structures, the energy and the integrated vorticity Ω_1 will be barely affected, but the other integrals, in particular the enstrophy Ω_2, can be drastically reduced because smoothing permits cancellation between neighboring vortical structures with opposite vorticities. In other words, a coarse-grained description of the vorticity may mimic the effect of viscous dissipation which removes enstrophy without much affecting the energy. Miller (1990) explicitly makes use of such a coarse-graining. Robert and Sommeria (1991) resort to a more radical approach by using 'Young measures': 'at each point we have a probability distribution of vorticity which gives, in some statistical sense, a local description of the small-scale oscillations of the microscopic vorticity function'. They actually give up, not only the deterministic description of the vorticity, but also its single-valuedness, although ordinary vorticity fields can be recovered as mean fields. The Euler equation is then extended to such Young measures and, by means of a suitable entropy, a steady-state solution is constructed and shown to characterize the 'most likely' behavior. Physical intuition (at least of a standard kind) may be somewhat lost in the intermediate steps. Fortunately, at the end Robert and Sommeria (1991) come up with a simple mean field relation, also found in Miller (1990). It differs from the Joyce and Montgomery relation (9.76), but

[68] See also Kuz'min (1982).

reduces to it in the limit of small initial vorticity patches with ± 1 circulation embedded in irrotational flow.

Specifically, let us assume that the dynamics is taking place in a bounded set Λ of area $|\Lambda|$ and let $P_0(\lambda)$ be the initial vorticity p.d.f., i.e. $P_0(\lambda)d\lambda$ is the fraction of the total area in which the initial vorticity is between λ and $\lambda + d\lambda$:

$$P_0(\lambda)d\lambda = \frac{1}{|\Lambda|}\int_{\lambda<\zeta(\boldsymbol{r},0)<\lambda+d\lambda} d\boldsymbol{r}. \tag{9.79}$$

Because vorticity and area are conserved, this quantity will be invariant under the evolution. The Robert–Sommeria–Miller equilibrium p.d.f. of the vorticity is a function of the point $\boldsymbol{r}$, given by

$$P(\boldsymbol{r},\lambda) = \frac{1}{Z[\psi(\boldsymbol{r})]}\, e^{[-\alpha(\lambda)-\beta\,\lambda\,\psi(\boldsymbol{r})]}, \tag{9.80}$$

where the parameter β (the inverse temperature) and the function $\alpha(\lambda)$ are Lagrange multipliers and $Z[\psi(\boldsymbol{r})]$, the partition function, is given by

$$Z(\psi) = \int d\lambda\, e^{[-\alpha(\lambda)-\beta\,\lambda\,\psi]}, \tag{9.81}$$

which ensures the normalization of $P(\boldsymbol{r},\lambda)$. The mean field result for point-vortices (9.76) is recovered in the limit $Z \to 1$ (with generalization to a distribution of vortex strengths λ). The vorticity is then so diluted in irrotational fluid that the local normalization constraint for $P(\boldsymbol{r},\lambda)$ is not effective. Note, however, that the distribution of vortex strengths is an arbitrary modeling choice for vortex statistics, while here it is completely determined by the global distribution of the vorticity levels in the initial condition.

For each $\boldsymbol{r}$, a mean (or macroscopic) vorticity is defined as

$$\zeta(\boldsymbol{r}) \equiv \int \lambda\, P(\boldsymbol{r},\lambda)\, d\lambda = -\frac{1}{\beta}\frac{\partial}{\partial\psi}\ln Z(\psi)\bigg|_{\psi=\psi(\boldsymbol{r})}. \tag{9.82}$$

It is in terms of this macroscopic vorticity that the stream function ψ is constructed via the Poisson equation:

$$-\nabla^2\psi = \zeta = F(\psi) \equiv -\frac{1}{\beta}\frac{\partial}{\partial\psi}\ln Z(\psi). \tag{9.83}$$

It follows that the macroscopic vorticity is a steady-state solution of the Euler equation. The Lagrange multipliers β and $\alpha(\lambda)$ are determined: (i) by the scalar constraint that the total energy (9.78) is equal to its initial value and (ii) by the one-parameter family of constraints that the space-averaged p.d.f. of the fluctuating (microscopic) vorticity is equal to the p.d.f. of the initial vorticity, i.e.

$$P_0(\lambda) = \frac{1}{|\Lambda|}\int_\Lambda P(r,\lambda)\,dr. \tag{9.84}$$

Eqs. (9.82) and (9.84) imply that the mean macroscopic vorticity is still equal to its initial value, but the mean square macroscopic vorticity is generally less than its initial value. Hence, some of the macroscopic enstrophy has been converted into microscopic enstrophy. Pomeau (1992) has shown that the mechanism of dissipation of a macroscopic invariant which gets transferred to very-small-scale fluctuations is quite general; it is also present, for example, in the (defocusing) cubic nonlinear Schrödinger equation, where the particle number and the energy play respectively the roles of the energy and the enstrophy.

A key issue in assessing the applicability of this sort of *quasi-inviscid*[69] equilibrium theory is whether the macroscopic to microscopic enstrophy conversion faithfully represents the effect of viscous dissipation. Does it have the right time scale? Is the final distribution of vorticity the same as predicted by the quasi-inviscid equilibrium theories? We have already quoted the numerical evidence in favor of applicability (Montgomery, Matthaeus, Stribling, Martinez and Oughton 1992). Some conceptual issues may be raised. First, it is obvious that a flow which is even slightly viscous will eventually decay to zero and not to the steady state predicted by quasi-inviscid equilibrium theory. This is, however, not an acceptable objection, since the flow may well remain for a very long time in a quasi-steady state where nonlinearities vanish, so that direct viscous decay (as opposed to cascade-enhanced decay) is exceedingly slow. It may be observed that, in their present state, quasi-inviscid equilibrium theories fail to take into account the *topological constraint* that vortex contours cannot cross (V. Zeitlin, private communication). Eyink and Spohn (1993) observe that strict ergodicity of a gas of point-vortices is: (i) unlikely and (ii) not really needed for applicability of quasi-inviscid equilibrium theories. As for the time scale, they observe that 'local equilibrium' of individual clusters of vortices is probably established rather quickly but global equilibrium may take much longer if the individual clusters are widely scattered, as in the example discussed in Section 9.7.3.

Our feeling is that much work remains to be done to delineate the *niche* of applicability of quasi-inviscid equilibrium theories in two dimensions. Already some attempts have been made to incorporate additional

[69] A term proposed by Pasmanter (1994).

elements such as the quasi-geostrophic extension of the two-dimensional Navier–Stokes equation, needed to describe planetary motion, to explain, for example, Jupiter's Great Red Spot (Sommeria, Nore, Dumont and Robert 1991; Michel and Robert 1994).

9.7.3 Conservative dynamics 'punctuated' by dissipative events

The quasi-inviscid equilibrium theory is clearly inapplicable as long as there are many well-separated vortices. In high-resolution numerical simulations with random initial conditions and a correlation length much smaller than the size of the numerical box, the formation of a large number of (monopolar) vortices is typically observed; their density $\rho(t)$, for example, the mean number of vortices per unit area, is found to decreases as

$$\rho(t) \propto t^{-\xi}, \tag{9.85}$$

with $\xi \approx 0.70 - 0.75$ (McWilliams 1990). A scaling theory of this phenomenon has been proposed by Carnevale, McWilliams, Pomeau, Weiss and Young (1991) and developed further by Weiss and McWilliams (1993); a similar approach may be found in Benzi, Colella, Briscolini and Santangelo (1992). The general idea is that most of the time the vortices are sufficiently far from each other that their motion can be described by the Hamiltonian dynamics of point-vortices. Occasionally, two vortices of like sign approach each other sufficiently closely to permit merger. Mergers are dissipative events which 'punctuate' the otherwise conservative dynamics.

It is assumed that during vortex-merger neither the energy nor the peak (absolute value of the) vorticity $\zeta_{\text{ext}}(t)$ of the vortices changes. However, the typical vortex size $a(t)$ and the enstrophy are allowed to change.[70] The conservation of energy is clearly a reasonable assumption for high-Reynolds-number two-dimensional turbulence. As for conservation of peak vorticity, it may be observed that it typically occurs at the center of the vortices while enstrophy dissipation during merger typically occurs at the periphery where filaments[71] are quickly destroyed by viscosity.

[70] In the simplest model, all the vortices are assumed to be of roughly equal size, but Weiss and McWilliams (1993) assume a distribution of sizes.

[71] It is a bit unfortunate that the same word *filament* is used: (i) in three dimensions for slender high-vorticity regions with approximately circular cross-section of the kind discussed in Section 8.9.1 and (ii) in two dimensions for very long and thin ribbons of quasi-uniform vorticity. Maybe the latter could be called 'hairs of vorticity'.

The energy $\mathscr{E}$ per unit area is approximately given by $\rho \zeta_{\text{ext}}^2 a^4$. (Note that ζa is the typical velocity induced within the vortex.) From the above assumptions, it follows that $\rho(t)a^4(t)$ should be constant. Hence, from (9.85), we have

$$a(t) \propto t^{\xi/4}. \tag{9.86}$$

There is no complete theoretical justification of the $t^{-\xi}$ law for the evolution of the density of vortices. The merger itself is amenable to rather detailed theoretical modeling, but finding the probability that point-vortices will come within the critical distance d_c needed for merging requires numerical calculations. However, a renormalization trick can be used to prevent the total number of vortices from dropping too quickly (Weiss and McWilliams 1993). This allows the determination of the exponent ξ with an accuracy of a few per cent.[72]

It is a tautology to state that numerical simulations support the scaling theory. Early experimental results by Tabeling, Burkhart, Cardoso and Willaime (1991) also gave some support to the theory/simulations. Their experiment uses a thin layer of electrolyte, a technique already used by Bondarenko, Gak and Dolzhansky (1979; see also Gledzer, Dolzhansky and Obukhov 1981) for studying the Kolmogorov flow. The initial eddy-motion is produced by forcing with an array of permanent magnets, acting on a current flowing through the electrolyte, which is then switched off to let the eddies decay freely. The measurements are made during a phase where vortex-merger proceeds faster than overall decay by friction against the bottom of the vessel, so that the latter may be safely ignored. Typically, the vortex size is found to grow as $a(t) \propto t^{0.22\pm0.03}$ which is close to being consistent with (9.86) when $\xi \approx 0.70 - 0.75$ is assumed. Further improvements in the technique with measurements of the peak vorticity (Cardoso, Marteau and Tabeling 1994) however reveal that it is far from being constant, in contrast to previously reported theoretical and numerical results. Furthermore, the density of vortices, instead of decreasing as $t^{-\xi}$, has a much smaller exponent, of about -0.44. Actually, in these experiments the vortices do not become sparse but remain packed like the molecules of a liquid. We suspect that the main reason for the discrepancy is not so much the bottom friction as the smallness of the Reynolds number R: based on the size of the cell, $R \simeq 2000$ and based on the size of individual vortices, $R \simeq 200$.

[72] With once more the *caveat* that dissipation in the simulations uses the squared Laplacian rather than the ordinary Laplacian.

9.7.4 From Flatland to three-dimensional turbulence

Doing physics in Flatland requires no apologies: many areas of theoretical and condensed matter physics have benefitted greatly from such studies. Indeed, two-dimensional models have often helped us in understanding new concepts, as in Onsager's theory of the the two-dimensional Ising model (see, e.g., Huang 1963). Furthermore, surprisingly many phenomena of our three-dimensional world are actually governed by two-dimensional equations. In this final section, we want to address two questions: (i) How far do we understand (high-Reynolds-number) two-dimensional turbulence? (ii) What have we learned that could be relevant in three dimensions?

The forced and unforced (decaying) problems differ much more in two than in three dimensions. For two-dimensional turbulence in an *unbounded* domain with forcing at a fixed scale, Kolmogorov's theory of a statistically self-similar flow with a $k^{-5/3}$ spectrum, as reworked by Kraichnan (1967a), constitutes a very good description of the inverse cascade; indeed, no coherent vortical structures (which would ruin this description) seem to be generated *at large scales*. At small scales, where vortices abound, their properties depend very much on the nature of the forcing, so that universality is questionable. It is conceivable that the background between the vortices has universal properties as predicted by the Batchelor (1969) and Kraichnan (1967a) theory of the enstrophy cascade or by the Polyakov (1993) theory of conformal turbulence, but this remains to be confirmed.

Decaying two-dimensional turbulence in an unbounded domain evolves into a dilute gas of vortices, governed by the kind of punctuated scaling régime discussed in Section 9.7.3 in which universality seems to hold. A complete theory is still lacking but there is good hope for one. In a bounded domain, the quasi-inviscid equilibrium theory discussed in Section 9.7.2 leads to the remarkable (and well-supported) prediction that the flow evolves[73] to a state with totally depleted nonlinearities; the detailed structure of the emerging vortex (or pair of vortices) may depend on the initial vorticity distribution.

Two-dimensional turbulence has thus a *Leonardo–Kolmogorov duality*, reminiscent of the particle–wave duality. Vortices are part of the Leonardian realm, while the near-Gaussian inverse cascade belongs to the Kolmogorovian realm. The latter may be analyzed indifferently in

[73] With possible transients described by the punctuated scaling régime when the initial correlation length is much smaller than the size of the domain.

the Fourier space or in the physical space,[74] while the former can become utterly obscure in the Fourier space.

Let us now try to assess what can be transposed from two- to three-dimensional turbulence. We are not aware of any successful attempt to extend quasi-inviscid equilibrium theories from two to three dimensions. The key physical difficulty is, of course, that in three dimensions a very small viscosity does not prevent the energy from decaying quickly.

The idea of conservative dynamics punctuated by dissipative events could be directly relevant in three dimensions. In two dimensions, it is intimately related to the generations of structures with strongly depleted nonlinearities, namely the vortices. In three dimensions, there are even more possibilities for depletion. The vortex filaments (Section 8.9.1) are an outstanding example: their interactions and self-interaction can be studied inviscidly as long as they remain well-separated; the dissipative events which punctuate such conservative dynamics include vortex breakdown and collisions of filaments leading to reconnection.

What is clear from the two-dimensional studies is that no meaningful statistical theory can be developed for flow dominated by coherent structures, until these have been identified and parametrized. They are in a sense the 'molecules' of the theory. In three dimensions, the analog could be strings, *viz* the vortex filaments, but there are many other possibilities, including vortex sheets, Burgers vortices (cf. p. 187) and other compact vortices and Fourier components.

The role of simulations and experiments needs some comment. In two dimensions we are now able to perform simulations which achieve very high Reynolds numbers and therefore can often be trusted to be truly asymptotic. Furthermore, simulated two-dimensional flow is very easy to visualize. Also, simulations clearly have an edge over experiments, since it is hard to set up genuine two-dimensional experiments, particularly at high Reynolds numbers. In three dimensions the opposite is true: computers have a long way to go before we can achieve inertial ranges of several decades, as needed for accurate measurements of exponents, while experiments can do this at relatively low cost. However, multidimensional reconstruction of fine-scale structures from computer simulations is considerably easier (but far from trivial) and von Neumann's (1949) statement about *'break(ing) the deadlock' by extensive, but well-planned, computational efforts* remains as true as ever. Actually, one observes an increasing synergy between experiments and computations

[74] And is well represented by closure.

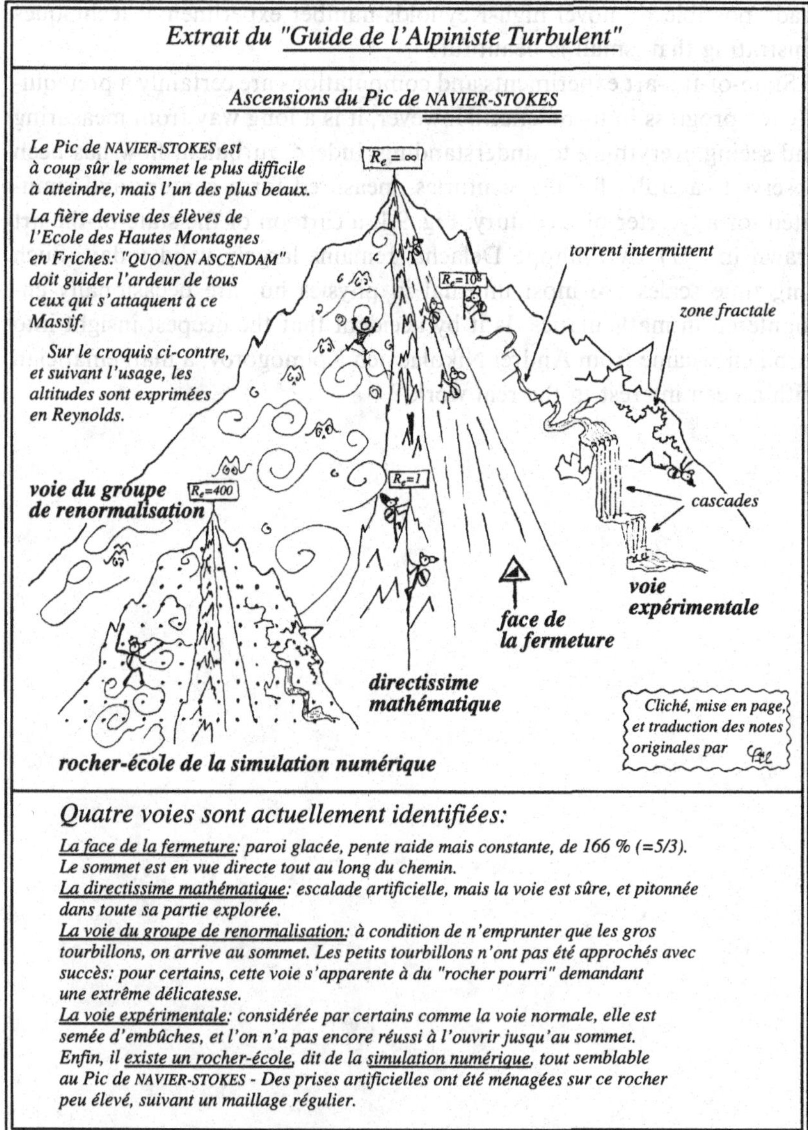

Fig. 9.1. Cartoon drawn in 1977 by the astronomer Philippe Delache, a penetrating observer of the turbulence community. He was the author's friend and died prematurely in 1994. The figure shows the 'Navier–Stokes peak' and four explored faces: experimentation, closure, mathematics and renormalization. It also shows a reduced model, the rock-climbing school of numerical simulation.

made possible by novel high-Reynolds-number experimental techniques illustrating that 'small is beautiful'.

State-of-the-art experiments and computations are certainly a prerequisite for progress in turbulence. However, it is a long way from measuring and seeing everything to understanding. Indeed, turbulent flow has been observed carefully for five centuries, measured for a century and simulated for a quarter of a century. Fig. 9.1, a cartoon of the state-of-the-art drawn in 1977 by Philippe Delache, remains largely valid today. Such long time scales are most unusual in physics but are occasionally encountered in mathematics. Is it by accident that the deepest insight into turbulence came from Andrei Nikolaevich Kolmogorov, a mathematician with a keen interest in the real world?

References

Abry, P., Fauve, S., Flandrin, P. & Laroche, C. 1994. Analysis of pressure fluctuations in swirling flows, *J. Phys. II France* **4**, 725–733.

Amit, D.J. 1984. *Field Theory, the Renormalization Group and Critical Phenomena*, 2nd edition. World Scientific, Singapore.

Anselmet, F., Gagne, Y. Hopfinger, E.J. & Antonia, R.A. 1984. High-order velocity structure functions in turbulent shear flow, *J. Fluid Mech.* **140**, 63–89.

Aref, H. 1983. Integrable, chaotic and turbulent vortex motion in two-dimensional flows, *Ann. Rev. Fluid Mech.* **15**, 345–389.

Aref, H. 1984. Stirring by chaotic advection, *J. Fluid Mech.* **143**, 1–21.

Arnold, V.I. 1963. Proof of A.N. Kolmogorov's theorem on the preservation of quasi-periodic motions under small pertubations of the Hamiltonian, *Usp. Math. Nauk* **18**, 13–40.

Arnold, V.I. 1965. Sur la topologie des écoulements stationnaires des fluides parfaits, *C. R. Acad. Sci. Paris* **261**, 17–20.

Arnold, V.I. 1966. Sur la géométrie différentielle des groupes de Lie de dimension infinie et ses applications à l'hydrodynamique des fluides parfaits, *Ann. Inst. Fourier* **16**, 319–361.

Arnold, V.I. 1972. Notes on the 3-D flow pattern of a perfect fluid in the presence of a small perturbation of the initial velocity field, *Appl. Math. Mech.* **36**, 236–242.

Arnold, V.I. 1978. *Mathematical Methods of Classical Mechanics*. Springer, Berlin.

Arnold, V.I. 1994. On A.N. Kolmogorov, in *Golden Year of Moscow Mathematics*, 129–153, vol. 6, in 'History of mathematics', eds. S. Zdravkovska & P.L. Duren. American Mathematical Society & London Mathematical Society.

Arnold, V.I. & Khesin, B.A. 1992. Topological methods in hydrodynamics, *Ann. Rev. Fluid Mech.* **24**, 145–166.

Arnold, V.I., Kozlov, V.V. & Neishtadt, A.I. 1988. Mathematical aspects of classical and celestial mechanics, in *Dynamical Systems III*, 1–291, in 'Encyclopaedia of Mathematical Sciences'. Springer, Berlin.

Arnold, V.I., Zeldovich, Ya.B., Ruzmaikin, A.A. & Sokoloff, D.D. 1981. A magnetic field in a stationary flow with stretching in Riemannian space, *Sov. Phys. JETP* **54**, 1083–1086.

Atten, P., Caputo, J.G., Malraison, B. & Gagne, Y. 1984. Determination of attractor dimension of various flows, in *Bifurcations and Chaotic Behaviours, J. Méc. Theor. Appl. Suppl.*, 133–156, eds. G. Iooss & R. Peyret.

Aubry, N., Holmes, P., Lumley, J.L. & Sone, E. 1988. The dynamics of coherent structures in the wall region of a turbulent boundary layer, *J. Fluid Mech.* **192**, 115–173.

Aurell, E., Frisch, U., Lutsko, J. & Vergassola, M. 1992. On the multifractal properties of the energy dissipation derived from turbulence data, *J. Fluid Mech.* **238**, 467–486.

Aurell, E., Boffetta, G., Crisanti, A., Frick, P., Paladin, G. & Vulpiani, A. 1994. Statistical mechanics of shell models for 2D-turbulence, *Phys. Rev.* E **50**, 4705–4715.

Avellaneda, M. & Majda, A.J. 1990. Mathematical models with exact renormalization for turbulent transport, *Commun. Math. Phys.* **131**, 381–429.

Bacry, E., Arnéodo, A., Frisch, U., Gagne, Y. & Hopfinger, E. 1990. Wavelet analysis of fully developed turbulence data and measurement of scaling exponents, in *Turbulence 89: Organized Structures and Turbulence in Fluid Mechanics*, 203–215, eds. O. Métais, & M. Lesieur. Kluwer, Dordrecht.

Barabási, A.-L. & Stanley, H.E. 1995. *Fractal Concepts in Surface Growth.* Cambridge University Press, Cambridge.

Bardos, C. & Benachour, S. 1977. Domaine d'analyticité des solutions de l'équation d'Euler dans un ouvert de $\mathbb{R}^n$, *Ann. Sci. Norm. Sup. Pisa* **4**, 647–687.

Bardos, C. & Frisch, U. 1976. Finite-time regularity for bounded and unbounded ideal incompressible fluids using Hölder estimates, in *Turbulence and the Navier–Stokes Equations, Lect. Notes in Math.* **565**, 1–13, ed. R. Temam. Springer, Berlin.

Basdevant, C., Legras, B., Sadourny, R. & Béland, M. 1981. A study of barotropic model flows: intermittency, waves and predictability, *J. Atmos. Sci.* **38**, 2305–2326.

Batchelor, G.K. 1946. Double velocity correlation function in turbulent motion, *Nature* **158**, 883–884.

Batchelor, G.K. 1947. Kolmogoroff's theory of locally isotropic turbulence, *Proc. Cambridge Phil. Soc.* **43**, 533–559.

Batchelor, G.K. 1951. Pressure fluctuations in isotropic turbulence, *Proc. Cambridge Phil. Soc.* **47**, 359–374.

Batchelor, G.K. 1953. *The Theory of Homogeneous Turbulence.* Cambridge University Press, Cambridge.

Batchelor, G.K. 1969. Computation of the energy spectrum in homogeneous two-dimensional turbulence, *Phys. Fluids Suppl. II* **12**, 233–239.

Batchelor, G.K. 1970. *An Introduction to Fluid Dynamics.* Cambridge University Press, Cambridge.

Batchelor, G.K. 1990. Kolmogorov's work in turbulence, in *Kolmogorov's obituary (organized by D. Kendall), Bull. London Math. Soc.* **22**, 47–51.

Batchelor, G.K. & Proudman, I. 1956. The large-scale structure of homogeneous turbulence, *Philos. Trans. Roy. Soc.* **268**, 369–405.

Batchelor, G.K. & Townsend, A.A. 1947. Decay of vorticity in isotropic turbulence, *Proc. R. Soc. Lond.* A **191**, 534–550.

Batchelor, G.K. & Townsend, A.A. 1949. The nature of turbulent motion at large wave-numbers, *Proc. R. Soc. Lond.* A **199**, 238–255.

Battimelli, G. & Vulpiani, A. 1982. Kolmogorov, Heisenberg, von Weizsäcker, Onsager: un caso di scoperta simultanea, in *Atti III congresso nazionale di storia della fisica, (Palermo 11–16 ottobre 1982)*, 169–175.

Bayly, B.J., Orszag, S.A. & Herbert, T. 1988. Instability mechanism in shear-flow transition, *Ann. Rev. Fluid Mech.* **20**, 359–391.

Beale, J.T., Kato, T. & Majda, A.J. 1989. Remarks on the breakdown of smooth solutions for the 3-D Euler equations, *Commun. Math. Phys.* **94**, 61–66.

Belin, F., Maurer, J., Tabeling, P. & Willaime, H. 1996. Observation of intense filaments in fully developed turbulence, *J. Phys. II France* **6**, 573–584.

Belinicher, V.I. & L'vov, V.S. 1987. A scale-invariant theory of fully developed hydrodynamic turbulence, *Sov. Phys. JETP* **66**, 303–313.

Bender, C. & Orszag, S.A. 1978. *Advanced Mathematical Methods for Scientists and Engineers*. McGraw-Hill, New York.

Benney, D.J. & Newell, A.C. 1969. Random wave closures, *Stud. Appl. Math.* **48**, 29–53.

Benney, D.J. & Saffman, P.G. 1966. Nonlinear interactions of random waves in a dispersive medium, *Proc. R. Soc. Lond.* A **289**, 301–320.

Bensoussan, A., Lions, J.L. & Papanicolaou, G. 1978. *Asymptotic Analysis for Periodic Structures*. North-Holland, Amsterdam.

Benzi, R., Biferale, L. & Parisi, G. 1993. On intermittency in a cascade model for turbulence, *Physica* D **65**, 163–171.

Benzi, R., Biferale, L., Crisanti, A., Paladin, G. Vergassola, M. & Vulpiani, A. 1993. A random process for the construction of multiaffine fields, *Physica* D **65**, 352–358.

Benzi, R., Biferale, L., Paladin, G., Vulpiani, A. & Vergassola, M. 1991. Multifractality in the statistics of velocity gradients, *Phys. Rev. Lett.* **67**, 2299–2302.

Benzi, R., Ciliberto, S., Baudet, C., Ruiz Chavarria, G. & Tripiccione, C. 1993. Extended self-similarity in the dissipation range of fully developed turbulence, *Europhys. Lett.* **24**, 275–279.

Benzi, R., Ciliberto, S., Baudet, C. & Ruiz Chavarria, G. 1995. On the scaling of three dimensional homogeneous and isotropic turbulence, *Physica* D **80**, 385–398.

Benzi, R., Ciliberto, S., Tripiccione, R., Baudet, C. Massaioli, F. & Succi, S. 1993. Extended self-similarity in turbulent flows, *Phys. Rev.* E **48**, R29–R32.

Benzi, R., Colella, M., Briscolini, M. & Santangelo, P. 1992. A simple point vortex model for two-dimensional decaying turbulence, *Phys. Fluids* A **4**, 1036–1039.

Benzi, R., Legras, B., Parisi, G. & Scardovelli, R. 1995. Conformal field theory and direct numerical simulation of two-dimensional turbulence, *Europhys. Lett.* **29**, 203–208.

Benzi, R, Paladin, G., Parisi, G. & Vulpiani, A. 1984. On the multifractal nature of fully developed turbulence and chaotic systems, *J. Phys.* **A17**, 3521-3531.

Benzi, R., Paladin, G., Paternello, S., Santangelo, P. & Vulpiani, A. 1986. Intermittency and coherent structures in two-dimensional turbulence, *J. Phys.* A **19**, 3771–3784.

Bergé, P., Pomeau, Y. & Vidal, C. 1984. *L'Ordre dans le Chaos: vers une Approche Déterministe de la Turbulence*. Hermann, Paris (English translation by Tuckerman, L.: *Order within Chaos: Towards a Deterministic Approach to Turbulence*, Wiley, New York, 1986).

Berry, M. 1977. Focusing and twinkling: critical exponents from catastrophes in non-Gaussian random short waves, *J. Phys.* A **10**, 2061–2081.

Berry, M. 1982. Universal power-law tails for singularity-dominated strong fluctuations, *J. Phys.* A **15**, 2735–2749.

Bertozzi, A. & Constantin, P. 1993. Global regularity for vortex patches, *Commun. Math. Phys.* **152**, 19–28.

Biferale, L., Blank, M. & Frisch, U. 1994. Chaotic cascades with Kolmogorov 1941 scaling, *J. Stat. Phys.* **75**, 781–795.

Biferale, L., Lambert, A., Lima, R. & Paladin, G. 1995. Transition to chaos in a shell model of turbulence, *Physica* D **80**, 105–119.

Blank, M. 1995. Geometric constructions in multifractality formalism, in *Lévy Flights and Applications in Physics, Lect. Notes in Phys.* **450**, 140–150, eds. M. Shlesinger, G.M. Zaslavsky & U. Frisch. Springer, Berlin.

Bondarenko, N.F., Gak, M.Z. & Dolzhansky, F.V. 1979. Laboratory and theoretical models of a plane periodic flow, *Atmos. Ocean. Phys.* **15**, 711–720.

Bonn, D., Couder, Y., van Dam, P.H.J. & Douady, S. 1993. From small scales to large scales in three-dimensional turbulence: the effect of diluted polymers, *Phys. Rev.* E **47**, R28–R31.

Boratav, O.N. & Pelz, R.B. 1994. Direct numerical simulation of transition to turbulence from a high-symmetry initial condition, *Phys. Fluids* A **6**, 2757–2784.

Borue, V. 1994. Inverse energy cascade in stationary two-dimensional homogeneous turbulence, *Phys. Rev. Lett.* **72**, 1475–1478.

Borue, V. & Orszag, S.A. 1995. Forced three-dimensional homogeneous turbulence with hyperviscosity, *Europhys. Lett.* **29**, 687–692.

Boussinesq, J. 1868. Mémoire sur l'influence des frottements dans les mouvements réguliers des fluides, *J. Math. Pures et Appliqu. (Paris)* **13 (2)**, 377–424.

Boussinesq, J. 1870. Essai théorique sur les lois trouvées expérimentalement par M. Bazin pour l'écoulement uniforme de l'eau dans les canaux découverts, *C. R. Acad. Sci. Paris* **71**, 389–393.

Boussinesq, J. 1877. Essai sur la théorie des eaux courantes, *Mém. prés. par div. savants à l'Acad. Sci.* **23**, 1–680.

Brachet, M.E. 1990. Géométrie des structures à petite échelle dans le vortex de Taylor–Green, *C. R. Acad. Sci. Paris, série II* **311**, 775–780.

Brachet, M.E. 1991. Direct simulation of three-dimensional turbulence in the Taylor–Green vortex, *Fluid Dyn. Res.* **8**, 1–8.

Brachet, M.E., Meiron, D.I., Orszag, S.A., Nickel, B.G., Morf, R.H. & Frisch, U. 1983. Small-scale structure of the Taylor–Green vortex, *J. Fluid Mech.* **130**, 411–452.

Brachet, M.E., Meneguzzi, M., Politano, H. & Sulem, P.-L. 1988. The dynamics of freely decaying two-dimensional turbulence, *J. Fluid Mech.* **194**, 333–349.

Brachet, M.E., Meneguzzi, M., Vincent, A., Politano, H. & Sulem, P.-L. 1992. Numerical evidence of smooth self-similar dynamics and possibility of subsequent collapse for three-dimensional ideal flow, *Phys. Fluids* A **4**, 2845–2854.

Bradshaw, P. 1971. *An Introduction to Turbulence and its Measurement.* Pergamon Press, New York.

Brissaud, A., Frisch, U., Léorat, J., Lesieur, M. & Mazure, A. 1973. Helicity cascades in fully developed isotropic turbulence, *Phys. Fluids* **16**, 1366–1367.

Brown, G.L. & Roshko, A. 1974. On density effects and large structures in turbulent mixing layers, *J. Fluid Mech.* **64**, 775–816.

Brush, S.G. 1976. *The Kind of Motion We Call Heat*. North-Holland, Amsterdam (2 vols.).

Burgers, J.M. 1939. Mathematical examples illustrating relations occuring in the theory of turbulent fluid motion, *Kon. Ned. Akad. Wet. Verh.* **17**, 1–53 (also in 'Selected Papers of J.M. Burgers', eds. F.T.M. Nieuwstadt & J.A. Steketee, pp. 281–334, Kluwer, 1995).

Burgers, J.M. 1948. A mathematical model illustrating the theory of turbulence, *Adv. in Appl. Mech.* **1**, 171–199.

Burgers, J.M. 1974. *The Nonlinear Diffusion Equation*. D. Reidel, Dordrecht.

Cafarelli, L., Kohn, R. & Nirenberg, L. 1982. Partial regularity of suitable weak solutions of the Navier–Stokes equations, *Commun. Pure Appl. Math.* **35**, 771-831.

Caglioti, E., Lions, P.-L., Marchioro, C. & Pulverenti, M. 1992. A special class of stationary flows for two-dimensional Euler equations: a statistical mechanics approach, *Commun. Math. Phys.* **143**, 501–525.

Cardoso, O., Marteau, D. & Tabeling, P. 1994. Quantitative experimental study of the free decay of quasi-two-dimensional turbulence, *Phys. Rev.* E **49**, 454–461.

Carnevale, G.F., McWilliams, J.C., Pomeau, Y., Weiss, J.B. & Young, W.R. 1991. Evolution of vortex statistics in two-dimensional turbulence, *Phys. Rev. Lett.* **66**, 2735–2738.

Castaing, B. 1989. Conséquences d'un principe d'extremum en turbulence, *J. Phys.* **50**, 147–156.

Castaing, B., Chabaud, B. & Hébral, B. 1992. Hot wire anemometry operating at cryogenic temperatures, *Rev. Sci. Instrum.* **63**, 4167–4173.

Castaing, B., Gagne, Y. & Hopfinger, E. 1990. Velocity probability density functions of high Reynolds number turbulence, *Physica* D **46**, 177–200.

Champagne, F.H. 1978. The fine-scale structure of the turbulent velocity field, *J. Fluid Mech.* **86**, 67–108.

Chandrasekhar, S. 1961. *Hydrodynamic and Hydromagnetic Stability*. Clarendon, Oxford.

Charney, J.G. 1947. The dynamics of long waves in a baroclinic westerly current, *J. Meteor.* **4**, 135–163.

Chemin, J.-Y. 1993. Persistence de structures géométriques dans les fluides incompressibles bidimensionnels, *Ann. Sci. École Norm. Sup (Paris)* **26**, 517–542.

Chertkov, M., Falkovich, G., Kolokolov, I. & Lebedev, V. 1995. Normal and anomalous scaling of the fourth-order correlation function of a randomly advected passive scalar, *Phys. Rev.* E **52**, 4924–4941.

Childress. S. & Gilbert, A. 1995. *Stretch, Twist, Fold – The Fast Dynamo*, in 'Springer Lecture Notes in Physics' **m37**. Springer, Berlin.

Childress, S., Collet, P., Frisch, U., Gilbert, A.D., Moffatt, H.K. & Zaslavsky, G.M. 1990. Small-diffusivity dynamos and dynamical systems, *Geophys. Astrophys. Fluid Dynam.* **52**, 263–270.

Chorin, A.J. 1988. Spectrum, dimension, and polymer analogies in fluid turbulence, *Phys. Rev. Lett.* **60**, 1947–1949.

Chorin, A.J. 1990. Constrained random walks and vortex filaments in turbulence theory, *Commun. Math. Phys.* **132**, 519–536.

Chorin, A.J. 1994. *Vorticity and Turbulence*. Springer, Berlin.

Chorin, A.J. & Akao, J.H. 1991. Vortex equilibria in turbulence theory and quantum analogues, *Physica* D **52**, 403–414.

Chorin, A.J. & Marsden, J.E. 1979. *A Mathematical Introduction to Fluid Mechanics*. Springer, Berlin.

Chossat, P. & Golubitsky, M. 1988. Symmetry-increasing bifurcations of chaotic attractors, *Physica* D **32**, 423–436.

Coleman, S. 1979. Lectures on Quantum Field Theory (Harvard University, Cambridge, MA, unpublished).

Collet P. & Koukiou, F. 1992. Large deviations for multiplicative chaos, *Commun. Math. Phys.* **47**, 329–342.

Constantin, P. 1991. Remarks on the Navier–Stokes equations, in *New Perpectives in Turbulence*, 229–261, ed. L. Sirovich. Springer, Berlin.

Constantin, P. 1994. Geometric statistics in turbulence, *SIAM Rev.* **36**, 73–98.

Constantin, P. & Fefferman, Ch. 1994. Scaling exponents in fluid turbulence: some analytic results, *Nonlinearity* **7**, 41–57.

Constantin, P. & Procaccia, I. 1993. Scaling in fluid turbulence: a geometric theory, *Phys. Rev.* E **47**, 3307–3315.

Constantin, P., E, W. & Titi, E.S. 1994. Onsager's conjecture on the energy conservation for solutions of Euler's equation, *Commun. Math. Phys.* **165**, 207–209.

Constantin, P., Fefferman, Ch. & Majda, A.J. 1995. Manuscript, quoted in Constantin (1994).

Constantin, P., Foias, C. & Temam, R. 1988. Dimension of the attractor of two-dimensional turbulence, *Physica* D **30**, 284–296.

Corrsin, S. 1959. Outline of some topics in homogeneous turbulent flow, *J. Geophys. Res.* **64**, 2134–2150.

Corrsin, S. 1963. Turbulence: experimental methods, in *Handbuch der Physik, Fluid Dynamics II*, 524–590, eds. S. Flügge & C. Truesdell. Springer, Berlin.

Couder, Y. 1984. Two-dimensional grid turbulence in a thin liquid film, *J. Phys. Lett. (Paris)* **45**, 353–360.

Coullet, P. & Fauve, S. 1985. Large scale oscillatory instability for sytems with translational and Galilean invariances, in *Macroscopic Modelling of Turbulent Flows, Lect. Notes in Phys.* **230**, 290–295, eds. U. Frisch, J.B. Keller, G. Papanicolaou & O. Pironneau. Springer, Berlin.

Cramér, H. 1938. Sur un nouveau théorème-limite de la théorie des probabilités, *Actualités Scientifiques et Industrielles* **736**, 5–23.

Cross, M.C. & Hohenberg, P.C. 1993. Pattern formation outside of equilibrium, *Rev. Mod. Phys.* **65**, 851–1111.

Crow, S.C. 1968. Viscoelastic properties of fine-grained incompressible turbulence, *J. Fluid Mech.* **33**, 1–20.

Crow, S.C. & Champagne, F.H. 1971. Orderly structures in jet turbulence, *J. Fluid Mech.* **48**, 547–591.

Cvitanović, P. 1989. *Universality in Chaos*, 2nd edition. Adam Hilger, Bristol and New York.

Dannevik, W.P., Yakhot, V. & Orszag, S.A. 1987. Analytic theories of turbulence and the ϵ expansion, *Phys. Fluids* **30**, 2021–2029.

de Dominicis, C. & Martin, P.C. 1979. Energy spectra of certain randomly-stirred fluids, *Phys. Rev.* A **19**, 419–422.

de Gennes, P.G. 1971. *Scaling Concepts in Polymer Physics*. Cornell Univ. Press, Ithaca.

Desnyansky, V.N. & Novikov, E.A. 1974. The evolution of turbulence spectra to the similarity regime, *Izv. Akad. Nauk SSSR Fiz. Atmos. Okeana* **10**(2), 127–136.

Devaney, R.L. 1989. *An Introduction to Chaotic Dynamical Systems*, 2nd edition. Addison-Wesley, Redwood City.

d'Humières, D., Pomeau, Y. & Lallemand, P. 1985. Simulation d'allées de von Kármán 2-D par méthode de gaz sur réseaux, *C. R. Acad. Sci. Paris, série II* **301**, 1391–1394.

Di Castro, C. & Jona-Lasinio, G. 1969. On the microscopic foundation of scaling laws, *Phys. Lett.* A **29**, 322–323.

Dombre, T., Frisch, U., Greene, J.M., Hénon, M., Mehr, A. & Soward, A.M. 1986. Chaotic streamlines in the ABC flows, *J. Fluid Mech.* **167**, 353–391.

Donnelly, R.J. & Swanson, C.E. 1986. Quantum turbulence, *J. Fluid Mech.* **173**, 387–429.

Donsker, M.D. 1964. On function space integrals, in *Analysis in Function Space*, 17–30, eds. W.T. Martin & I. Segal. MIT Press, Cambridge, MA.

Douady, S. & Couder, Y. 1993. On the dynamical structures observed in 3D turbulence, in *Turbulence in Spatially Extended Systems, Les Houches 1992*, 3–17, eds. R. Benzi, C. Basdevant & S. Cilberto. Nova Science, Commack, New York.

Douady, S., Couder, Y. & Brachet, M.E. 1991. Direct observation of the intermittency of intense vorticity filaments in turbulence, *Phys. Rev. Lett.* **67**, 983–986.

Dracos, Th. & Tsinober, A. 1993. *New Appoaches and Concepts in Turbulence*, eds. Th. Dracos & A. Tsinober. Birkhäuser, Basel (Proceedings Monte Verita, 1991).

Drazin, P.G. & Reid, W.H. 1981. *Hydrodynamic Stability.* Cambridge University Press, Cambridge.

Dritschel, D.G. 1989. Contour dynamics and contour surgery: numerical algorithms for extended high resolution modelling of vortex dynamics in two-dimensional inviscid, incompressible flows, *Computer Phys. Rep.* **10**, 7–146.

Dritschel, D.G. & Legras, B. 1993. Modeling oceanic and atmospheric vortices, *Physics Today* **46**, 44–51.

Du, Y., & Ott, E. 1993. Fractal dimensions of fast dynamo magnetic fields, *Physica* D **67**, 387–417.

Dubrulle, B. 1994. Intermittency in fully developed turbulence: log–Poisson statistics and generalized scale-covariance, *Phys. Rev. Lett.* **73**, 959–962.

Dubrulle, B. & Frisch, U. 1991. The eddy viscosity of parity-invariant flow, *Phys. Rev.* A **43**, 5355–5364.

E, W., & Shu, C.-W. 1993. Effective equations and the inverse cascade theory for Kolmogorov flows, *Phys. Fluids* A **5**, 998–1010.

E, W., & Shu, C.-W. 1994. Small-scale structures in Boussinesq convection, *Phys. Fluids* **6**, 49–58.

Ebin, D., & Marsden, J. 1970. Groups of diffeomorphisms and the motion of an incompressible fluid, *Ann. Math.* **92**, 102–163.

Edwards, S.F. 1964. The statistical dynamics of homogeneous turbulence, *J. Fluid Mech.* **18**, 239–273.

Edwards, S.F. 1965. Turbulence in hydrodynamics and plasma physics, in *Int. Conf. on Plasma Physics*, 595–623. International Atomic Energy Agency, Vienna.

Einstein, A. 1914. Méthode pour la détermination de valeurs statistiques d'observations concernant des grandeurs soumise à des fluctuations irrégulières, *Arch. Sci. Phys. et Natur.* **37**, 254–255.

Ellis, R.S. 1985. *Entropy, Large Deviations and Statistical Mechanics.* Springer, Berlin.

Eneva, M. 1994. Monofractal or multifractal: a case study of spatial distribution of mining induced seismic activity, *Nonlinear Processes in Geophysics* **1**, 182–190.

Evertsz, C.J.G. & Mandelbrot, B.B. 1992. Multifractal measures, in *Chaos and Fractals; New Frontiers of Science*, 921–953, eds. H.-O. Peitgen, H. Jürgens & D. Saupe. Springer, Berlin.

Eyink, G.L. 1994a. Energy dissipation without viscosity in ideal hydrodyamics, I. Fourier analysis and local transfer, *Physica* D **78**, 222–240.

Eyink, G.L. 1994b. The renormalization group method in statistical hydrodynamics, *Phys. Fluids* **6**, 3063–3078.

Eyink, G.L. & Goldenfeld, N. 1994. Analogies between scaling in turbulence, field theory and critical phenomena, *Phys. Rev.* E **50**, 4638–4679.

Eyink, G.L. & Spohn, H. 1993. Negative temperature states and large-scale long-lived vortices in two-dimensional turbulence, *J. Stat. Phys.* **70**, 833–886.

Fairhall, A.L., Gat, O., L'vov, V.S. & Procaccia, I. 1996. Anomalous scaling in a model of passive scalar advection: exact results, *Phys. Rev.* E, **53**, 3518–3535.

Falconer, K. 1990. *Fractal Geometry: Mathematical Foundations and Applications.* Wiley, Chichester.

Falkovich, G. & Hanany, A. 1993. Is 2D turbulence a conformal turbulence?, *Phys. Rev. Lett.* **71**, 3454–3457.

Falkovich, G. & Lebedev, V. 1994. Universal direct cascade in two-dimensional turbulence, *Phys. Rev.* E **50**, 3883–3899.

Fannjiang, A. & Papanicolaou, G. 1994. Convection enhanced diffusion for periodic flows, *SIAM J. Appl. Math.* **54**, 333–408.

Farge, M. 1992. Wavelet transforms and their applications to turbulence, *Ann. Rev. Fluid Mech.* **24**, 395–407.

Fauve, S. 1983. Instabilité convective et équation d'amplitude, in *Instabilités Hydrodynamiques et Applications Astrophysiques. Proceedings Goutelas 1983*, 83–115, ed. A. Baglin. Soc. Franç. Spéc. Astron., Observatoire de Nice.

Fauve, S. 1991. Lectures on patterns, cellular flow, etc., in *Patterns in Fluid Flow. Proceedings GFD 1991*, 2–101. Woods-Hole Oceanographic Institute, Woods Hole (WHOI-92-16).

Fauve, S., Laroche, C. & Castaing, B. 1993. Pressure fluctuations in swirling turbulent flows, *J. Phys. II France* **3**, 271–278.

Favre, A., Kovasznay, L.S.G., Dumas, R., Gaviglio, J. & Coantic, M. 1976. *La Turbulence en Mécanique des Fluides.* Gauthiers-Villars, Paris.

Feller, W. 1968a. *An Introduction to Probability Theory and its Applications*, 3rd edition, vol. 1. Wiley, New York.

Feller, W. 1968b. *An Introduction to Probability Theory and its Applications*, 3rd edition, vol. 2. Wiley, New York.

Feynman, R.P. 1950. Mathematical formulation of the quantum theory of electromagnetic interaction, *Phys. Rev.* **80**, 440–457.

Feynman, R.P. 1964. *The Feynman Lectures on Physics/Feynman, Leighton, Sands.* Addison-Wesley, Redwood City.

Fornberg, B. 1977. A numerical study of two-dimensional turbulence, *J. Comput. Phys.* **25**, 1–31.

Forster, D., Nelson, D.R. & Stephen, M.J. 1977. Large distance and long time properties of a randomly stirred fluid, *Phys. Rev.* A **16**, 732–749.

Fournier, J.D. & Frisch, U. 1983a. Remarks on the renormalization group in statistical fluid dynamics, *Phys. Rev.* A **28**, 1000–1002.

Fournier, J.D. & Frisch, U. 1983b. L'équation de Burgers déterministe et statistique, *J. Méc. Théor. Appl. (Paris)* **2**, 699–750.

Fournier, J.D., Sulem, P.-L. & Pouquet, A. 1982. Infrared properties of forced MHD turbulence, *J. Phys.* A **15**, 1393–1420.

Frisch, U. 1968. Wave propagation in random media, in *Probabilistic Methods in Applied Mathematics* **1**, 75-198, ed. A.T. Bharucha-Reid. Academic Press, New York (see also *Ann. Astrophys. (Paris)*, **29**, 645–682 (1966) and **30**, 565–601 (1967)).

Frisch, U. 1983. Fully developed turbulence and singularities, in *Chaotic Behavior of Deterministic Systems, Les Houches 1981*, 665–704, eds. G. Iooss, R. Hellemann & R. Stora. North-Holland, Amsterdam.

Frisch, U. 1989. Lectures on turbulence and lattice gas hydrodynamics, in *Lecture Notes on Turbulence, NCAR-GTP Summer School June 1987*, 219–371, eds. J.R. Herring & J.C. McWilliams. World Scientific, Singapore.

Frisch, U. 1991. From global scaling, à la Kolmogorov, to local multifractal scaling in fully developed turbulence, in *Kolmogorov's Ideas 50 Years on*, *Proc. R. Soc. Lond.* A **434**, 89–99, eds. J.C.R. Hunt, O.M. Phillips & D. Williams.

Frisch, U. & Fournier J.D. 1978. d-dimensional turbulence, *Phys. Rev.* A **17**, 747–762.

Frisch, U. & Morf, R. 1981. Intermittency in nonlinear dynamics and singularities at complex times, *Phys. Rev.* A **23**, 2673–2705.

Frisch, U. & Orszag, S.A. 1990. Turbulence: challenges for theory and experiments, *Phys. Today*, 24–32 (January 1990).

Frisch, U. & She, Z.S. 1991. On the probability density function of velocity gradients in fully developed turbulence, *Fluid Dyn. Res.* **8**, 139–142.

Frisch, U. & Sulem P.-L. 1984. Numerical simulation of the inverse cascade in two-dimensional turbulence, *Phys. Fluids* **27**, 1921–1923.

Frisch, U. & Vergassola, M. 1991. A prediction of the multifractal model: the intermediate dissipation range, *Europhys. Lett.* **14**, 439–444.

Frisch, U., Fournier, J.D. & Rose, H. 1978. Infinite-dimensional turbulence, *J. Phys.* A **11**, 187–198.

Frisch, U., Legras, B. & Villone, B. 1996. Large-scale Kolmogorov flow on the beta-plane and resonant wave interactions, *Physica* D **94**, 36–56.

Frisch, U., Lesieur, M. & Schertzer, D. 1980. Comments on the quasi-normal Markovian approximation for fully-developed turbulence, *J. Fluid Mech.* **97**, 181–192.

Frisch, U., Lesieur, M. & Sulem, P-.L. 1976. On crossover dimensions for fully developed turbulence, *Phys. Rev. Lett.* **37**, 895–897.

Frisch, U., She, Z.S. & Sulem, P.-L. 1987. Large-scale flow driven by the anisotropic kinetic alpha effect, *Physica* D **28**, 382–392.

Frisch, U., She, Z.S. & Thual, O. 1986. Viscoelastic behavior of cellular solutions to the Kuramoto–Sivashinsky model, *J. Fluid Mech.* **168**, 221–240.

Frisch, U., Sulem, P.-L. & Nelkin, M. 1978. A simple dynamical model of intermittent fully developed turbulence, *J. Fluid Mech.* **87**, 719–736.

Frisch, U., Pouquet, A., Sulem, P.-L. & Meneguzzi, M. 1983. The dynamics of two-dimensional ideal MHD, *J. Méc Théor. Appliqu. (Paris)*, 191–216 (Special issue on two-dimensional turbulence).

Fröhlich, J. & Ruelle, D. 1982. Statistical mechanics of vortices in an inviscid two-dimensional fluid, *Commun. Math. Phys.* **87**, 1–36.

Frostman, O. 1935. Potentiel d'équilibre et capacité des ensembles, Thesis, Lund, Sweden.

Fung, J.C.H. & Vassilicos, J.C. 1991. Fractal dimensions of lines in chaotic advection, *Phys. Fluids* A **3**, 2725–2733.

Furutsu, K. 1963. On the statistical theory of electromagnetic waves in a fluctuating medium, *J. Res. Nat. Bur. Standards* D **67**, 303–323.

Gagne, Y. 1980. Contribution à l'étude expérimentale de l'intermittence de la turbulence à petite échelle, Thèse de Docteur-Ingénieur, Université de Grenoble.

Gagne, Y. 1987. Etude expérimentale de l'intermittence et des singularités dans le plan complexe en turbulence développée, Thèse de Docteur ès-Sciences Physiques, Université de Grenoble.

Gagne, Y. & Castaing, B. 1991. A universal representation without global scaling invariance of energy spectra in developed turbulence, *C. R. Acad. Sci. Paris, série II* **312**, 441–445.

Gagne, Y., Hopfinger, E. & Frisch, U. 1990. A new universal scaling for fully developed turbulence: the distribution of velocity increments, in *New Trends in Nonlinear Dynamics and Pattern-Forming Phenomena, NATO ASI* **237**, 315–319, eds. P. Coullet & P. Huerre. Plenum Press, New York.

Gallavotti, G. 1993. Some rigorous results about 3D Navier–Stokes, in *Turbulence in Spatially Extended Systems, Les Houches 1992*, 45–74, eds. R. Benzi, C. Basdevant & S. Cilberto. Nova Science, Commack, New York.

Gama, S. & Frisch, U. 1993. Local helicity, a material invariant for the odd-dimensional Euler equations, in *NATO-ASI: Solar and Planetary Dynamos*, 115–119, eds. M.R.E. Proctor, P.C. Mathews & A.M. Rucklidge. Cambridge University Press, Cambridge.

Gama, S., Vergassola, M. & Frisch, U. 1994. Negative eddy viscosity in isotropically forced two-dimensional flow: linear and nonlinear dynamics, *J. Fluid Mech.* **260**, 95–126.

Gat, O., Procaccia, I. & Zeitak, R. 1994. Breakdown of dynamical scaling and intermittency in a cascade model of turbulence, *Phys. Rev.* E **51**, 1148–1154.

Gawędzki, K. & Kupiainen, A. 1995. Anomalous scaling of the passive scalar, *Phys. Rev. Lett.* **75**, 3834–3837.

Gharib, M. & Derango, P. 1989. A liquid film (soap film) tunnel to study laminar and turbulent shear flow, *Physica* D **37**, 406–416.

Gibson, C.H. & Schwarz, W.H. 1963. The universal equilibrium spectra of turbulent velocity and scalar fields, *J. Fluid Mech.* **16**, 365–384.

Gilbert, A., Frisch, U. & Pouquet, A. 1988. Helicity is unnecessary for alpha-effect dynamos, but it helps, *Geophys. Astrophys. Fluid Dynam.* **42**, 151–161.

Gledzer, E.B. 1973. System of hydrodynamic type admitting two quadratic integrals of motion, *Sov. Phys. Dokl.* **18**, 216–217.

Gledzer, E.B., Dolzhansky, F.V. & Obukhov, A.M. 1981. *Systems of Hydrodynamic Type and Applications.* Nauka, Moscow (in Russian).

Goldstein, H. 1980. *Classical Mechanics*, 2nd edition. Addison-Wesley, Reading, MA.

Goldstein, S. 1969. Fluid mechanics in the first half of this century, *Ann. Rev. Fuid Mech.* **1**, 1–28.

Golitsyn, G.S. 1962. Dissipation fluctuations in locally isotropic flow, *Dokl. Akad. Nauk SSSR* **144**, 520–523.

Gollub, J., Clarke, J., Gharib, M., Lane, B. & Mesquita, O. 1991. Fluctuations and transport in a stirred fluid with a mean gradient, *Phys. Rev. Lett.* **67**, 3507–3510.

Gottlieb, D. & Orszag, S.A. 1977. *Numerical Analysis of Spectral Methods.* SIAM, Philadelphia.

Grant, H.L., Stewart, R.W. & Moilliet, A. 1962. Turbulent spectra from a tidal channel, *J. Fluid Mech.* **12**, 241–268.

Grassberger, P. 1983. Generalized dimensions of strange attractors, *Phys. Lett.* A **97**, 227–230.

Grassberger, P. & Procaccia, I. 1983. Characterization of strange attractors, *Phys. Rev. Lett.* **50**, 346–349.

Grassberger, P. & Procaccia, I. 1984. Dimensions and entropies of strange attractors from a fluctuating dynamics approach, *Physica* D **13**, 34–54.

Grauer, R. & Sideris, T. 1991. Numerical computations of 3-D incompressible ideal fluids with swirl, *Phys. Rev. Lett.* **25**, 3511–3514.

Guckenheimer, J. & Holmes, P. 1986. *Nonlinear Oscillations, Dynamical Systems and Bifurcations of Vector Fields.* Springer, Berlin.

Gurbatov, S.N., Malakhov, A.N., & Saichev, A.I. 1991. *Nonlinear Random Waves and Turbulence in Nondispersive media: Waves, Rays, Particles.* Manchester University Press, Manchester.

Gurvitch, A.S. 1960. Experimental research on frequency spectra of atmospheric turbulence, *Izv. Akad. Nauk SSSR, geofiz. ser.*, 1042–1055.

Guyon, E., Hulin, J.-P. & Petit, L. 1991. *Hydrodynamique Physique.* InterEditions/Editions du CNRS, Paris.

Halmos, P.R. 1956. *Lectures on Ergodic Theory.* Chelsea, New York.

Halsey, T.C., Jensen, M.H., Kadanoff, L., Procaccia, I. & Shraiman, B.I. 1986. Fractal measures and their singularities: the characterization of strange sets, *Phys. Rev.* A **33**, 1141–1151.

Heisenberg, W. 1948. Zur statistischen Theorie der Turbulenz, *Zeit. f. Phys.* **124**, 628–657.

Hénon, M. 1966. Sur la topologie des lignes de courant dans un cas particulier, *C. R. Acad. Sci. Paris* **262**, 312–314.

Hénon, M. 1976. A two-dimensional mapping with a strange attractor, *Commun. Math. Phys.* **50**, 69–77.

Hentschel, H.G.E. 1994. Stochastic multifractality and universal scaling distributions, *Phys. Rev.* E **50**, 243–261.

Hentschel, H.G.E. & Procaccia, I. 1983. The infinite number of generalized dimensions of fractals and strange attractors, *Physica* D **8**, 435–444.

Herring, J.R. & Kerr, R.M. 1993. Development of enstrophy and spectra in numerical turbulence, *Phys. Fluids* A **5**, 2792–2798.

Herring, J.R. & McWilliams, J.C. 1985. Comparison of direct-numerical simulations of two-dimensional turbulence with two-point closure: the effect of intermittency, *J. Fluid Mech.* **153**, 229–242.

Hinze, J.O. 1959. *Turbulence.* McGraw-Hill, New York.

Hölder, E. 1933. Über die unbeschränkte Fortsetzbarkeit einer stetigen ebenen Bewegung in einer unbegrezten inkompressiblen Flüssigkeit, *Math. Z.* **37**, 727–738.

Holmes, P. 1990. Can dynamical systems approach turbulence?, in *Wither Turbulence, Lect. Notes in Physics* **357**, 195–249, ed. J.L. Lumley. Springer, Berlin.

Holzer, M. & Pumir, A. 1993. Simple models of non-Gaussian statistics for a turbulently advected passive scalar, *Phys. Rev.* E **47**, 202–219.

Hopf, E. 1952. Statistical hydrodynamics and functional calculus, *J. Ratl. Mech. Anal.* **1**, 87–123.

Hosokawa, I. & Yamamoto, K. 1990. Intermittency of dissipation in directly simulated fully developed turbulence, *J. Phys. Soc. Japan* **59**, 401–404.

Hosokawa, I. & Yamamoto, K. 1992. Evidence against the Kolmogorov refined similarity hypothesis, *Phys. Fluids* A **4**, 457–459.

Huang, K. 1963. *Statistical Mechanics.* Wiley, New York.

Iooss, G. & Adelmeyer, M. 1992. *Topics in Bifurcation Theory and Applications,* vol. 3, in 'Advanced Series in Nonlinear Dynamics'. World Scientific, Singapore.

Isichenko, M.B. 1992. Percolation, statistical topography and transport in random media, *Rev. Mod. Phys.* **64**, 961–1043.

Isserlis, L. 1918. On a formula for the product-moment coefficient in any number of variables, *Biometrika* **12**, 134–139.

Itzykson, C. & Drouffe, J.-M. 1989. *Statistical Field Theory.* Cambridge University Press, Cambridge.

Itzykson, C. & Zuber, J.-B. 1985. *Quantum Field Theory.* McGraw-Hill, New York.

Jensen, M.H. 1995. Multifractals and multiscaling, *Phys. Reports* (to appear).

Jensen, M.H., Paladin, G. & Vulpiani, A. 1991. *Phys. Rev.* A **43**, 798–805.

Jiménez, J. 1993. Small-scale vortices in turbulent flows, in *New Approaches and Concepts in Turbulence,* 95–110, eds. T. Dracos & A. Tsinober. Birkhäuser, Basel.

Jiménez, J., Wray, A.A., Saffman, P.G. & Rogallo, R.S. 1993. The structure of intense vorticity in isotropic turbulence, *J. Fluid Mech.* **255**, 65–90.

Joyce, G. & Montgomery, D. 1973. Negative temperature states for the two-dimensional guiding center plasma, *J. Plasma Phys.* **10**, 107–121.

Kac, M. 1959. *Probability and Related Topics in Physical Sciences.* Wiley (Interscience), New York.

Kadanoff, L. 1966. Scalings laws for Ising models near T_c, *Physics (USA)* **2**, 263–272.

Kadanoff, L., Lohse, D., Wang, J. & Benzi, R. 1995. Scaling and dissipation in the GOY shell model, *Phys. Fluids* **7**, 617–629.

Kahane, J.P. 1985. *Some Random Series of Functions,* 2nd edition. Cambridge University Press, Cambridge.

Kahane, J.P. 1993. Fractals and random measures, *Bull. Sc. math.* **117**, 13–159.

Kahane, J.P. & Peyrière, J. 1976. Sur certaines martingales de B. Mandelbrot, *Advances in Math.* **2**, 131–145.

Kahane, J.P. & Salem, R. 1994. *Ensembles Parfaits et Séries Trigonométriques,* 2nd edition. Hermann, Paris.

Kampé de Fériet, J. 1953. Fonctions aléatoires et théorie statistique de la turbulence, in *Théorie des Fonctions Aléatoires*, 568–623, eds. A. Blanc-Lapierre & R. Fortet. Masson, Paris.

Kardar, M., Parisi, G. & Zhang, Y.C. 1986. Dynamical scaling of growing interfaces, *Phys. Rev. Lett.* **56**, 889–892.

Kármán, T. von 1921. Über laminare und turbulente Reibung, *Z. f. angew. Math. u. Mech.* **1**, 233–252.

Kármán, T. von 1930. Mechanische Änlichkeit und Turbulenz, *Nachr. Ges. Wiss. Göttingen, Math.-Phys. Kl.*, 58–76.

Kármán, T. von & Howarth, L. 1938. On the statistical theory of isotropic turbulence, *Proc. R. Soc. Lond.* A **164**, 192–215.

Kato, T. 1967. On classical solutions of the two-dimensional non stationary Euler equation, *Arch. Rat. Mech. Analys.* **25**, 188–200.

Kato, T. 1972. Nonstationary flows of viscous and ideal fluids in $\mathbb{R}^3$, *J. Funct. Anal.* **9**, 296–305.

Keller, B.S. & Yaglom, A.M. 1970. Effect of fluctuations in energy dissipation rate on the functional form of the energy spectrum in the far short-wave range, *Izv. Akad. Nauk SSSR ser. Mekh. Zhidk. i Gaza* (3), 70-79.

Kelvin, Lord (Sir W. Thomson) 1887. On the propagation of laminar motion through a turbulently moving inviscid liquid, *Phil. Mag.* **24**, 342–353.

Kendall, D.G. 1990. Obituary Andrei Nikolaevich Kolmogorov, *Bull. London Math. Soc.* **22**, 31–100.

Kerr, R. 1985. Higher-order derivative correlations and the alignement of small-scale structures in isotropic numerical turbulence, *J. Fluid Mech.* **153**, 31–58.

Kida, S. & Ohkitani, K. 1992. Spatio-temporal intermittency and instability of a forced turbulence, *Phys. Fluids* A **4**, 1018–1027.

Kirchhoff, G. 1877. *Vorlesungen über mathematische Physik*. B.G. Teubener, Leipzig.

Kolmogorov, A.N. 1933. *Grundbegriffe der Wahrscheinlichkeitsrechnung*. Springer, Berlin.

Kolmogorov, A.N. 1935. La transformation de Laplace dans les espaces linéaires, *C. R. Acad. Sci. Paris* **200**, 1717–1718.

Kolmogorov, A.N. 1940. Wiener's spiral and some interesting curves in Hilbert space, *Dokl. Akad. Nauk SSSR*, 115–118.

Kolmogorov, A.N. 1941a. The local structure of turbulence in incompressible viscous fluid for very large Reynolds number, *Dokl. Akad. Nauk SSSR* **30**, 299–303 (reprinted in *Proc. R. Soc. Lond.* A **434**, 9–13 (1991)).

Kolmogorov, A.N. 1941b. On degeneration (decay) of isotropic turbulence in an incompressible viscous liquid, *Dokl. Akad. Nauk SSSR* **31**, 538–540.

Kolmogorov, A.N. 1941c. Dissipation of energy in locally isotropic turbulence, *Dokl. Akad. Nauk SSSR* **32**, 16–18 (reprinted in *Proc. R. Soc. Lond.* A **434**, 15–17 (1991)).

Kolmogorov, A.N. 1941d. On the logarithmically normal law of distribution of the size of particles under pulverization, *Dokl. Akad. Nauk SSSR* **31**, 99–101.

Kolmogorov, A.N. 1942. The equation of turbulent motion in an incompressible viscous fluid, *Izv. Akad. Nauk SSSR, Ser.Fiz.* **VI**(1–2), 56–58.

Kolmogorov, A.N. 1954. On conservation of conditional periodic motions under small perturbations of the Hamiltonian, *Dokl. Akad. Nauk SSSR* **98**, 527–530.

Kolmogorov, A.N. 1961. Précisions sur la structure locale de la turbulence dans un fluide visqueux aux nombres de Reynolds élevés, in *La Turbulence en Mécanique des Fluides*, 447–451, eds. A. Favre, L.S.G. Kovasznay, R. Dumas, J. Gaviglio & M. Coantic. Gauthier-Villars, Paris.

Kolmogorov, A.N. 1962. A refinement of previous hypotheses concerning the local structure of turbulence in a viscous incompressible fluid at high Reynolds number, *J. Fluid Mech.* **13**, 82–85.

Kraichnan, R.H. 1958. A theory of turbulence dynamics, in *Second Symposium on Naval Hydrodynamics*, 29–44. Office of Naval Research, Washington, DC (Ref. ACR-38).

Kraichnan, R.H. 1959. The structure of isotropic turbulence at very high Reynolds numbers, *J. Fluid Mech.* **5**, 497–543.

Kraichnan, R.H. 1961. Dynamics of nonlinear stochastic systems, *J. Math. Phys.* **2**, 124–148 (erratum *ibid.* **3**, p. 205).

Kraichnan, R.H. 1964. Kolmogorov's hypotheses and Eulerian turbulence theory, *Phys. Fluids* **7**, 1723–1734.

Kraichnan, R.H. 1965. Lagrangian-history closure approximation for turbulence, *Phys. Fluids* **8**, 575–598 (erratum 1966 *ibid.*, p. 1884).

Kraichnan, R.H. 1966. Isotropic turbulence and inertial-range structure, *Phys. Fluids* **9**, 1728–1752.

Kraichnan, R.H. 1967a. Inertial ranges in two-dimensional turbulence, *Phys. Fluids* **10**, 1417–1423.

Kraichnan, R.H. 1967b. Intermittency in the very small scales of turbulence, *Phys. Fluids* **10**, 2080–2082.

Kraichnan, R.H. 1968a. Lagrangian-history statistical theory for Burgers' equation, *Phys. Fluids* **11**, 266–277.

Kraichnan, R.H. 1968b. Small-scale structure of a scalar field convected by turbulence, *Phys. Fluids* **11**, 945–953.

Kraichnan, R.H. 1971a. An almost-Markovian Galilean-invariant turbulence model, *J. Fluid Mech.* **47**, 513–524.

Kraichnan, R.H. 1971b. Inertial-range transfer in two and three-dimensional turbulence, *J. Fluid Mech.* **47**, 525–535.

Kraichnan, R.H. 1972. Some modern developments in the statistical theory of turbulence, in *Statistical Mechanics: New Concepts, New Problems, New Applications*, 201–228, eds. S.A. Rice, K.F. Freed & J.C. Light. University of Chicago, Chicago.

Kraichnan, R.H. 1974. On Kolmogorov's inertial-range theories, *J. Fluid Mech.* **62**, 305–330.

Kraichnan, R.H. 1975. Remarks on turbulence theory, *Advanc. Math.* **16**, 305–331.

Kraichnan, R.H. 1976. Eddy viscosity in two and three dimensions, *J. Atmos. Sci.* **33**, 1521–1536.

Kraichnan, R.H. 1987. An interpretation of the Yakhot–Orszag turbulence theory, *Phys. Fluids* **30**, 2400-2405.

Kraichnan, R.H. 1990. Models of intermittency in hydrodynamic turbulence, *Phys. Rev. Lett.* **65**, 575–578.

Kraichnan, R.H. 1994. Anomalous scaling of a randomly advected passive scalar, *Phys. Rev. Lett.* **72**, 1016–1019.

Kraichnan, R.H. & Montgomery, D. 1980. Two-dimensional turbulence, *Reports Progr. Phys.* **43**, 547–619.

Kraichnan, R.H., Yakhot, V. & Chen, S. 1995. Scaling relations for a randomly advected passive scalar, *Phys. Rev. Lett.* **75**, 240–243.

Kuo, A.Y. & Corrsin, S. 1971. Experiments on internal intermittency and fine-structure distribution function in fully turbulent fluid, *J. Fluid Mech.* **50**, 285–320.

Kuz'min, G.A. 1982. Statistical mechanics of the organisation into two-dimensional coherent structures, in *Structural Turbulence*, 103–114, ed. M.A. Goldshtik. Acad. Nauk USSR, Institute of Thermophysics, Novosibirsk (in Russian).

Kuz'min, G.A. 1983. Ideal incompressible hydrodynamics in terms of the vortex momentum density, *Phys. Lett.* A **96**, 88–90.

Kuznetsov, V.R., Praskovsky, A.A. & Sabelnikov, V.A. 1992. Fine-scale turbulence structure of intermittent shear flows, *J. Fluid Mech.* **243**, 595–622.

Ladyzhenskaya, O.A. 1969. *The Mathematical Theory of Viscous Incompressible Flow*, 2nd edition. Gordon and Breach, New York.

Lamb, H. 1932. *Hydrodynamics*, 6th edition. Dover, New York.

Landau, L.D. & Lifshitz, E.M. 1987. *Fluid Mechanics*, 2nd edition. Pergamon Press, Oxford.

Lanford, O.E. 1973. Entropy and equilibrium states in classical mechanics, in *Statistical Mechanics and Mathematical Problems, Lect. Notes in Physics* **20**, 1–113, ed. A. Lenard. Springer, Berlin.

Le Bellac, M. 1992. *Quantum and Statistical Field Theory*. Oxford University Press (translated from French *Des Phénomènes Critiques aux Champs de Jauge*, published by Interéditions/éditions du CNRS, Paris, 1988).

Lebedev, V.V. & L'vov, V.S. 1994. Scaling of correlation functions of velocity gradients in hydrodynamic turbulence, *JETP Lett.* **59**, 577–583.

Legras, B., Santangelo, P. & Benzi, R. 1988. High-resolution numerical experiments for forced two-dimensional turbulence, *Europhys. Lett.* **5**, 37–42.

Leibovich, S. 1978. The structure of vortex breakdown, *Ann. Rev. Fluid Mech.* **10**, 221–246.

Leray, J. 1934. Sur le mouvement d'un liquide visqueux emplissant l'espace, *Acta Math.* **63**, 193–248.

Lesieur, M. 1990. *Turbulence in Fluids*, 2nd edition. Kluwer, Dordrecht.

Lesieur, M. 1994. *La Turbulence*. Presses Universitaires de Grenoble, Grenoble.

Lesieur, M. & Schertzer, D. 1978. Amortissement autosimilaire d'une turbulence à grand nombre de Reynolds, *J. Méc. (Paris)* **17**, 610–646.

Lesieur, M., Frisch, U. & Brissaud, A. 1971. Théorie de Kraichnan de la turbulence. Application à l'étude d'une turbulence possédant de l'hélicité, *Ann. Géophys. (Paris)* **27**, 151–165.

Leslie, D.C. 1973. *Developments in the Theory of Turbulence*. Clarendon Press, Oxford.

Lévêque, E. & She, Z.S. 1995. Viscous effects on inertial-range scaling in a dynamical model of turbulence, *Phys. Rev. Lett.* **75**, 2690–2693.

Lévy, P. 1965. *Processus Stochastiques et Mouvement Brownien*. Gauthiers-Villars, Paris.

Lewis, R.M.& Kraichnan, R.H. 1962. A space-time functional formalism for turbulence, *Comm. Pure Appl. Math.* **15**, 397–411.

Lichtenstein, L. 1925. Über einige Existenzsätze der Hydrodynamik homogener, unzusammendrückbarer, reibungsloser Flüssigkeiten und die Helmholtzschen Wirbelsätze, *Math. Zeitschr.* **23**, 89–154; 309–316.

Lilly, D.K. 1971. Numerical simulation of two-dimensional turbulence, *Phys. Fluids Suppl. II* **12**, 240–249.

Lilly, D.K. & Peterson, E.L. 1983. Aircraft measurements of atmospheric kinetic energy spectra, *Tellus* **35A**, 379–382.

Lin, S.J. & Corcos, G. 1984. The mixing layer: deterministic models of a turbulent flow: part 3. The effect of plain strain on the dynamics of streamwise vortices, *J. Fluid Mech.* **246**, 569–591.

Lin, C.C. & Reid, W.H. 1963. Turbulent flow, theoretical aspects, in *Handbuch der Physik, Fluid Dynamics II*, 438–523, eds. S. Flügge & C. Truesdell. Springer, Berlin.

Lions, J.L. 1969. *Quelques Méthodes de Résolution des Problèmes aux Limites non Linéaires.* Gauthier-Villars, Paris.

Loitsyansky, L.G. 1939. Some basic laws for isotropic turbulent flow, *Trudy Tsentr. Aero.-Gidrodin. Inst*, 3–23.

Lorenz, E.N. 1963. Deterministic nonperiodic flow, *J. Atmos. Sci.* **20**, 130–141.

Lorenz, E.N. 1972. Low order models representing realizations of turbulence, *J. Fluid Mech.* **55**, 545–563.

Lumley, J.L. 1965. On the interpretation of time-spectra measured in high intensity shear flows, *Phys. Fluids* **8**, 1056–1062.

Lumley, J.L. 1970. *Stochastic Tools in Turbulence.* Academic Press, New York.

Lumley, J.L. 1972. Application of central limit theorems to turbulence problems, in *Statistical Models and Turbulence, Lect. Notes in Phys.* **12**, 1–26, eds. M. Rosenblatt & C.W. van Atta. Springer, Berlin.

Lumley, J.L. 1990. *Wither Turbulence? Turbulence at Crossroads, Lect. Notes in Phys.* **357**, ed. J.L. Lumley. Springer, Berlin.

Lumley, J.L. 1992. Some comments on turbulence, *Phys. Fluids* A **4**, 203–211.

Lundgren, T.S. 1982. Strained spiral vortex model for turbulent fine structure, *Phys. Fluids* **25**, 2193–2203.

Lundgren, T.S. 1993. A small-scale turbulence model, *Phys. Fluids* A **5**, 1472–1483.

L'vov, V.S. 1991. Scale invariant theory of fully developed hydrodynamic turbulence – Hamiltonian approach, *Phys. Rep.* **207**, 1–47.

L'vov, V.S. 1994. *Wave Turbulence under Parametric Excitation.* Springer, Berlin.

L'vov, V.S. & Procaccia, I. 1995. "Intermittency" in hydrodynamic turbulence as intermediate asymptotics to Kolmogorov scaling, *Phys. Rev. Lett.* **74**, 2690–2693.

L'vov, V.S., Procaccia, I. & Fairhall, A.L. 1994. Anomalous scaling in fluid mechanics: the case of the passive scalar, *Phys. Rev.* E **50**, 4684–4704.

Lynden-Bell, D. 1967. Statistical mechanics of violent relaxation in stellar systems, *Mon. Not. R. astr. Soc.* **136**, 101–121.

Ma, S.K. & Mazenko, G.F. 1975. Critical dynamics of ferromagnets in $6 - \epsilon$ dimensions: general discussion and detailed calculation, *Phys. Rev.* B **11**, 4077–4100.

Machlup, S. & Onsager, L. 1953. Fluctuations and irreversible processes. II. Systems with kinetic energy, *Phys. Rev.* **91**, 1512–1515.

Majda, A.J. 1986. Vorticity and the mathematical theory of incompressible flow, *Comm. Pure Appl. Math.* **S39**, 187–220.

Mandelbrot, B. 1968. On intermittent free turbulence, in *Turbulence of Fluids and Plasmas, New York, April 16–18, 1968.* Brooklyn Polytechnic Inst., Brooklyn, New York (abstract).

Mandelbrot, B. 1972. Possible refinement of the lognormal hypothesis concerning the distribution of energy dissipation in intermittent turbulence, in *Statistical Models and Turbulence, Lect. Notes in Phys.* **12**, 333–351, eds. M. Rosenblatt & C.W. van Atta. Springer, Berlin.

Mandelbrot, B. 1974a. Multiplications aléaoires itérées et distributions invariantes par moyenne aléatoire: quelques extensions, *C. R. Acad. Sci. Paris* **278**, 355–358.

Mandelbrot, B. 1974b. Intermittent turbulence in self-similar cascades: divergence of high moments and dimension of the carrier, *J. Fluid Mech.* **62**, 331–358.

Mandelbrot, B. 1977. *Fractals: Form, Chance and Dimension.* Freeman & Co., San Francisco.

Mandelbrot, B. 1989. Multifractal measures, especially for the geophysicist, *Pure Appl. Geophys.* **131**, 5–42.

Mandelbrot, B. 1990. Negative fractal dimensions and multifractals, *Physica A* **163**, 306–315.

Mandelbrot, B. 1991. Random multifractals: negative dimensions and the resulting limitations of the thermodynamic formalism, *Proc. R. Soc. Lond.* A **434**, 79–88.

Manneville, P. 1990. *Dissipative Structures and Weak Turbulence.* Academic Press, Boston.

Marsden, J. & Weinstein, A. 1983. Coadjoint orbits, vortices and Clebsch variables for incompressible fluids, *Physica* D **7**, 305–323.

Martin, P.C., Siggia, E.D. & Rose, H.A. 1973. Statistical dynamics of classical systems, *Phys. Rev.* A **8**, 423–437.

Maurer, J., Tabeling, P. & Zocchi, G. 1994. Statistics of turbulence between two counter-rotating disks in low temperature helium gas, *Europhys. Lett.* **26**, 31–36.

McWilliams, J.C. 1984. The emergence of isolated coherent vortices in turbulent flow, *J. Fluid Mech.* **146**, 21–43.

McWilliams, J.C. 1990. The vortices of two-dimensional turbulence, *J. Fluid Mech.* **219**, 361–385.

Meneveau, C.M. & Sreenivasan, K.R. 1987. The multifractal spectrum of the dissipation field in turbulent flows, *Nucl. Phys. B Proc. Suppl.* **2**, 49–76.

Meneveau, C.M. & Sreenivasan, K.R. 1991. The multifractal nature of turbulent energy dissipation, *J. Fluid Mech.* **224**, 429–484.

Meshalkin, L.D. & Sinai, Ya.G. 1961. Investigation of the stability of a stationary solution of a system of equations for the plane movement of an incompressible viscous liquid, *Appl. Math. Mech.*. **25**, 1700–1705.

Michel, J. & Robert, R. 1994. Statistical mechanical theory of the Great Red Spot of Jupiter, *J. Stat. Phys.* **77**, 645–666.

Migdal, A.A. 1994. Loop equations and area law in turbulence, *Int. J. Mod. Phys.* **9**, 1197–1238.

Miles, R.B., Lempert, W., Zhang, B. & Zhang, L. 1991. Turbulent structure measurements by RELIEF flow tagging, *Fluid Dyn. Res.* **8**, 9–17.

Miller, J. 1990. Statistical mechanics of Euler equations in two dimensions, *Phys. Rev. Lett.* **65**, 2137–2140.

Millionshchikov, M.D. 1941a. Theory of homogeneous isotropic turbulence, *Dokl. Acad. Nauk. SSSR* **22**, 236–240.

Millionshchikov, M.D. 1941b. Theory of homogeneous isotropic turbulence, *Izv. Akad. Nauk SSSR Ser. Geogr. Geofiz.* **5**, 433–446.

Moffatt, H.K. 1969. The degree of knotedness of tangled vortex lines, *J. Fluid Mech.* **35**, 117–129.

Moffatt, H.K. 1978. *Magnetic Field Generation in Electrically Conducting Fluids*. Cambridge University Press, Cambridge.

Moffatt, H.K. 1984. Simple topological aspects of turbulent vorticity dynamics, in *Turbulence and Chaotic Phenomena in Fluids*, 223–230, ed. T. Tatsumi. North-Holland, Amsterdam.

Moffatt, H.K. 1994. *Tourbillons et Turbulence*, École Polytechnique, Palaiseau, France (vol. 2 of lecture notes 'Dynamique des fluides').

Moffatt, H.K. & Tsinober, A. 1992. Helicity in laminar and turbulent flow, *Ann. Rev. Fluid Mech.* **24**, 281–312.

Moffatt, H.K., Kida, S. & Ohkitani, K. 1994. Stretched vortices — the sinews of turbulence; high Reynolds number asymptotics, *J. Fluid Mech.* **259**, 241–264.

Monin, A.S. 1959. Theory of locally isotropic turbulence, *Dokl. Akad. Nauk SSSR* **125**, 515–518.

Monin, A.S. & Yaglom, A.M. 1971. *Statistical Fluid Mechanics*, vol. 1, ed. J. Lumley. MIT Press, Cambridge, MA.

Monin, A.S. & Yaglom, A.M. 1975. *Statistical Fluid Mechanics*, vol. 2, ed. J. Lumley. MIT Press, Cambridge, MA.

Montgomery, D. & Joyce, G. 1974. Statistical mechanics of negative temperature states, *Phys. Fluids* **17**, 1139–1145.

Montgomery, D., Matthaeus, W.H., Stribling, W.T., Martinez, D. & Oughton, S. 1992. Relaxation in two dimensions and the "sinh–Poisson" equation, *Phys. Fluids* A **4**, 3–6.

Moreau, J.J. 1961. Constantes d'un îlot tourbillonnaire en fluide parfait barotrope, *C. R. Acad. Sci. Paris* **252**, 2810–2812.

Morf, R.H., Orszag, S.A. & Frisch, U. 1980. Spontaneous singularity in three-dimensional, inviscid incompressible flow, *Phys. Rev. Lett.* **44**, 572–575.

Morrison, P.J. 1993. Hamiltonian description of the ideal fluid, in *Geometrical Methods in Fluid Dynamics. Proceedings GFD 1993*, 17–110. Woods-Hole Oceanographic Institute, Woods Hole (WHOI-94-12).

Moser, J. 1962. On invariant curves of area-preserving mappings of an annulus, *Nachr. Akad. Wiss. Gött., II, Math.-Phys. Kl.* **11a**, 1–20.

Muzy, J.F., Bacry, E. & Arnéodo, A. 1993. Multifractal formalism for fractal signals: the structure-function approach versus the wavelet-transform modulus-maximum method, *Phys. Rev.* E **47**, 875–884.

Navier, C.L.M.H. 1823. Mémoire sur les lois du mouvement des fluides, *Mém. Acad. Roy. Sci.* **6**, 389–440.

Nelkin, M. 1974. Turbulence, critical phenomena and intermittency, *Phys. Rev.* A **9**, 388–395.

Nelkin, M. 1990. Multifractal scaling of velocity derivatives in turbulence, *Phys. Rev.* A **42**, 7226–7229.

Nelkin, M. 1994. Universality and scaling in fully developed turbulence, *Advances in Physics* **43**, 143–181.

Nepomnyashchy, A.A. 1976. On the stability of the secondary flow of a viscous fluid in an infinite domain, *Appl. Math. Mech.* **40**, 886–891.

Neu, J.C. 1984. The dynamics of stretched vortices, *J. Fluid Mech.* **143**, 253–276.

Neumann, J. von 1949. Recent theories of turbulence, in *Collected works (1949–1963)* **6**, 437–472, ed. A.H. Taub. Pergamon Press, New York (1963).

Noether, E. 1918. Invariante Variationsprobleme, *Nachr. v. d. Ges. d. Wiss. zu Göttingen*, 235–257.

Noullez, A., Wallace, G., Lempert, W., Miles, R.B. & Frisch, U. 1996. Transverse velocity increments in turbulent flow using the RELIEF technique, *J. Fluid Mech* (submitted).

Novikov, E.A. 1964. Functionals and the method of random forces in turbulence theory, *Zh. Exper. Teor. Fiz.* **47**, 1919-1926.

Novikov, E.A. 1969. Scale similarity for random fields, *Sov. Phys. Dokl.* **14**, 104–107.

Novikov, E.A. 1970. Intermittency and scale similarity of the structure of turbulent flow, *Prikl. Math. Mekh.* **35**(2), 266–277.

Novikov, E.A. & Stewart, R.W. 1964. The intermittency of turbulence and the spectrum of energy dissipation, *Izv., Akad. Nauk SSSR, Ser. Geoffiz.*, 408–413.

Obukhov, A.M. 1941a. On the distribution of energy in the spectrum of turbulent flow, *Dokl. Akad. Nauk SSSR* **32**(1), 22–24.

Obukhov, A.M. 1941b. Spectral energy distribution in a turbulent flow, *Izv. Akad. Nauk SSSR Ser. Geogr. Geofiz.* **5**(4–5), 453–466.

Obukhov, A.M. 1962. Some specific features of atmospheric turbulence, *J. Fluid Mech.* **13**, 77–81.

Ogura, Y. 1963. A consequence of the zero-fourth-cumulant approximation in the decay of isotropic turbulence, *J. Fluid Mech.* **16**, 33–40.

Ohkitani, K. & Kishiba, S. 1995. Nonlocal nature of vortex stretching in an inviscid fluid, *Phys. Fluids* **7**, 411–421.

Ohkitani, K. & Yamada, M. 1989. Temporal intermittency in the energy cascade process and local Lyapunov analysis in fully-developed model turbulence, *Progr. Theoret. Phys.* **89**, 329–341.

Onsager, L. 1945a. Letter to C.C. Lin, June 1945 *T. von Kármán papers*, National Air and Space Museum, Smithsonian Institution, Washington, DC (copy of microfiches of the originals at the CALTECH Archives), box 18, folder 22.

Onsager, L. 1945b. The distribution of energy in turbulence, *Phys. Rev.* **68**, p. 285 (Abstract of talk presented at the meeting of the Metropolitan Section of the American Physical Society, Columbia University, New York, November 9-10, 1945).

Onsager, L. 1949. Statistical hydrodynamics, *Nuovo Cimento* **6**(2), 279–287 (Suppl. Ser. IX).

Oono, Y. 1989. Large deviations and statistical physics, *Progr. Theor. Phys. Suppl.* **99**, 165–205.

Orszag, S.A. 1966. Dynamics of fluid turbulence, Princeton Plasma Physics Laboratory, report PPL-AF-13.

Orszag, S.A. 1970. Indeterminacy of the moment problem for intermittent turbulence, *Phys. Fluids* **13**, 2211–2212.

Orszag, S.A. 1974. Numerical Simulation of the Taylor–Green Vortex, in *Computing Methods in Applied Sciences and Engineering, Part 2, Lect. Notes in Computer Science* **11**, 50–64, eds. R. Glowinski & J.L. Lions. Springer, Berlin.

Orszag, S.A. 1977. Statistical Theory of Turbulence, in *Fluid Dynamics, Les Houches 1973*, 237–374, eds. R. Balian & J.L. Peube. Gordon and Breach, New York.

Orszag, S.A. & Kruskal, M.D. 1966. Theory of turbulence, *Phys. Rev. Lett.* **16**, 441–444.

Orszag, S.A. & Kruskal, M.D. 1968. Formulation of the theory of turbulence, *Phys. Fluids* **11**, 43–60.

Orszag, S.A. & Patterson, G.S. 1972. Numerical simulation of turbulence, in *Statistical Models and Turbulence, Lect. Notes in Phys.* **12**, 127–147, eds. M. Rosenblatt & C.W. Van Atta. Springer, Berlin.

Orszag, S.A., Yakhot, V., Flannery, W.S., Boysan, F., Choudhury, D., Maruzewski, J. & Patel, B. 1993. Renormalization Group modeling and turbulence simulations, in *Proc. Int. Conf. on Near-Wall Turbulent Flows*, 1031–1046, eds. R.M.C. So, C.G. Speziale & B.E. Launde. Elsevier, Amsterdam.

Oseledets, V.I. 1989. On a new way of writing the Navier–Stokes equation. Hamiltonian formalism, *Russian Math. Surveys* **44**, 210–211.

Ott, E., Du, Y., Sreenivasan, K.R., Juneja, A. & Suri, A.K. 1992. Sign-singular measures: fast magnetic dynamos, and high-Reynolds-number fluid turbulence, *Phys. Rev. Lett.* **69**, 2654–2657.

Ottino, J.M. 1989. *The Kinematics of Mixing: Stretching, Chaos, and Transport.* Cambridge University Press, Cambridge.

Paladin, G. & Vulpiani, A. 1987a. Degrees of freedom of turbulence, *Phys. Rev.* A **35**, 1971–1973.

Paladin, G. & Vulpiani, A. 1987b. Anomalous scaling laws in multifractal objects, *Phys. Rep.* **156**, 147–225.

Panchev, S. 1971. *Random Functions and Turbulence.* Pergamon Press, New York.

Papoulis, A. 1991. *Probability, Random Variables, and Stochastic Processes*, 3rd edition. McGraw-Hill, New York.

Parisi, G. 1988. *Statistical Field Theory.* Addison-Wesley, Redwood City.

Parisi, G. 1990. A mechanism for intermittency in a cascade model for turbulence, University of Rome, preprint ROM2F-90/37.

Parisi, G. & Frisch, U. 1985. On the singularity structure of fully developed turbulence, in *Turbulence and Predictability in Geophysical Fluid Dynamics, Proceed. Intern. School of Physics 'E. Fermi', 1983, Varenna, Italy*, 84–87, eds. M. Ghil, R. Benzi & G. Parisi. North–Holland, Amsterdam.

Parisi, G. & Sourlas, N. 1979. Random magnetic fields, supersymmetry, and negative dimensions, *Phys. Rev. Lett.* **43**, 744–745.

Pasmanter, R.A. 1994. On long lived vortices in 2-D viscous flows, most probable states of inviscid 2-D flows and a soliton solution, *Phys. Fluids* **6**, 1236–1241.

Passot, T., Politano, H., Sulem, P.-L., Angilella, J.R. & Meneguzzi, M. 1995. Instability of strained vortex layers and vortex tube formation in homogeneous turbulence, *J. Fluid Mech.* **282**, 313–338.

Pedlosky, J. 1979. *Geophysical Fluid Dynamics.* Springer, Berlin.

Phythian, R. 1977. The functional formalism of classical statistical dynamics, *J. Phys.* A **10**, 777–789.

Pinton, J.-F. & Labbé, R. 1994. Correction to the Taylor hypothesis in swirling flows, *J. Phys. II France* **4**, 1461–1468.

Pisarenko, D., Biferale, L., Courvoisier, D., Frisch, U. & Vergassola, M. 1993. Further results on multifractality in shell models, *Phys. Fluids* A **5**, 2533–2538.

Piumati, G. 1894. *Il Codice Atlantico di Leonardo da Vinci.* Reproduced and published by Academia dei Lincei, Ulrico Hoepli, Milan.

Polyakov, A.M. 1972. Scale invariance of strong interactions and its application to lepton–hadron reactions, in *International School of High Energy Physics in Erevan 23 November – 4 December 1971 (Chernogolovka 1972).* Academy Sci. USSR, Moscow.

Polyakov, A.M. 1993. The theory of turbulence in two dimensions, *Nucl. Phys.* **B396**, 367–385.

Pomeau, Y. 1992. Asymptotic time behaviour of nonlinear classical field equations, *Nonlinearity* **5**, 707–720.

Pomeau, Y. 1994. Statistical approach to 2D turbulence, in *Turbulence: a Tentative Dictionary*, 117–123, eds. P. Tabeling & O. Cardoso. Plenum Press, New York.

Pomeau, Y. & Manneville, P. 1980. Intermittent transition to turbulence in dissipative dynamical systems, *Commun. Math. Phys.* **74**, 189–197.

Pomeau, Y., Pumir, A. & Young, W.R. 1988. Transitoires dans l'advection-diffusion d'impuretés, *C. R. Acad. Sci. Paris, série II* **306**, 741–746.

Pouquet, A. 1978. On two-dimensional magnetohydrodynamic turbulence, *J. Fluid Mech.* **88**, 1–16.

Pouquet, A., Gloaguen, C., Léorat, J. & Grappin, R. 1984. A scalar model of MHD turbulence, in *Chaos and statistical methods Sept. 1983, Kyoto*, 206–210, ed. Y. Kuramoto. Springer, Berlin.

Pouquet, A., Lesieur, M., André, J.C. & Basdevant, C. 1975. Evolution of high Reynolds number two-dimensional turbulence, *J. Fluid Mech.* **72**, 305–319.

Prandtl, L. 1925. Bericht über Untersuchungen zur ausgebildeten Turbulenz, *Zs. angew. Math. Mech.* **5**, 136–139.

Prandtl, L. 1932. Zur turbulenten Strömung in Röhren und längs Platten, *Ergebn. Aerodyn. Versuchsanst.* **4**, 18–29.

Praskovsky, A.A. 1992. Experimental verification of the Kolmogorov refined similarity hypothesis, *Phys. Fluids* A **4**, 2589–2591.

Praskovsky, A.A., Foss, J.F., Kleiss, S.J. & Karyakin, M.Yu. 1993. Fractal properties of isovelocity surfaces in high Reynolds number laboratory shear flows, *Phys. Fluids* A **5**, 2038–2042.

Proudman, I. & Reid, W.H. 1954. On the decay of a normally distributed and homogeneous turbulent velocity field, *Phil. Trans. R. Soc. Lond.* A **247**, 163–189.

Pullin, D.I. & Saffman, P.G. 1993. On the Lundgren–Townsend model of turbulent fine scales, *Phys. Fluids* A **5**, 126–145.

Pumir, A. 1994. A numerical study of the mixing of a passive scalar in three dimensions in the presence of a mean gradient, *Phys. Fluids* **6**, 2118–2132.

Pumir, A. & Siggia, E.D. 1990. Collapsing solutions in the 3-D Euler equations, in *Topological Fluid Mechanics*, 469–477, eds. H.K. Moffat & A. Tsinober. Cambridge University Press, Cambridge.

Pumir, A. & Siggia, E.D. 1992. Development of singular solutions to the axisymmetric Euler equations, *Phys. Fluids* A **4**, 1472–1491.

Pumir, A., Shraiman, B.I. & Siggia, E.D. 1991. Exponential tails and random advection, *Phys. Rev. Lett.* **66**, 2984–2987.

Reynolds, O. 1894. On the dynamical theory of turbulent incompressible viscous fluids and the determination of the criterion, *Phil. Trans. R. Soc. London* A **186**, 123–161.

Richardson, L.F. 1922. *Weather Prediction by Numerical Process.* Cambridge University Press, Cambridge.

Richardson, L.F. 1926. Atmospheric diffusion shown on a distance–neighbour graph, *Proc. R. Soc. Lond.* A **110**, 709–737 (also in *Collected Papers of L.F. Richardson*, vol. 1, ed. P.G. Drazin, pp. 523–551, Cambridge University Press 1993).

Rivet, J.P. 1991. Brisure spontanée de symétrie dans le sillage tri-dimensionnel d'un cylindre allongé, simulé par la méthode des gaz sur réseaux., *C. R. Acad. Sci. Paris, série II* **313**, 151-157.

Robert, R. 1990. État d'équilibre statistique pour l'écoulement bidimensionnel d'un fluide parfait, *C. R. Acad. Sci. Paris, série I* **311**, 575–578.

Robert, R. 1991. A maximum-entropy principle for two-dimensional perfect fluid dynamics, *J. Stat. Phys.* **65**, 531–553.

Robert, R. & Sommeria, J. 1991. Statistical equilibrium states for two-dimensional flows, *J. Fluid Mech.* **229**, 291–310.

Rose, H.A. 1977. Eddy diffusivity, eddy noise and subgrid-scale modelling, *J. Fluid Mech.* **81**, 719–734.

Rose, H.A. & Sulem, P.-L. 1978. Fully developed turbulence and statistical mechanics, *J. Phys. France*, 441–484.

Rosen, G. 1960. Turbulence theory and functional integration. I, II, *Phys. Fluids* **3**, 519–524, 525–528.

Ruelle, D. 1989. *Chaotic Evolution and Strange Attractors.* Cambridge University Press, Cambridge.

Ruelle, D. 1990. Is there screening in turbulence?, *J. Stat. Phys.* **61**, 865–868.

Ruelle, D. 1991. The turbulent fluid as a dynamical system, in *New Perpectives in Turbulence*, 123–138, ed. L. Sirovich. Springer, Berlin.

Saffman, P.G. 1960. On the effect of the molecular diffusivity on turbulent diffusion, *J. Fluid Mech.* **8**, 273–283.

Saffman, P.G. 1967a. The large scale structure of homogeneous turbulence, *J. Fluid Mech.* **27**, 581–593.

Saffman, P.G. 1967b. Note on the decay of homogeneous turbulence, *Phys. Fluids* **10**, p. 1349.

Saffman, P.G. 1968. Lectures on homogeneous turbulence, in *Topics in Nonlinear Physics*, 485–614, ed. N.J. Zabusky. Springer, Berlin.

Saffman, P.G. 1981. Dynamics of vorticity, *J. Fluid Mech.* **106**, 49–58.

Saffman, P.G. 1992. *Vortex Dynamics.* Cambridge University Press, Cambridge.

Saint-Venant, A.J.C. (Barré) de 1843. Note à joindre au Mémoire sur la dynamique des fluides, *C. R. Acad. Sci. Paris* **17**, 1240–1243.

Saint-Venant, A.J.C. (Barré) de 1850. Mémoire sur des formules nouvelles pour la solution des problèmes relatifs aux eaux courantes, *C. R. Acad. Sci. Paris* **31**, 283–286.

Saint-Venant, A.J.C. (Barré) de 1851. Formules et tables nouvelles pour les eaux courantes, *Ann. Mines* **20**, p. 49.

Sanada, T. 1991. Cluster statistics of homogeneous fluid turbulence, *Phys. Rev.* A **44**, 6480–6489.

Santangelo, P., Benzi, R. & Legras, B. 1989. The generation of vortices in high-resolution, two-dimensional decaying turbulence and the influence of initial conditions on the breaking of self-similarity, *Phys. Fluids* A **1**, 1027–1034.

Scheffer, V. 1976. Turbulence and Hausdorff dimension, in *Turbulence and the Navier–Stokes equations, Lect. Notes in Math.* **565**, 94–112, ed. R. Temam. Springer, Berlin.

Scheffer, V. 1977. Hausdorff measure and the Navier–Stokes equation, *Commun. Math. Phys.* **55**, 97–112.

Schertzer, D. & Lovejoy, S. 1984. On the dimension of atmospheric motion, in *Turbulence and Chaotic Phenomena in Fluids*, 505–512, ed. T. Tatsumi. North-Holland, Amsterdam.

Schertzer, D. & Lovejoy, S. 1985. The dimension and intermittency of atmospheric dynamics, in *Turbulent Shear Flow* **4**, 7–33, eds. L.J.S. Bradbury, F. Durst, B. Launder, F.W. Schmidt & J.H. Whitelaw. Springer, Berlin.

Schewe, G. 1983. On the force fluctuations acting on a circular cylinder in crossflow from subcritical up to transcritical Reynolds numbers, *J. Fluid Mech.* **133**, 265–285.

Schmitt, F., Lavallée, D., Schertzer, D. & Lovejoy, S. 1992. Empirical determination of universal multifractal exponents in turbulent velocity fields, *Phys. Rev. Lett.* **68**, 305–308.

Schuster, H.G. 1988. *Deterministic Chaos: an Introduction*, 2nd edition. VCH, Weinheim, Germany.

Schwarz, K.W. 1988. Three-dimensional vortex dynamics in superfluid ^{4}He: Homogeneous superfluid turbulence, *Phys. Rev.* B **38**, 2398–2417.

Schwinger, J. 1951a. On gauge invariance and vacuum polarization, *Phys. Rev.* **82**, 664–679.

Schwinger, J. 1951b. On the Green's functions of quantized fields. I, II, *Proc. Natl. Acad. Sci.* **37**, 452–459.

Schwinger, J. 1958. *Selected Papers on Quantum Electrodynamics*. Dover, New York.

Serrin, J. 1959. Mathematical principles of classical fluid mechanics, in *Handbuch der Physik, Fluid Dynamics I*, 125–263, eds. S. Flügge & C. Truesdell. Springer, Berlin.

Sewell, M.J. 1987. *Maximum and Minimum Principles*. Cambridge University Press, Cambridge.

She, Z.S. 1987. Metastability and vortex pairing in the Kolmogorov flow, *Phys. Lett.* A **124**, 161–164.

She, Z.S. 1991. Physical model of intermittency in turbulence: near-dissipation-range non-Gaussian statistics, *Phys. Rev. Lett.* **66**, 600–603.

She, Z.S. & Jackson, E. 1993. On the universal form of energy spectra in fully developed turbulence, *Phys. Fluids* A **5**, 1526–1528.

She, Z.S. & Lévêque, E. 1994. Universal scaling laws in fully developed turbulence, *Phys. Rev. Lett.* **72**, 336–339.

She, Z.S. & Orszag, S.A. 1991. Physical model of intermittency: inertial-range non-Gaussian statistics, *Phys. Rev. Lett.* **66**, 1701–1704.

She, Z.S. & Waymire, E.C. 1995. Quantized energy cascade and log–Poisson statistics in fully developed turbulence, *Phys. Rev. Lett.* **74**, 262–265.

She, Z.S., Aurell, E. & Frisch, U. 1992. The inviscid Burgers equation with initial data of Brownian type, *Commun. Math. Phys.* **148**, 623–641.

She, Z.S., Jackson, E. & Orszag, S.A. 1990. Intermittent vortex structures in homogeneous isotropic turbulence, *Nature* **344**, 226–228.

She, Z.S., Jackson, E. & Orszag, S.A. 1991. Structure and dynamics of homogeneous turbulence: models and simulations, *Proc. R. Soc. Lond.* A **434**, 101–124.

Shraiman, B.I. & Siggia, E.D. 1995. Anomalous scaling of a passive scalar in turblent flow, *C. R. Acad. Sci. Paris, série II* **321**, 279–284.

Siggia, E.D. 1978. Model of intermittency in three-dimensional turbulence, *Phys. Rev.* A **17**, 1166–1176.

Siggia, E.D. 1981. Numerical study of small scale intermittency in three-dimensional turbulence, *J. Fluid Mech.* **107**, 375–406.

Siggia, E.D. 1994. High Rayleigh number convection, *Ann. Rev. Fluid Mech.* **26**, 137–168.

Siggia, E.D. & Aref, H. 1981. Point-vortex simulation of the inverse cascade in two-dimensional turbulence, *Phys. Fluids* **24**, 171–173.

Sinai, Ya.G. 1992. Statistics of shocks in solutions of inviscid Burgers equation, *Commun. Math. Phys.* **148**, 601–622.

Sivashinsky, G.I. 1985. Weak turbulence in periodic flows, *Physica* D **170**, 243–255.

Sivashinsky, G.I. & Frenkel, A.L. 1992. On negative eddy viscosity under conditions of isotropy, *Phys. Fluids* A **4**, 1608–1610.

Slutsky, E.E. 1938. Sur les fonctions aléatoires presque périodiques et sur la décomposition des fonctions aléatoires stationnaires en composantes, in *Actualités scientifiques et industrielles* **738**, 33–55. Hermann, Paris.

Smith, L.A., Fournier, J.D & Spiegel, E.A. 1986. Lacunarity and intermittency in fluid turbulence, *Phys. Lett.* A **114**, 465–468.

Smith, L.M. & Yakhot, V. 1993. Bose condensation and small-scale structure generation in a random force driven 2D turbulence, *Phys. Rev. Lett.* **71**, 352–355.

Smith, L.M., & Yakhot, V. 1994. Finite-size effects in forced, two-dimensional turbulence, *J. Fluid Mech.* **274**, 115–138.

Sommeria, J. 1986. Experimental study of the inverse energy cascade in a square box, *J. Fluid Mech.* **170**, 139–168.

Sommeria, J., Nore, C., Dumont, T. & Robert, R. 1991. Théorie statistique de la Tache Rouge de Jupiter, *C. R. Acad. Sci. Paris, série II* **312**, 999–1005.

Sornette, D. & Sammis, C.G. 1995. Complex critical exponents from renormalization group theory of earthquakes: implications for earthquake predictions, *J. Phys. I France* **5**, 607–619.

Spalding, D.B. 1991. Kolmogorov's two-equation model of turbulence, *Proc. R. Soc. Lond.* A **434**, 211–216.

Sreenivasan, K.R. 1984. On the scaling of the turbulent energy dissipation rate, *Phys. Fluids* **27**, 1048–1051.

Sreenivasan, K.R. 1991. On local isotropy of passive scalars in turbulent shear flows, *Proc. R. Soc. Lond.* A **434**, 165–182.

Sreenivasan, K.R. 1995. On the universality of the Kolmogorov constant, *Phys. Fluids* **7**, 2778–2784.

Starr, V.P. 1968. *Physics of Negative Viscosity Phenomena.* McGraw-Hill, New York.

Steenbeck, M., Krause, F. & Rädler, K.H. 1966. Zur Berechnung de mittlerer Lorentz-FeldStärke $v \times B$ für ein elektrisch leitendes Medium in turbulenter, durch Coriolis-Kräfte beeinflusster Bewegung, *Z. Naturforsch.* **21**, 369–376.

Stenflo, J.O. 1973. Magnetic-field structure of the photospheric network, *Solar Phys.* **32**, 41–63.

Stenflo, J.O. 1994. *Solar Magnetic Fields*. Kluwer, Dordrecht.

Stewart, R.W. 1972. Turbulence, in *Illustrated Experiments in Fluid Mechanics*, 82–88, ed. National Committee for Fluid Mechanics Films (illustrates a film available from International Division, Encyclopedia Britanica, 425 N. Michigan Avenue, Chicago).

Stewart, R.W. & Townsend, A.A. 1951. Similarity and self-preservation in isotropic turbulence, *Phil. Trans. R. Soc. Lond.* A **243**, 359–386.

Stokes, G.G. 1843. On some cases of fluid motion, *Trans. Camb. Phil. Soc.* **8**, p. 105 (also in *Mathematical and Physical Papers by G.G. Stokes* **1**, 17–68. Cambridge University Press, Cambridge, 1880. Republished by Johnson Reprint Corporation, New York, 1966).

Stolovitzky, G. & Sreenivasan, K.R. 1994. Kolmogorov's refined similarity hypotheses for turbulence and general stochastic processes, *Rev. Mod. Phys.* **66**, 229–240.

Stolovitzky, G., Kailasnath, P. & Sreenivasan, K.R. 1992. Kolmogorov's refined similarity hypothesis, *Phys. Rev. Lett.* **69**, 1178–1181.

Sulem, P.-L. & Frisch, U. 1975. Bounds on energy flux for finite energy turbulence, *J. Fluid Mech.* **72**, 417–423.

Sulem, C. & Sulem, P.-L. 1983. The well-posedeness of two-dimensional ideal flow, *J. Méc. Théor. Appliqu. (Paris)*, 217–242 (Special issue on two-dimensional turbulence).

Sulem, C., Sulem, P.-L. & Frisch, H. 1983. Tracing complex singularities with spectral methods, *J. Comput. Phys.* **50**, 138–161.

Sulem, P.-L., She, Z.S., Scholl, H. & Frisch, U. 1989. Generation of large-scale structures in three-dimensional flow lacking parity-invariance, *J. Fluid Mech.* **205**, 341–358.

Swift, J. 1983. On Poetry, in *The Complete Poems*, ed. P. Rogers. Yale University Press, New Haven.

Tabeling, P. & Cardoso, O. 1994. *Turbulence: a Tentative Dictionary*. Plenum Press, New York.

Tabeling, P., Burkhart, S., Cardoso, O. & Willaime, H. 1991. Experimental study of freely decaying two-dimensional turbulence, *Phys. Rev. Lett.* **67**, 3772–3775.

Tabeling, P., Zocchi, G., Belin, F., Maurer, J. & Willaime, H. 1996. Probability density functions, skewness and flatness in large Reynolds number turbulence, *Phys. Rev.* E **55**, 1613–1621.

Tatarskii, V.I. 1962. Application of quantum field theory to the problem of decay of homogeneous turbulence, *Sov. Phys. JETP* **15**, 961–971.

Tatsumi, T. 1957. The theory of decay process of incompressible isotropic turbulence, *Proc. R. Soc. Lond.* A **239**, 16–45.

Taylor, G.I. 1915. Eddy motion in the atmosphere, *Phil. Trans. R. Soc.* A **215**, 1–26.

Taylor, G.I. 1921. Diffusion by continuous movements, *Proc. London Math. Soc.* **20**, 196–211.

Taylor, G.I. 1935. Statistical theory of turbulence, *Proc. R. Soc. Lond.* A **151**, 421–478.

Taylor, G.I. 1938. The spectrum of turbulence, *Proc. R. Soc. Lond.* A **164**, 476–490.

Temam, R. 1977. *Navier–Stokes Equations.* North-Holland, Amsterdam (revised edition 1984).

Tennekes, H. 1989. Two and three-dimensional turbulence, in *Lecture Notes on Turbulence, NCAR-GTP 1987*, 1–73, eds. J.R. Herring & J.C. McWilliams. World Scientific, Singapore.

Tennekes, H. & Lumley, J.L. 1972. *A First Course in Turbulence.* MIT Press, Cambridge, MA.

Tikhomirov, V.M. 1991. *Selected works of A.N. Kolmogorov.* Dordrecht, Boston.

Tong. C. & Warhaft, Z. 1994. On passive scalar derivative statistics in grid turbulence, *Phys. Fluids* **6**, 2165–2176.

Toulouse, G. & Pfeuty, P. 1975. *Introduction au Groupe de Renormalisation et à ses Applications.* Presses Universitaires de Grenoble, Grenoble.

Townsend, A.A. 1951. On the fine-scale structure of turbulence, *Proc. R. Soc. Lond.* A **208**, 534–542.

Townsend, A.A. 1976. *The Structure of Turbulent Shear Flow.* Cambridge University Press, Cambridge.

Tritton, D.J. 1988. *Physical Fluid Dynamics*, 2nd edition. Clarendon, Oxford.

Van Atta, C.W. & Chen, W.Y. 1970. Structure functions of turbulence in the atmospheric boundary layer over the ocean, *J. Fluid Mech.* **44**, 145–159.

Van Atta, C.W. & Park, J. 1972. Statistical self-similarity and inertial subrange turbulence, in *Statistical Models and Turbulence, Lect. Notes in Phys.* **12**, 402–426, eds. M. Rosenblatt & C.W. Van Atta. Springer, Berlin.

van de Water, W., van der Vorst, B. & van de Wetering, E. 1991. Multiscaling of turbulent structure functions, *Europhys. Lett.* **16**, 443–448.

Van Dyke, M. 1982. *An Album of Fluid Motion.* The Parabolic Press, Stanford, CA.

Varadhan, S.R.S. 1984. *Large Deviations and Applications.* SIAM, Philadelphia.

Vergassola, M. 1993. Chiral nonlinearities in forced 2-D Navier–Stokes flows, *Europhys. Lett.* **24**, 41–45.

Vergassola, M. 1996. Anomalous scaling for passively advected magnetic fields, *Phys. Rev.* E, **53**, R3021–R3024.

Vergassola, M. & Gama, S. 1994. Slow-down of nonlinearity in 2-D Navier–Stokes flow, *Physica* D **76**, 291–296.

Vergassola, M., Gama, S. & Frisch, U. 1993. Proving the existence of negative isotropic eddy viscosity, in *NATO-ASI: Solar and Planetary Dynamos*, 321–327, eds. M.R.E. Proctor, P.C. Mathews & A.M. Rucklidge. Cambridge University Press, Cambridge.

Vergassola, M., Benzi, R., Biferale, L. & Pisarenko, D. 1993. Wavelet analysis of a Gaussian Kolmogorov signal, *J. Phys.* A **26**, 6093–6099.

Vergassola, M., Dubrulle, B., Frisch, U. & Noullez, A. 1994. Burgers' equation, Devil's staircases and the mass distribution for large-scale structures, *Astron. Astrophys.* **289**, 325–356.

Vincent, A. & Meneguzzi, M. 1991. The spatial structure and statistical properties of homogeneous turbulence, *J. Fluid Mech.* **225**, 1–25.

Vishik, M.J. & Fursikov, A.V. 1988. *Mathematical Problems of Statistical Hydrodynamics.* Kluwer, Dordrecht.

Vogel, S. 1981. *Life in Moving Fluids.* Princeton University Press, Princeton.

Waugh, D.W., Plumb, R.A., Atkinson, R.J., Schoeberl, M.R., Lait, L.R., Newman, P.A., Lowenstein, M., Toohey, D.W., Avallone, L.M.,

Webster, C.R. & May, R.D. 1994. Transport of material out of the stratospheric Arctic vortex by Rossby wave breaking, *J. Geophys. Res.* **99**, 1071–1088.

Wax, N. 1954. *Noise and Stochastic Processes (Selected Papers on).* Dover, New York.

Waymire, E.C. & Williams, S.C. 1994. A general decomposition theory for random cascades, *Bull. Amer. Math. Soc.* **31**, 216–222.

Weiss, J.B. & McWillams, J.C. 1993. Temporal scaling behavior of decaying two-dimensional turbulence, *Phys. Fluids* A **5**, 608–621.

Weizsäcker, C.F. von 1948. Das Spektrum der Turbulenz bei großen Reynoldschen Zahlen, *Zeit. f. Phys.* **124**, 614–627.

Wiener, N. 1930. Generalized harmonic analysis, *Acta Math.* **55**, 117–258.

Wilson, K. 1972. Feynman-graph expansion for critical exponents, *Phys. Rev. Lett.* **28**, 548–551.

Wilson, K. & Fisher, M.E. 1972. Critical exponents in 3.99 dimensions, *Phys. Rev. Lett.* **28**, 240–243.

Wilson, K. & Kogut, J. 1974. The renormalization group and the ϵ expansion, *Phys. Rep.* **12C**, 75–200.

Wirth, A., Gama, S. & Frisch, U. 1995. Eddy viscosity of three-dimensional flow, *J. Fluid Mech.* **288**, 249–264.

Wolibner, W. 1933. Un théorème sur l'existence du mouvement plan d'un fluide parfait, homogène, incompressible, pendant un temps infiniment long, *Math. Z.* **37**, 698–726.

Wu, X.Z., Kadanoff, L., Libchaber, A. & Sano, M. 1990. *Phys. Rev. Lett.* **64**, 2140–2143.

Wyld, H.W. 1961. Formulation of the theory of turbulence in an incompressible fluid, *Ann. Phys.* **14**, 143–165.

Wyngaard, J.C. & Tennekes, H. 1970. Measurements of the small-scale structure of turbulence at moderate Reynolds numbers, *Phys. Fluids* **13**, 1962–1969.

Yaglom, A.M. 1966. Effect of fluctuations in energy dissipation rate on the form of turbulence characteristics in the inertial subrange, *Dokl. Akad. Nauk SSSR* **166**, 49–52.

Yaglom, A.M. 1987. Einstein's 1914 paper on the theory of irregularly fluctuating series of observations, *IEEE ASSP Magazine* **4**(4), 6–11.

Yaglom, A.M. 1994. A.N. Kolmogorov as a fluid mechanician and founder of a school in turbulence research, *Ann. Rev. Fluid Mech.* **26**, 1–22.

Yakhot, V. & Orszag, S.A. 1986a. Renormalization group analysis of turbulence. I. Basic theory, *J. Sci. Comput.* **1**, 3–52.

Yakhot, V. & Orszag,S.A. 1986b. Renormalization group analysis of turbulence, *Phys. Rev. Lett.* **57**, 1722–1724.

Yamada, M., Kida, S. & Ohkitani, K. 1993. Wavelet analysis of PDF's in turbulence – a power law distribution in the dissipation range, in *Unstable and Turbulent Motion of Fluid*, 188–199, ed. S. Kida. World Scientific, Singapore.

Zabusky, N.J., Hughes, M. & Roberts, K.V. 1979. Contour dynamics of the Euler equations in two dimensions, *J. Comput. Phys.* **30**, 96–106.

Zakharov, V.E., L'vov, V.S. & Falkovich, G. 1992. *Kolmogorov Spectra of Turbulence I.* Springer, Berlin.

Zeitlin, V. 1991. Finite-mode analogs of 2-D ideal hydrodynamics: coadjoint orbits and local canonical structure, *Physica* D **49**, 353–362.

Zeitlin, V. 1992. On the structure of phase-space, Hamiltonian variables and statistical approach to the description of two-dimensional hydrodynamics and magnetohydrodynamics, *J. Phys.* A **25**, L171–L175.
Zeldovich, Ya.B., Ruzmaikin, A.A. & Sokoloff, D.D. 1983. *Magnetic Fields in Astrophysics*. Gordon and Breach, New York.
Zeldovich, Ya.B., Ruzmaikin, A.A. & Sokoloff, D.D. 1990. *The Almighty Chance*. World Scientific, Singapore.

Author index

D

E

F

G

H

I

J

K

L

T

V

W

Y

Z

Subject index

S

T

U

V